深入浅出 USB 系统开发
——基于 ARM Cortex - M3

主　编　王川北　刘　强
副主编　屈召贵　孙　活　蔡德洋

北京航空航天大学出版社

内 容 简 介

本书系统地阐述了 USB 协议、Stellaris USB 处理器的体系结构、工作原理和设计方法，并通过多个 USB 开发实例，详细介绍了 USB 开发思路、流程及编程方法，并在此基础上讲解了嵌入式 USB 主机、USB OTG 开发。全书共分 15 章：第 1 章介绍 USB 系统基础知识、基本术语、USB 基本结构、开发流程、USB 枚举、USB 描述符格式、主机和设备开发过程等；第 2 章介绍 Cortex-M3 内核的 USB 处理器，包括 USB 基本模块、工作方式、USB 寄存器操作、寄存器级编程等；第 3 章介绍使用设备驱动库函数进行 Cortex-M3 编程，包括内核操作、中断控制、GPIO 编程、USB 基本编程等；第 4 章介绍 TI 的 USB 库使用及编程；第 5～10 章介绍 USB 设备开发；第 11 章介绍 USB 主机开发；第 12 章介绍 USB OTG 开发；第 13 章介绍 USB 设备开发总结及注意事项；第 14 章介绍 USB 主机开发总结及注意事项；第 15 章是 USB 系统开发总结，包括常见概念性问题、开发问题等，阐述其产生的基本原因，并提供了解决此类问题的方案。

本书可作为高等院校电子类、仪器仪表类、控制类等专业的 USB 系统开发教材或参考用书，也可供广大从事 USB 系统开发的工程技术人员参考。

图书在版编目(CIP)数据

深入浅出 USB 系统开发：基于 ARM Cortex-M3 / 王川北，刘强主编. --北京：北京航空航天大学出版社，2012.8

ISBN 978-7-5124-0872-2

Ⅰ. ①深… Ⅱ. ①王… ②刘… Ⅲ. ①USB 总线—串行接口 ②微控制器 Ⅳ. ①TP334.7 ②TP332.3

中国版本图书馆 CIP 数据核字(2012)第 157759 号

版权所有，侵权必究。

深入浅出 USB 系统开发——基于 ARM Cortex-M3

主　编　王川北　刘　强
副主编　屈召贵　孙　活　蔡德洋
责任编辑　陈　旭

*

北京航空航天大学出版社出版发行

北京市海淀区学院路 37 号(邮编 100191)　http://www.buaapress.com.cn
发行部电话：(010)82317024　传真：(010)82328026
读者信箱：emsbook@gmail.com　邮购电话：(010)82316936
涿州市新华印刷有限公司印装　各地书店经销

开本：710×1000　1/16　印张：27.25　字数：597 千字
2012 年 7 月第 1 版　2012 年 7 月第 1 次印刷　印数：4 000 册
ISBN 978-7-5124-0872-2　定价：52.00 元

若本书有倒页、脱页、缺页等印装质量问题，请与本社发行部联系调换。联系电话：(010)82317024

前 言

　　USB最初是由英特尔与微软公司倡导发起的,其最大的特点是支持热插拔和即插即用。当设备插入时,主机侦测此设备并加载所需的驱动程序,因此使用远比PCI和ISA总线方便。USB规范第一次是于1995年,由Intel、IBM、Compaq、Microsoft、NEC、Digital、North Telecom等7家公司组成的USBIF(USB Implement Forum)共同提出,USBIF于1996年1月正式提出USB 1.0规范,频宽为12 Mbps,不过因为当时支持USB的外围设备很少,所以主机板厂商不太把USB Port直接设计在主机板上。1998年9月,USBIF提出USB 1.1规范来修正USB 1.0,主要修正了技术上的小细节,但传输的频宽不变,仍为12 Mbps。USB 1.1向下兼容USB 1.0,因此对于一般使用者而言,感受不到USB 1.1与USB 1.0的规范差异。

　　2000年4月,广泛使用的USB 2.0推出,速度达到了480 Mbps,是USB 1.1的40倍;如今8个半年头过去了,USB 2.0的速度早已无法满足应用的需要,USB 3.0也就应运而生了,最大传输带宽高达5.0 Gbps,也就是625 MB/s,同时在使用A型的接口时向下兼容。

　　随着电子信息科学的飞速发展,近年来嵌入式技术的应用越来越广泛,生活中无所不在,无处不用。面对当前嵌入式行业的优越形势,业界人士掀起了学习嵌入式系统理论及应用开发的热潮,选择合适的嵌入式系统学习至关重要,嵌入式系统涉及嵌入式处理器、实时操作系统等内容。在众多的嵌入式处理器中,基于ARM系列的处理器得到了行业人士的认可,其中Cortex-M3是ARM公司推出的针对微控制器应用的内核,它提供了适用于众多高性能和低成本需求的嵌入式应用,成为时下MCU应用的热点。

　　本书选用Stellaris的USB控制器为控制芯片介绍USB系统开发,该类USB处理器支持USB主机/设备/OTG功能,在点对点通信过程中可运行在全速和低速模式。它符合USB 2.0标准,包含挂起和恢复信号。它包含32个端点,其中包含两个用于控制传输的专用连接端点(一个用于输入,一个用于输出)以及30个由固件定义的端点,并带有一个大小可动态变化的FIFO,以支持多包队列。可通过μDMA来访问FIFO,将对系统软件的依赖降至最低。USB设备启动方式灵活,可软件控制是否在启动时连接。USB控制器遵从OTG标准的会话请求协议(SRP)和主机协商协议(HNP)。本书将重点讲解Luminary(TI)的Stellaris USB控制器原理及应用。

　　本书深入浅出地讲解USB系统的基础知识,在内容上力求精简,快速上手,书中

前言

大量实例来自科研实践和电子竞赛优秀作品,可作为高等院校本、专科的 USB 系统开发的教材,也可作为对 USB 系统自学和工程技术人员的参考书。

全书共分 15 章。

第 1 章:介绍 USB 系统基础知识、基本术语、USB 基本结构、开发流程、USB 枚举、USB 描述符格式、主机和设备开发过程等。

第 2 章:介绍 Cortex-M3 的内核 USB 处理器,包括 USB 基本模块、工作方式、USB 寄存器操作、寄存器级编程等。

第 3 章:介绍使用设备驱动库函数进行 Cortex-M3 编程,包括内核操作、中断控制、GPIO 编程、USB 基本编程等。

第 4 章:TI 的 USB 库使用及编程。

第 5~10 章:介绍 USB 设备开发。

第 11 章:介绍 USB 主机开发。

第 12 章:介绍 USB OTG 开发。

第 13 章:介绍 USB 设备开发总结及注意事项。

第 14 章:介绍 USB 主机开发总结及注意事项。

第 15 章:USB 系统开发总结,包括常见的概念性问题、开发问题等,阐述其产生的基本原因,并提供了解决此类问题的方案。

参加本书编写的有王川北、屈召贵、刘强、孙活、蔡德洋等。王川北负责全书的统稿工作。其中第 1~8 章由王川北和孙活编写;第 8~15 章由屈召贵和刘强编写。本书配备有实验开发装置,由王川北和蔡德洋设计。蔡德洋和孙活参与了本书各章节的审阅。

本书在编写过程得到了四川师范大学成都学院的大力支持,还得到了学院胡增江老师、汪光宅高级工程师、林信元教授、鲁顺昌教授、吴自恒教授、杨肇基教授、郭群扬工程师的大力帮助与指导。在此表示衷心感谢。

这是在非计算机、电子专业嵌入式系统课程教学改革的一个尝试,因编者水平所限,难免有不尽如人意之处,敬请读者提出宝贵意见和建议!

笔者的通信地址:四川成都郫县团结镇学院街 65 号,邮编:611745。

有兴趣的读者可以发送电子邮件到:paulhyde@126.com 或登录:www.isjtag.com 与笔者联系;也可发送邮件到:xdhydcd5@sina.com,与本书策划编辑沟通。

编 者

2012.5

目 录

第1章 USB 基础 ... 1
1.1 USB 介绍 ... 1
1.2 USB 常用术语 ... 2
1.3 USB 设备开发流程 ... 6
1.4 USB 设备枚举 ... 6
1.4.1 USB 设备请求 ... 7
1.4.2 描述符 ... 9
1.4.3 设备枚举过程 ... 20
1.5 USB 主机开发流程 ... 29
1.6 USB OTG 介绍 ... 30
1.7 小 结 ... 31

第2章 Stellaris 的 USB 处理器 ... 32
2.1 Stellaris 处理器简介 ... 32
2.2 Stellaris USB 模块 ... 42
2.2.1 功能描述 ... 43
2.2.2 USB 控制器作为 USB 设备 ... 44
2.2.3 USB 控制器作为主机 ... 49
2.2.4 OTG 模式 ... 51
2.3 寄存器描述 ... 52
2.3.1 控制状态寄存器 ... 54
2.3.2 中断控制 ... 61
2.3.3 端点寄存器 ... 69
2.4 USB 处理器配置使用 ... 84
2.5 小 结 ... 86

目 录

第 3 章　底层库函数 87

3.1　底层库函数 87
3.2　通用库函数 88
　3.2.1　内核操作 88
　3.2.2　系统中断控制 91
　3.2.3　GPIO 控制 92
3.3　USB 基本操作 97
3.4　设备库函数 111
3.5　主机库函数 114
3.6　小　结 122

第 4 章　USB 库介绍 123

4.1　USB 库函数简介 123
4.2　USBlib 介绍 126
4.3　使用底层驱动开发 130
4.4　使用 USB 库开发 146
4.5　小　结 148

第 5 章　HID 设备 149

5.1　HID 介绍 149
5.2　HID 类描述符 149
5.3　USB 键盘 155
　5.3.1　数据类型 155
　5.3.2　API 函数 161
　5.3.3　USB 键盘开发 162
5.4　USB 鼠标 174
　5.4.1　数据类型 174
　5.4.2　API 函数 177
　5.4.3　USB 鼠标开发 178
5.5　小　结 190

第 6 章　Audio 设备 191

6.1　Audio 设备介绍 191
6.2　Audio 描述符 192
6.3　Audio 数据类型 198
6.4　API 函数 201

6.5 Audio 设备开发 ………………………………………… 202
6.6 小　结 …………………………………………………… 219

第 7 章　Bulk 设备 …………………………………………… 220

7.1 Bulk 设备介绍 …………………………………………… 220
7.2 Bulk 数据类型 …………………………………………… 220
7.3 API 函数 ………………………………………………… 223
7.4 Bulk 设备开发 …………………………………………… 228
7.5 小　结 …………………………………………………… 253

第 8 章　CDC 设备 …………………………………………… 254

8.1 CDC 设备介绍 …………………………………………… 254
8.2 CDC 数据类型 …………………………………………… 254
8.3 API 函数 ………………………………………………… 257
8.4 CDC 设备开发 …………………………………………… 260
8.5 小　结 …………………………………………………… 294

第 9 章　Mass Storage 设备 ………………………………… 295

9.1 Mass Storage 设备介绍 ………………………………… 295
9.2 MSC 数据类型 …………………………………………… 295
9.3 API 函数 ………………………………………………… 298
9.4 MSC 设备开发 …………………………………………… 299
9.5 小　结 …………………………………………………… 314

第 10 章　Composite 设备 …………………………………… 315

10.1 Composite 设备介绍 …………………………………… 315
10.2 Composite 数据类型 …………………………………… 315
10.3 API 函数 ………………………………………………… 316
10.4 Composite 设备开发 …………………………………… 317
10.5 小　结 …………………………………………………… 334

第 11 章　USB 主机开发 …………………………………… 335

11.1 USB 主机开发介绍 ……………………………………… 335
11.2 USB 主机开发过程 ……………………………………… 337
　11.2.1 主机配置 …………………………………………… 338
　11.2.2 注册驱动 …………………………………………… 340
　11.2.3 运行主机 …………………………………………… 344

目录

11.3 主机开发实例 ·········· 350
   11.3.1 鼠　标 ·········· 350
   11.3.2 键　盘 ·········· 356
   11.3.3 U　盘 ·········· 365
11.4 小　结 ·········· 372

第 12 章　USB OTG 开发 ·········· 373

12.1 OTG 介绍 ·········· 373
   12.1.1 主机通信协议与对话请求协议 ·········· 374
   12.1.2 OTG 功能的构建 ·········· 374
   12.1.3 LM3S 的 OTG 功能 ·········· 375
   12.1.4 OTG 函数 ·········· 376
12.2 OTG B 开发 ·········· 381
12.3 OTG A 开发 ·········· 381
12.4 OTG 开发实例 ·········· 381
12.5 OTG 开发小结 ·········· 385

第 13 章　USB 设备工程实例 ·········· 386

13.1 USB 设备开发流程 ·········· 386
13.2 USB 设备之 USB BootLoader ·········· 387
13.3 USB 设备开发总结 ·········· 393

第 14 章　USB 主机开发实例 ·········· 396

14.1 USB 主机开发流程 ·········· 396
14.2 USB 主机之音频输入输出 ·········· 399
14.3 USB 主机开发总结 ·········· 402

第 15 章　USB 系统开发总结 ·········· 403

15.1 常见问题 ·········· 403
   15.1.1 概念问题 ·········· 403
   15.1.2 开发问题 ·········· 409
15.2 本章小结 ·········· 415

附录 A　LM3S5749 应用电路图 ·········· 416

附录 B　LM-Link 下载器原理图 ·········· 420

附录 C　USB 常见术语及缩略词 ·········· 422

参考文献 ·········· 426

第 1 章

USB 基础

　　USB 最初是由英特尔与微软公司倡导发起的,其最大的特点是支持热插拔和即插即用。当设备插入时,主机侦测此设备并自动加载所需的驱动程序。随着便携式设备的发展,USB 的优越性更加明显,没有哪一种手机没有 USB 接口;没有哪一种通信接口能有 USB 那么方便可靠;没有哪一种通信方式的传输速度有 USB 那么快且用线少。本章主要介绍 USB 的基本理论、基础知识、基本术语、基本结构等,同时也简单介绍 USB 的开发流程、枚举过程等。通过本章的学习,读者可以掌握 USB 的基础知识,为后面 USB 系统开发打好基础。

1.1　USB 介绍

　　USB,Universal Serial BUS(通用串行总线)的缩写,其中文简称为"通用串行总线",是一个外部总线标准,用于规范计算机与外部设备的连接和通信,是应用在 PC 和嵌入式领域的接口技术。USB 接口支持设备的即插即用和热插拔功能。USB 通信方式为串行通信。在 1994 年底由英特尔、康柏、IBM、Microsoft 等多家公司联合提出,目前使用最多的是 USB 1.1 和 USB 2.0,几乎每一台计算机上都有 4~8 个 USB 接口头。USB 3.0 已经"崛起",并向下兼容;USB 3.0 芯片已经大量生产,其相关产品已经大量投入使用。USB 具有传输速度快(USB 1.1 是 12 Mbps,USB 2.0 是 480 Mbps,USB 3.0 是 5 Gbps)、使用方便、支持热插拔、连接灵活、独立供电等优点,可以连接鼠标、键盘、打印机、扫描仪、摄像头、闪存盘、MP3 机、手机、数码相机、移动硬盘、外置光软驱、USB 网卡、ADSL Modem、Cable Modem 等几乎所有的外部设备。所以,掌握 USB 系统开发是很重要的,就像 10 年前,电子工程师必须掌握 UART 串行通信一样,不懂 USB 不能称其为真正的电子工程师。

　　常用 USB 接口如图 1-1 所示。有方型头(D 型头)、扁型公头、扁型母头等,其中公头主要用于 USB 设备中,比如 U 盘使用 USB 扁型公头;母头主要用在 USB 主机上,其中 USB 扁型母头用得最多,比如计算机主板上全使用 USB 扁型母头做 USB 主机接头。

图 1-1　USB 常用接口图

1.2　USB 常用术语

学习 USB 系统开发之前,有必要先了解 USB 开发中可能会遇到的一些常用术语,有助于理解 USB 的工作模式和编程模型。USB 系统开发分为 USB 主机开发和 USB 设备(从机)开发。在一个 USB 系统中,某一个时刻,只有一个 USB 主机,其余均为 USB 设备。但是为了让一个 USB 系统既有 USB 主机功能,又有从机功能,便出现了 USB OTG。因此,USB 开发主要包括 USB 主机、USB 设备、USB OTG 系统开发。下面就一些常用术语进行详细解释:

① USB 主机:在任何一个 USB 系统中,只有一个主机。USB 和主机系统的接口称作主机控制器,主机控制器可由硬件、固件和软件综合实现。USB 主机主要与 USB 设备进行通信、交换数据、设备控制。

② USB 设备:主机的"下行"设备,为系统提供具体功能,可以是多个功能并受主机控制的外部 USB 设备。也称作 USB 外设(从机),使用 USB B 型连接器连接。USB 主机最多可以支持 127 个 USB 设备。

③ USB OTG:简单地说,OTG 就是 On The Go,正在进行中的意思。是近几年发展起来的技术,2001 年 12 月 18 日由 USB Implementers Forum(USB IF)公布,主要应用于各种不同的设备或移动设备间的连接,进行数据交换,特别是 PDA、移动电话、消费类设备中运用得最多。改变如数码照相机、摄像机、打印机等设备间多种不同制式连接器,多达 7 种制式的存储卡间数据交换的不便。USB OTG 标准在完全兼容 USB 2.0 标准的基础上,增添了电源管理(节省功耗)功能,它允许设备既可作为主机,也可作为外设操作(两用 OTG)。OTG 两用设备完全符合 USB 2.0 标准,并可提供一定的主机检测能力,支持主机协商协议(HNP)和对话请求协议(SRP)。

在 OTG 中,初始主机称为 A 设备,设备称为 B 设备。简单地说,USB OTG 既是 USB 主机也是 USB 设备。但是在任意时刻,只能有一个 A 设备(主机)。

④ 集线器:集线器(Hub)扩展 USB 主机所能连接设备的数量,主要用于扩展。图 1-2 为 USB Hub 使用图,从图中可以看出,一个 USB Hub 将一个 USB 接口扩展为 4 个,增加了 USB 主机连接设备的数量。Hub 在 USB 系统中的功能就像交换机在以太网通信中的功能一样,起到数据交换、增加设备连接数量的作用。

图 1-2 USB Hub 使用图

⑤ SIE:串行接口引擎,USB 控制器内部的"核心",将低级信号转换成字节,以供控制器使用,并负责处理底层协议,如填充位、CRC 生成和校验,且可发出错误报告。物理层的数据功能是根据 USB 协议规定把 USB 物理接口上的电信号解码为 USB 数据。一个 USB 设备系统由以下几部分组成:进行数据处理的微处理器部分; USB SIE(串行接口引擎)部分;用于数据存储的存储器部分;进行数据编码与外部相连的 USB 收发器。串行接口引擎完成发送和接收,不对数据内容进行分析。SIE 的主要功能是对 USB 总线上传输的数据进行编码与解码;检测与产生信息包的起始 SOP 标志和信息包的结束 EOP 标志;对数据进行串/并转换;检测接收信息包、传送信息包;串行接口引擎是 USB 设备系统必不可少的接口模块,它的可靠性是 USB 设备正确执行后续操作的前提。

⑥ 端点:是主机与设备之间通信的目的或来源。控制端点可以双向传输数据,而其他端点只能在单方向传输数据。主机和设备的通信最终作用于设备的各个端点上,是主机与设备间通信流的一个逻辑终端。每个 USB 设备有一个唯一的地址,这个地址是在设备连上主机时,由主机分配的,而设备中的每个端点在设备内部有唯一

第 1 章 USB 基础

的端点号。这个端点号是在设计设备时给定的。每个端点都是一个简单的连接点，或者支持数据流进设备，或者支持其流出设备，两者不可兼得。基于 PnP 机制，设备被枚举时，它必须向主机报告各个端点的特性，包括端点号、通信方向、端点支持的最大包大小、带宽要求等（其中端点支持的最大包大小叫作数据有效负载）。每个设备必须有端点 0，用于设备枚举和对设备进行一些基本的控制功能。除了端点 0，其余的端点在设备配置之前不能与主机通信，只有向主机报告这些端点的特性并被确认后才能被激活。端点位于 USB 系统内部，是一个可寻址的 FIFO 空间。类似于高速公路收费口的入口或出口，一个端点地址对应一个方向。所以，端点 2-IN 与端点 2-OUT 完全不同。端点 0 默认双向控制传输，共享一个 FIFO 空间。

⑦ 管道：是 USB 通信设备上的一个端点和主机上软件之间的联系，体现了主机缓存和端点间传送数据的能力，有流和消息两种不同且互斥的管道通信格式。流指不具有 USB 定义格式的数据流。流通道中的数据是流的形式，也就是该数据的内容不具有 USB 要求的结构。数据从流通道一端流进的顺序与它们从流通道另一端流出时的顺序是一样的，流通道中的通信流总是单方向的。对于在流通道中传送的数据，USB 认为它来自同一个客户。USB 系统软件不能够提供使用同一流通道的多个客户的同步控制。在流通道中传送的数据遵循先进先出原则。流管道只能连到一个固定号码的端点上，或者流进，或者流出。而具有这个号码的另一个方向的端点可以被分配给其他流通道。流通道支持同步传送、中断传送和批量传送；消息指具有某种 USB 定义的格式的数据流。消息通道与端点的关系同流通道与端点的关系是不同的。首先，主机向 USB 设备发出一个请求；接着，就是数据的传送；最后，是一个状态阶段。为了能够容纳请求/数据/状态的变化，消息通道要求数据有一个格式，此格式保证了命令能够被可靠地传送和确认。消息通道允许双方向的信息流，虽然大多数的通信流是单方向的。特别地，默认控制通道也是一个消息通道。在某一时刻，一个端点只能为单个信息请求服务。主机上的多个客户软件通过默认管道能产生请求，但它们以先进先出的顺序被送到端点上。在响应主机事务的数据和状态阶段，端点能控制信息的流动。只有当端点上的当前信息处理完成后，端点才会正常地发送下一个信息。一个设备信息管道在两个方向（IN 或 OUT 标记）要求有一个单一的设备端点号。对于每个方向，USB 不允许信息管道与不同的端点号相联系。

⑧ 设备接口：此接口非物理接口，是一种与主机通信的信道管理。设备接口中包含多个端点，与主机进行通信。此接口相当于一个 USB 系统功能，这个功能可能会使用到多个端点。比如，一个 USB 设备既有 USB 键盘功能，又有 U 盘功能，这种设备就具备两个接口功能，每一个接口都要使用多个端口。

⑨ 描述符：是一个数据结构，使主机了解设备的格式化信息。打个比方，就像平时填写申请表一样，格式与内容都有规定，必须按规定格式和字数编写，否则不予理睬。每一个描述符可能包含整个设备的信息，或是设备中的一个组件（接口）。标准 USB 设备有 5 种描述符：设备描述符、配置描述符、字符串描述符、接口描述符及端

点描述符。

⑩ 枚举：USB主机通过一系列命令要求USB设备发送描述符信息，从而知道设备具有什么功能、属于哪一类设备、要占用多少带宽、使用哪类传输方式及数据量的大小，只有主机确定了这些信息之后，设备才能真正地开始工作。打个比方，这就相当于申请处理过程，让你提交什么资料，就必须得提交什么资料，并且必须按申请表规定的格式与字符数进行填写，否则，申请无效。所以枚举对主机与设备进行通信是非常重要的，枚举不成功，主机就无法了解设备，USB设备也不能称其为USB设备。

⑪ 3种传输速率：低速模式传输速率为1.5 Mbps，多用于键盘和鼠标；全速模式传输速率为12 Mbps；高速模式传输速率为480 Mbps。

⑫ 输入：相对主机而言，如果设备输入端点发送的数据为设备发送数据到主机。设备发送数据，主机接收数据。

⑬ 输出：相对主机而言，如果设备输出端点接收的数据为主机发送到设备的数据。设备接收数据，主机发送数据。

⑭ HCD：包含主机控制器和根HUB的硬件，为程序员提供了由硬件实现定义的接口主机控制器设备（HCD）。而实际上HCD在计算机上表现为端口和内存的映射。USB 1.0和USB 1.1的标准都有两个HCD标准，分别是康柏公司的开放主机控制器接口（OHCI）和Intel公司的通用主机控制器接口（UHCI），而VIA公司的USB芯片采纳了Intel的UHCI标准；其他USB芯片大多使用康柏公司的OHCI标准。它们的主要区别是UHCI更加依赖软件驱动，因此对CPU要求更高，但是自身的硬件会更廉价。它们的并存导致操作系统开发和硬件厂商都必须在两个方案上开发和测试，从而导致费用上升。因此USB IF在USB 2.0的设计阶段坚持只能有一个实现规范，这就是扩展主机控制器接口（EHCI）。因为EHCI只支持全速传输，所以EHCI控制器包括4个虚拟的全速或者慢速控制器。这里同样是Intel和VIA使用虚拟UHCI，其他一般使用OHCI控制器。某些版本的Windows上，打开设备管理器，如果设备说明中有"增强"("Enhanced")，就能够确认它是USB 2.0版的。而在Linux系统中，命令lspci能够列出所有的PCI设备，而USB会分别命名为OHCI、UHCI或者EHCI。列出为16位地址的为EHCI，32位的为OHCI。命令lsusb能够显示所有USB设备的信息，命令dmesg能够显示OS启动时关于USB设备的信息。

⑮ USB电源：USB接头提供一组5 V的电压，可作为USB设备的电源。实际上，设备接收到的电源可能会低于5 V，只略高于4 V。USB规范要求在任何情形下，电压均不能超过5.25 V；在最坏情形下（经由USB供电HUB所连接的LOW POWER设备）电压均不能低于4.375 V，一般情形下电压会接近5 V。一个USB的根集线器最多只能提供500 mA的电流。如此的电流已足以驱动许多电子设备，不过连接在总线供电Hub的所有设备需要共享500 mA的电流额度。一个由总线供电的设备可以使用到它所连接到Hub的所有电能。总线供电的Hub可以将电源供

给连接在 Hub 上的所有设备,不过 USB 的规范只允许总线供电的 Hub 下游串接一层总线供电的设备,因此,总线供电的 Hub 下游不允许再串接另一个由总线供电的 Hub。许多 Hub 有外加电源,因此可以提供电源给下游的设备,不会消耗总线上的电源。若设备需要的电压超过 5 V,或是需要电流超过 500 mA,都需要使用外加电源。相对于之前其他沟通介面仅能传递信息资料,高电压 USB 插槽本身还能提供 5 V 的主动电压及 0.5 A 的电流,因此对于一些小型设备而言,可以不必再外接电源供应装置,就能利用来自 USB 插槽的电力顺利运作。利用这个特点,也有厂商开发出适当的排线,将 USB 拿来当作供电插座使用,例如作为移动电话的充电器;或是供给小型电灯的电力需要,而与原本用来连接计算机的主要用途无关。

1.3　USB 设备开发流程

图 1-3 为 USB 设备开发流程图。USB 设备开发流程相当简单,①首先确认 USB 系统的开发类型,是 USB 主机、USB 设备还是 OTG;②如果确定是 USB 设备,必须明确该设备的类型:HID、UDIO、CDC、HUB、IMAGE 等;③查找相关设备手册,确定其描述符;④完成描述符后,编写 USB 枚举程序,观察是否枚举成功,如果枚举成功,此设备开发已经完成大部分;⑤编写应用程序,在枚举成功后,主要是进行数据处理,编写应用程序。以上就是 USB 设备开发的主要流程,其中最主要的是枚举过程,枚举不成功,该设备就不能称为 USB 设备,更不能完成 USB 设备所赋予的任务。

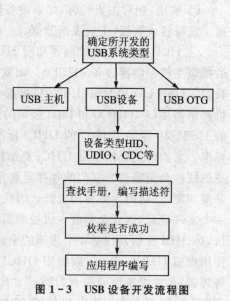

图 1-3　USB 设备开发流程图

1.4　USB 设备枚举

对于 USB 设备开发来说,最重要的是枚举,即让主机知道设备的相关信息。若枚举不成功,则设备无法识别、更不能使用。本节主要讲枚举过程,有关设备的其他信息,请参阅 USB 官方协议手册。USB 设备的属性通过一组描述符来反映它们,这些描述符是具有一定格式的数据结构,主机软件可通过 GET_DESCRIPTOR 请求获取这些描述符。每一个描述符的第一个字节表明本描述符的长度,其后是一个字节的描述符类型信息。如果描述符中的长度域值小于描述符的定义长度,此描述符被认为是非法的,不能被主机接收;如果返回描述符中的长度域值大于描述符的定义长度,则过长部分被忽略。

1.4.1 USB 设备请求

设备描述符是表征对该设备及所有设备配置起全程作用的信息。在设备枚举时,主机使用 GET_DESCRIPTOR 控制指令直接从设备端点 0 读取该描述符,一个 USB 设备只能有一个设备描述符。USB 设备与主机连接时会发出 USB 请求命令。每个请求命令数据包由 8 个字节(5 个字段)组成,具有相同的数据结构。表 1-1 为 USB 设备请求数据包格式表:

表 1-1 USB 设备请求数据包格式表

偏移量	域	大小	值	描述
0	bmRequestType	1	位	请求特征
1	bRequest	1	值	请求命令
2	wValue	2	值	请求不同,含义不同
4	wIndex	2	索引	请求不同,含义不同
6	wLength	2	值	数据传输阶段,为数据字节数

标准的 USB 设备请求命令是用在控制传输中"初始设置步骤"里的数据包阶段(即 DATA0,由 8 个字节构成)。标准 USB 设备请求命令共有 11 个,大小都是 8 个字节,具有相同的结构,由 5 个字段构成(字段是标准请求命令的数据部分),结构如下(括号中的数字表示字节数,首字母 bm、b、w 分别表示位图、字节和双字节):

bmRequestType(1)+bRequest(1)+wvalue(2)+wIndex(2)+wLength(2)

在 USB 2.0 规范里定义的标准 USB 设备请求结构体数据头如下:

```
typedef struct
{
    //定义请求的方向和类型
    unsigned char bmRequestType;
    //请求类型
    unsigned char bRequest;
    //根据请求而定其实际含义
    unsigned short wValue;
    //根据请求而定,通常提供索引或偏移
    unsigned short wIndex;
    //在数据传输阶段时,指示传输数据的字节数
    unsigned short wLength;
}
tUSBRequest;
```

第1章 USB 基础

(1) bmRequestType 参数

```
//位 7,说明请求的传输方向
//输入
#define USB_RTYPE_DIR_IN        0x80
//输出
#define USB_RTYPE_DIR_OUT       0x00
//位 6:5,定义请求的类型
#define USB_RTYPE_TYPE_M        0x60
#define USB_RTYPE_VENDOR        0x40
#define USB_RTYPE_CLASS         0x20
#define USB_RTYPE_STANDARD      0x00
```

USB_RTYPE_TYPE_M 用于提取请求中的 6:5 位,并通过 6:5 位判断请求类型,大多数为 0x00,即标准请求(USB_RTYPE_STANDARD)。

```
// 位 4:0,定义接收者
#define USB_RTYPE_RECIPIENT_M   0x1f
#define USB_RTYPE_OTHER         0x03
#define USB_RTYPE_ENDPOINT      0x02
#define USB_RTYPE_INTERFACE     0x01
#define USB_RTYPE_DEVICE        0x00
```

USB_RTYPE_RECIPIENT_M 用于提取请求中的 4:0 位,并能通过 4:0 位判断请求的接收者。0x00,设备;0x01,接口;0x02,端点;0x03 为其他请求。

(2) bRequest 参数

```
//标准请求的请求类型
#define USBREQ_GET_STATUS       0x00
#define USBREQ_CLEAR_FEATURE    0x01
#define USBREQ_SET_FEATURE      0x03
#define USBREQ_SET_ADDRESS      0x05
#define USBREQ_GET_DESCRIPTOR   0x06
#define USBREQ_SET_DESCRIPTOR   0x07
#define USBREQ_GET_CONFIG       0x08
#define USBREQ_SET_CONFIG       0x09
#define USBREQ_GET_INTERFACE    0x0a
#define USBREQ_SET_INTERFACE    0x0b
#define USBREQ_SYNC_FRAME       0x0c
```

当 bmRequestType 为标准请求(即设备请求的第 7 位为 0)时,bRequest(第 6:5 位)参数提示当前请求类型,以上请求类型经常在枚举时使用。

(3) wValue 参数

① USBREQ_CLEAR_FEATURE 和 USBREQ_SET_FEATURE 请求命令时:

```
#define USB_FEATURE_EP_HALT          0x0000    // Endpoint halt feature
#define USB_FEATURE_REMOTE_WAKE      0x0001    // Remote wake feature, device only
#define USB_FEATURE_TEST_MODE        0x0002    // Test mode
```

② USBREQ_GET_DESCRIPTOR 请求命令时：

```
#define USB_DTYPE_DEVICE             1
#define USB_DTYPE_CONFIGURATION      2
#define USB_DTYPE_STRING             3
#define USB_DTYPE_INTERFACE          4
#define USB_DTYPE_ENDPOINT           5
#define USB_DTYPE_DEVICE_QUAL        6
#define USB_DTYPE_OSPEED_CONF        7
#define USB_DTYPE_INTERFACE_PWR      8
#define USB_DTYPE_OTG                9
#define USB_DTYPE_INTERFACE_ASC      11
#define USB_DTYPE_CS_INTERFACE       36
```

请求命令数据包由主机通过端点 0 发送,当设备端点为接收到请求命令数据包时,会发出端点 0 中断,控制器可以读取端点 0 中的数据,此数据格式由 tUSBRequest 定义。根据 bmRequestType 判断是不是标准请求,bRequest 获得具体请求类型,wValue 指定更具体的对像,wIndex 指出索引和偏移。在数据传输阶段时,wLength 为传输数据的字节数。整个控制传输都依靠这 11 个标准请求命令,非标准请求通过 callback 函数返回给用户处理。

1.4.2 描述符

USB 采用 USB 标准描述符说明一个 USB 设备,这些描述符包括设备描述符、配置描述符、接口描述符、端点描述符和字符串描述符。对于高速设备还包括设备限定描述符和其他速率配置描述符。设备类和供应商也可以自己定义其设备专用描述符,分别称为设备类定义描述符和供应商自定义描述符。每一个设备都有自己一套完整的描述符,包括设备描述符、配置描述符、接口描述符、端点描述符和字符串描述符。这些描述符反映设备特性,由特定格式组合的一个数据结构体。在 USB 设备枚举过程中,主机通过标准命令 USBREQ_GET_DESCRIPTOR 来获取描述符,从而得知设备的各种特性。

1. 设备描述符

设备描述符给出了 USB 设备的一般信息。一个 USB 设备只能有一个设备描述符。主机发出 USBREQ_GET_DESCRIPTOR 请求命令,且 wValue 值为 USB_DTYPE_DEVICE 时,获取设备描述符,如表 1-2 所列,为 USB 标准设备描述符表。

表1-2 标准设备描述符表

偏移量	域	大小	值	描述
0	bLength	1	数字	此描述表的字节数
1	bDecriptorType	1	常量	描述表种类为设备
2	bcdUSB	2	BCD码	USB设备版本号(BCD码)
4	bDeviceClass	1	类	设备类码
5	bDeviceSubClass	1	子类	子类码
6	bDevicePortocol	1	协议	协议码
7	bMaxPacketSize0	1	数字	端点0最大包大小(8,16,32,64)
8	idVendor	2	ID	厂商标志(VID)
10	idProduct	2	ID	产品标志(PID)
12	bcdDevice	2	BCD码	设备发行号(BCD码)
14	iManufacturer	1	索引	厂商信息字串索引
15	iProduct	1	索引	产品信息字串索引
16	iSerialNumber	1	索引	设备序列号信息字串索引
17	bNumConfigurations	1	数字	配置描述符数目

C语言设备描述符结构体为：

```
typedef struct
{
    //本描述符字节数,设备描述符为18字节
    unsigned char bLength;
    //本描述符类型,设备描述符为1,USB_DTYPE_DEVICE
    unsigned char bDescriptorType;
    // USB规格版本号格式的BCD码。对于USB 2.0是0x0200
    unsigned short bcdUSB;
    //设备类代码
    unsigned char bDeviceClass;
    //子类代码
    unsigned char bDeviceSubClass;
    //协议代码
    unsigned char bDeviceProtocol;
    //端点0最大包长
    unsigned char bMaxPacketSize0;
    //VIID
    unsigned short idVendor;
    //PID
    unsigned short idProduct;
```

```c
// BCD 格式的设备的版本号
unsigned short bcdDevice;
//厂商字符描述索引
unsigned char iManufacturer;
//产品字符描述索引
unsigned char iProduct;
//设备序列号信息字串索引
unsigned char iSerialNumber;
//该设备提供可能的配置数量
unsigned char bNumConfigurations;
}
tDeviceDescriptor;
```

① bDescriptorType 表示本描述符类型，不同描述符，取值不一样。

```c
#define USB_DTYPE_DEVICE            1
#define USB_DTYPE_CONFIGURATION     2
#define USB_DTYPE_STRING            3
#define USB_DTYPE_INTERFACE         4
#define USB_DTYPE_ENDPOINT          5
#define USB_DTYPE_DEVICE_QUAL       6
```

② bDeviceClass 表示本设备使用类的代码。

```c
#define USB_CLASS_DEVICE            0x00
#define USB_CLASS_AUDIO             0x01
#define USB_CLASS_CDC               0x02
#define USB_CLASS_HID               0x03
#define USB_CLASS_PHYSICAL          0x05
#define USB_CLASS_IMAGE             0x06
#define USB_CLASS_PRINTER           0x07
#define USB_CLASS_MASS_STORAGE      0x08
#define USB_CLASS_HUB               0x09
#define USB_CLASS_CDC_DATA          0x0a
#define USB_CLASS_SMART_CARD        0x0b
#define USB_CLASS_SECURITY          0x0d
#define USB_CLASS_VIDEO             0x0e
#define USB_CLASS_HEALTHCARE        0x0f
#define USB_CLASS_DIAG_DEVICE       0xdc
#define USB_CLASS_WIRELESS          0xe0
#define USB_CLASS_MISC              0xef
#define USB_CLASS_APP_SPECIFIC      0xfe
#define USB_CLASS_VEND_SPECIFIC     0xff
#define USB_CLASS_EVENTS            0xffffffff
```

如果设备为鼠标,则 bDeviceClass 选为 USB_CLASS_HID 人机接口。

③ bDeviceSubClass 表示子类码、bDevicePortocol 表示设备协议码,根据不同的设备类 bDeviceClass 来选择。例如,bDeviceClass＝USB_CLASS_HID,即为人机接口,则 bDeviceSubClass 有以下参数可选择:

```
#define USB_HID_SCLASS_NONE       0x00
#define USB_HID_SCLASS_BOOT       0x01
```

bDevicePortocol 有以下参数可选择:

```
#define USB_HID_PROTOCOL_NONE     0
#define USB_HID_PROTOCOL_KEYB     1
#define USB_HID_PROTOCOL_MOUSE    2
```

配置一个键盘设备的描述符如下:

```
tDeviceDescriptor    * HIDKEYDeviceDescriptor;
unsigned char g_pHIDDeviceDescriptor[] =
{
    18,                         // Size of this structure
    USB_DTYPE_DEVICE,           // Type of this structure
    USBShort(0x110),            // USB version 1.1
    USB_CLASS_HID,              // USB Device Class
    USB_HID_SCLASS_BOOT,        // USB Device Sub-class
    USB_HID_PROTOCOL_KEYB,      // USB Device protocol
    64,                         // Maximum packet size for default pipe
    USBShort(0x1234),           // Vendor ID (VID)
    USBShort(0x5678),           // Product ID (PID)
    USBShort(0x100),            // Device Version BCD
    1,                          // Manufacturer string identifier
    2,                          // Product string identifier
    3,                          // Product serial number
    1                           // Number of configurations
};
HIDKEYDeviceDescriptor = (tDeviceDescriptor    *)g_pHIDDeviceDescriptor;
```

2. 配置描述符

配置描述符是一个 USB 设备配置相关的数据结构体,包括该描述符的字符长度、供电方式、最大耗电量等。主机发出 USBREQ_GET_DESCRIPTOR 请求,且 wValue 值为 USB_DTYPE_CONFIGURATION 时,获取设备描述符,那么此配置包含的所有接口描述符与端点描述符都将发送给 USB 主机。如表 1-3 所列,为 USB 标准配置描述符表。

表 1-3 标准配置描述符表

偏移量	域	大 小	值	描 述
0	bLength	1	数字	字节数
1	bDescriptorType	1	常量	配置描述符类型
2	wTotalLength	2	数字	配置总长（配置、接口、端点、设备类）
4	bNumInterfaces	1	数字	此配置所支持的接口个数
5	bCongfigurationValue	1	数字	USBREQ_SET_CONFIG 请求选定此配置
6	iConfiguration	1	索引	配置的字串索引
7	bmAttributes	1	位图	实际电源模式
8	MaxPower	1	mA	总线电源耗费量，以 2 mA 为一个单位

C 语言配置描述符结构体为：

```c
typedef struct
{
    //配置描述符,长度为 9
    unsigned char bLength;
    //本描述符类型,2,USB_DTYPE_CONFIGURATION
    unsigned char bDescriptorType;
    //配置总长,包括配置、接口、端点、设备类描述符
    unsigned short wTotalLength;
    //支持接口个数
    unsigned char bNumInterfaces;

    // USBREQ_SET_CONFIG 请求选定此配置
    unsigned char bConfigurationValue;
    //配置描述符的字串索引
    unsigned char iConfiguration;
    //电源模式
    unsigned char bmAttributes;
    //最大耗电量,2 mA 为单位
    unsigned char bMaxPower;
}
tConfigDescriptor;
```

bmAttributes 电源模式参数如下：

```
#define USB_CONF_ATTR_PWR_M       0xC0    //方便使用,定义了一个屏蔽参数
#define USB_CONF_ATTR_SELF_PWR    0xC0    //自身供电
#define USB_CONF_ATTR_BUS_PWR     0x80    //总线取电
#define USB_CONF_ATTR_RWAKE       0xA0    //Remove Wakeup
```

配置一个键盘配置描述符如下：

```
tConfigDescriptor * HIDKEYConfigDescriptor;
unsigned char g_pHIDDescriptor[] =
{
    9,                              // Size of the configuration descriptor
    USB_DTYPE_CONFIGURATION,        // Type of this descriptor
    USBShort(34),                   // The total size of this full structure
    1,                              // The number of interfaces in this
                                    // configuration
    1,                              // The unique value for this configuration
    5,                              // The string identifier that describes this
                                    // configuration.
    USB_CONF_ATTR_SELF_PWR,         // Bus Powered, Self Powered, remote wake up
    250,                            // The maximum power is 500 mA
};
HIDKEYConfigDescriptor = (tConfigDescriptor *)pHIDDescriptor;
```

3. 接口描述符

配置描述符中包含了一个或多个接口描述符，接口描述符中的"接口"理解为"功能"的意思，例如一个设备既有鼠标功能又有键盘功能，则这个设备至少就有两个"接口"。

如果配置描述符不止支持一个接口描述符，并且每个接口描述符都有一个或多个端点描述符，那么在响应 USB 主机的配置描述符命令时，USB 设备的端点描述符总是紧跟在相关的接口描述符后面，作为配置描述符的一部分。接口描述符不可用 Set_Descriptor 和 Get_Descriptor 来存取。如果一个接口仅使用端点 0，则接口描述符以后就不再返回端点描述符，并且此接口表现的是一个控制接口的特性，在这种情况下 bNumberEndpoints 域应被设置成 0。接口描述符在说明端点个数时并不把端点 0 计算在内。如表 1-4 所列，为标准接口描述符表。

表 1-4 标准接口描述符表

偏移量	域	大小	值	说明
0	bLength	1	数字	此表的字节数
1	bDescriptorType	1	常量	接口描述表类
2	bInterfaceNumber	1	数字	接口号，从零开始
3	bAlternateSetting	1	数字	索引值
4	bNumEndpoints	1	数字	接口端点数量
5	bInterfaceClass	1	类	类值
6	bInterfaceSubClass	1	子类	子类码
7	bInterfaceProtocol	1	协议	协议码
8	iInterface	1	索引	接口字串描述索引

C 语言接口描述符结构体为：

```c
typedef struct
{
    //接口描述符长度,9 字节
    unsigned char bLength;
    //本描述符类型,4,USB_DTYPE_INTERFACE
    unsigned char bDescriptorType;
    //接口号,从 0 开始编排
    unsigned char bInterfaceNumber;
    //接口索引值
    unsigned char bAlternateSetting;
    //本接口使用除端点 0 外的端点数
    unsigned char bNumEndpoints;
    //USB 接口类码
    unsigned char bInterfaceClass;
    //子类码
    unsigned char bInterfaceSubClass;
    //接口协议
    unsigned char bInterfaceProtocol;
    //描述本接口的字符串索引
    unsigned char iInterface;
}
tInterfaceDescriptor;
```

bInterfaceClass 接口使用类的代码如下：

```c
#define USB_CLASS_DEVICE          0x00
#define USB_CLASS_AUDIO           0x01
#define USB_CLASS_CDC             0x02
#define USB_CLASS_HID             0x03
#define USB_CLASS_PHYSICAL        0x05
#define USB_CLASS_IMAGE           0x06
#define USB_CLASS_PRINTER         0x07
#define USB_CLASS_MASS_STORAGE    0x08
#define USB_CLASS_HUB             0x09
#define USB_CLASS_CDC_DATA        0x0a
#define USB_CLASS_SMART_CARD      0x0b
#define USB_CLASS_SECURITY        0x0d
#define USB_CLASS_VIDEO           0x0e
#define USB_CLASS_HEALTHCARE      0x0f
#define USB_CLASS_DIAG_DEVICE     0xdc
#define USB_CLASS_WIRELESS        0xe0
```

第1章 USB 基础

```
#define USB_CLASS_MISC              0xef
#define USB_CLASS_APP_SPECIFIC      0xfe
#define USB_CLASS_VEND_SPECIFIC     0xff
#define USB_CLASS_EVENTS            0xffffffff
```

例如：定义 USB 键盘接口描述符。

```
tInterfaceDescriptor   * HIDKEYIntDescriptor;
unsigned char g_pHIDInterface[] =
{
    9,                              // Size of the interface descriptor
    USB_DTYPE_INTERFACE,            // Type of this descriptor
    0,                              // The index for this interface
    0,                              // The alternate setting for this interface
    2,                              // The number of endpoints used by this
                                    // interface
    USB_CLASS_HID,                  // The interface class
    USB_HID_SCLASS_BOOT,            // The interface sub-class
    USB_HID_PROTOCOL_KEYB,          // The interface protocol for the sub-class
                                    // specified above
    4,                              // The string index for this interface
};
```

4. 字符串描述符

字符串描述符是可有可无的，如果一个设备无字符串描述符，则所有其他描述符中有关字符串描述表的索引都必须为 0。字符串描述符使用 UNICODE 编码。如表 1-5 所列，为标准字符串描述符表。

表 1-5　标准字符串描述符表

偏移量	域	大小	值	描述
0	bLength	1	数字	此描述表的字节数
1	bDescriptorType	1	常量	字串描述表类型
2	bString	N	数字	UNICODE 编码的字串

C 语言字符串描述符结构体为：

```
typedef struct
{
    //字符串描述总长度
    unsigned char bLength;
    //本描述符类型 USB_DTYPE_STRING (3)
    unsigned char bDescriptorType;
```

```c
    //unicode 字符串
    unsigned char bString;
}
tStringDescriptor;
```

例如:一设备的所有字符串描述符。

```c
// ****************************************************************
// USB 键盘语言描述
// ****************************************************************
const unsigned char g_pLangDescriptor[] =
{
    4,
    USB_DTYPE_STRING,
    USBShort(USB_LANG_EN_US)
};
// ****************************************************************
// 制造商 字符串 描述
// ****************************************************************
const unsigned char g_pManufacturerString[] =
{
    (11 + 1) * 2,
    USB_DTYPE_STRING,
    'O', 0, 'g', 0, 'a', 0, 'w', 0, 'a', 0, 's', 0, 't', 0, 'u', 0, 'd', 0,
    'i', 0, 'o', 0,
};
// ****************************************************************
//产品 字符串 描述
// ****************************************************************
const unsigned char g_pProductString[] =
{
    (17 + 1) * 2,
    USB_DTYPE_STRING,
    'K', 0, 'e', 0, 'y', 0, 'b', 0, 'o', 0, 'a', 0, 'r', 0, 'd', 0, ' ', 0,
    'F', 0, 'o', 0, 'r', 0, ' ', 0, 'p', 0, 'a', 0, 'u', 0, 'l', 0
};
// ****************************************************************
//产品 序列号 描述
// ****************************************************************
const unsigned char g_pSerialNumberString[] =
{
    (7 + 1) * 2,
    USB_DTYPE_STRING,
```

```c
    '6',0,'6',0,'7',0,'2',0,'1',0,'1',0,'5',0
};
//***************************************************************
//设备接口字符串描述
//***************************************************************
const unsigned char g_pHIDInterfaceString[] =
{
    (22 + 1) * 2,
    USB_DTYPE_STRING,
    'H',0,'I',0,'D',0,' ',0,'K',0,'e',0,'y',0,'b',0,
    'o',0,'a',0,'r',0,'d',0,' ',0,'I',0,'n',0,'t',0,
    'e',0,'r',0,'f',0,'a',0,'c',0,'e',0
};
//***************************************************************
//设备配置字符串描述
//***************************************************************
const unsigned char g_pConfigString[] =
{
    (26 + 1) * 2,
    USB_DTYPE_STRING,
    'H',0,'I',0,'D',0,' ',0,'K',0,'e',0,'y',0,'b',0,
    'o',0,'a',0,'r',0,'d',0,' ',0,'C',0,'o',0,'n',0,
    'f',0,'i',0,'g',0,'u',0,'r',0,'a',0,'t',0,'i',0,
    'o',0,'n',0
};
//***************************************************************
//字符串描述集合
//***************************************************************
const unsigned char * const g_pStringDescriptors[] =
{
    g_pLangDescriptor,
    g_pManufacturerString,
    g_pProductString,
    g_pSerialNumberString,
    g_pHIDInterfaceString,
    g_pConfigString
};
```

也可以使用专用的软件生成 USB 字符串描述符,从而简化字符串描述符的生成,特别是在使用中文字符串描述符时,使用专用的软件生成更简洁、更方便。

5. 端点描述符

每个接口使用的端点都有自己的描述符,此描述符被主机用来决定每个端点的带宽需求。每个端点的描述符总是作为配置描述符的一部分，端点 0 无描述符。如表 1-6 所列，为标准端点描述表。

表 1-6 标准端点描述符表

偏移量	域	大小	值	说明
0	bLength	1	数字	字节数
1	bDescriptorType	1	常量	端点描述符类型
2	bEndpointAddress	1	端点	端点的地址及方向
3	bmAttributes	1	位图	传送类型
4	wMaxPacketSize	2	数字	端点能够接收或发送的最大数据包的大小
6	bInterval	1	数字	数据传送端点的时间间隙

C 语言端点描述符结构体为：

```
typedef struct
{
    //本描述符总长度
    unsigned char bLength;
    //描述符类型 USB_DTYPE_ENDPOINT (5)
    unsigned char bDescriptorType;
    //端点地址及方向
    unsigned char bEndpointAddress;
    //传输类型
    unsigned char bmAttributes;
    //传输包最大长度
    unsigned short wMaxPacketSize;
    //时间间隔
    unsigned char bInterval;
}
tEndpointDescriptor;
```

(1) bEndpointAddress 定义端点地址及方向，方向参数如下：

```
#define USB_EP_DESC_OUT         0x00
#define USB_EP_DESC_IN          0x80
```

(2) bmAttributes 定义端点传输类型：

```
#define USB_EP_ATTR_CONTROL     0x00
#define USB_EP_ATTR_ISOC        0x01
#define USB_EP_ATTR_BULK        0x02
#define USB_EP_ATTR_INT         0x03
#define USB_EP_ATTR_TYPE_M      0x03
```

第 1 章 USB 基础

```
#define USB_EP_ATTR_ISOC_M            0x0c
#define USB_EP_ATTR_ISOC_NOSYNC       0x00
#define USB_EP_ATTR_ISOC_ASYNC        0x04
#define USB_EP_ATTR_ISOC_ADAPT        0x08
#define USB_EP_ATTR_ISOC_SYNC         0x0c
#define USB_EP_ATTR_USAGE_M           0x30
#define USB_EP_ATTR_USAGE_DATA        0x00
#define USB_EP_ATTR_USAGE_FEEDBACK    0x10
#define USB_EP_ATTR_USAGE_IMPFEEDBACK 0x20
```

键盘端点描述符为:

```
const unsigned char g_pHIDInEndpoint[] =
{
    7,                        // The size of the endpoint descriptor
    USB_DTYPE_ENDPOINT,       // Descriptor type is an endpoint.
    USB_EP_DESC_IN | 3,       // Endpoint 3→Input
    USB_EP_ATTR_INT,          // Endpoint is an interrupt endpoint
    USBShort(0x8),            // The maximum packet size
    16,                       // The polling interval for this endpoint
};
const unsigned char g_pHIDOutEndpoint[] =
{
    7,                        // The size of the endpoint descriptor
    USB_DTYPE_ENDPOINT,       // Descriptor type is an endpoint
    USB_EP_DESC_OUT | 3,      // Endpoint 3→Output
    USB_EP_ATTR_INT,          // Endpoint is an interrupt endpoint
    USBShort(0x8),            // The maximum packet size
    16,                       // The polling interval for this endpoint
};
```

g_pHIDInEndpoint 定义了端点 3 方向为输入,g_pHIDOutEndpoint 定义了端点 3 方向为输出,虽然都是"端点 3",但使用的 FIFO 不一样,实际为两个端点。尤如高速路出口,一进一出,但是两个路口。注意,每个接口中使用的端点,相应端点描述符必须紧跟接口描述符返回给主机。

由此可以看出,USB 的描述符之间的关系是一层一层的,最上一层是设备描述符,紧接着是配置描述符,下面是接口描述符,再下面是端点描述符。在获取描述符时,先获取设备描述符,然后再获取配置描述符,根据配置描述符中的配置集合长度,将配置描述符、接口描述符、端点描述符一起一次读回。其中可能还会获取设备序列号、厂商字符串和产品字符串等。

1.4.3 设备枚举过程

枚举是 USB 主机通过一系列命令要求设备发送描述符信息,标准 USB 设备有

5 种描述符:设备描述符、配置描述符、字符串描述符、接口描述符以及端点描述符。枚举过程就是对设备识别的过程。

① 首先,USB 主机检测到有 USB 设备插入,会对设备复位。USB 设备复位后其地址都为 0,主机就可以和刚刚插入的设备通过 0 地址的端点 0 进行通信。

② USB 主机对设备发送获取设备描述符的标准请求,设备收到请求后,将设备描述符发给主机。

③ 主机对总线进行复位,之后发送 USBREQ_SET_ADDRESS 请求,设置设备地址。

④ 主机发送请求到新的 USB 地址,并再获取设备描述符的标准请求。

⑤ 主机获取 USB 设备配置信息,包括配置描述符、接口描述符、设备类描述符(如 HID 设备)、端点描述符等。

⑥ 主机根据已获得的描述符,请求相关字符串描述符。

对于 HID 设备,主机还会请求 HID 报告描述符。此时枚举完成,同时 USB 设备初始化工作完成。可以通过端点与主机通信。

例如:C 语言枚举程序。

定义 USB 标准请求调用函数,方便函数管理:

```
typedef void ( * tStdRequest)(void * pvInstance, tUSBRequest * pUSBRequest);
static const tStdRequest g_psUSBDStdRequests[] =
{
    USBDGetStatus,
    USBDClearFeature,
    0,
    USBDSetFeature,
    0,
    USBDSetAddress,
    USBDGetDescriptor,
    USBDSetDescriptor,
    USBDGetConfiguration,
    USBDSetConfiguration,
    USBDGetInterface,
    USBDSetInterface,
    USBDSyncFrame
};
```

枚举处理主函数如下:

```
void USBDeviceEnumHandler(tDeviceInstance * pDevInstance)
{
    unsigned long ulEPStatus;
    //获取端点 0 的中断情况,并清 0
```

```c
ulEPStatus = USBEndpointStatus(USB0_BASE, USB_EP_0);
//判断端点 0 的当前状态
switch(pDevInstance->eEP0State)
{
    //如果处理于等待状态
    case USB_STATE_STATUS:
    {
        //修改端点 0 为空闲状态
        pDevInstance->eEP0State = USB_STATE_IDLE;
        // 检查地址是否改变
        if(pDevInstance->ulDevAddress & DEV_ADDR_PENDING)
        {
            //设置设备地址
            pDevInstance->ulDevAddress &= ~DEV_ADDR_PENDING;
            USBDevAddrSet(USB0_BASE, pDevInstance->ulDevAddress);
        }
        //判断端点 0 是否有数据要接收
        if(ulEPStatus & USB_DEV_EP0_OUT_PKTRDY)
        {
            //接收并处理
            USBDReadAndDispatchRequest(0);
        }
        break;
    }
    //端点 0 空闲状态
    case USB_STATE_IDLE:
    {
        //判断端点 0 是否有数据要接收
        if(ulEPStatus & USB_DEV_EP0_OUT_PKTRDY)
        {
            //接收并处理
            USBDReadAndDispatchRequest(0);
        }
        break;
    }
    //端点 0 数据发送阶段
    case USB_STATE_TX:
    {
        USBDEP0StateTx(0);
        break;
    }
    // 发送端点 0 配置
```

```c
case USB_STATE_TX_CONFIG:
{
    USBDEP0StateTxConfig(0);
    break;
}
//接收阶段
case USB_STATE_RX:
{
    unsigned long ulDataSize;
    if(pDevInstance->ulEP0DataRemain > EP0_MAX_PACKET_SIZE)
    {
        ulDataSize = EP0_MAX_PACKET_SIZE;
    }
    else
    {
        ulDataSize = pDevInstance->ulEP0DataRemain;
    }
    USBEndpointDataGet(USB0_BASE, USB_EP_0, pDevInstance->pEP0Data,
                    &ulDataSize);
    if(pDevInstance->ulEP0DataRemain < EP0_MAX_PACKET_SIZE)
    {
        USBDevEndpointDataAck(USB0_BASE, USB_EP_0, true);
        pDevInstance->eEP0State =  USB_STATE_IDLE;
        if((pDevInstance->psInfo->sCallbacks.pfnDataReceived) &&
            (pDevInstance->ulOUTDataSize != 0))
        {
            pDevInstance->psInfo->sCallbacks.pfnDataReceived(
                pDevInstance->pvInstance,
                pDevInstance->ulOUTDataSize);
            pDevInstance->ulOUTDataSize = 0;
        }
    }
    else
    {
        USBDevEndpointDataAck(USB0_BASE, USB_EP_0, false);
    }
    pDevInstance->pEP0Data += ulDataSize;
    pDevInstance->ulEP0DataRemain -= ulDataSize;

    break;
}
```

```c
            // STALL 状态
            case USB_STATE_STALL:
            {
                //发送 STALL 包
                if(ulEPStatus & USB_DEV_EP0_SENT_STALL)
                {
                    USBDevEndpointStatusClear(USB0_BASE, USB_EP_0,
                                        USB_DEV_EP0_SENT_STALL);
                    pDevInstance->eEP0State = USB_STATE_IDLE;
                }
                break;
            }
            default:
            {
                ASSERT(0);
            }
        }
    }
```

枚举主要发生在端点 0 处于 USB_STATE_STATUS 和 USB_STATE_IDLE 阶段，USBDReadAndDispatchRequest(0)获取主机发送的请求，在内部进行主机请求解析并响应。USBDReadAndDispatchRequest()函数如下：

```c
USBDReadAndDispatchRequest(unsigned long ulIndex)
{
    unsigned long ulSize;
    tUSBRequest * pRequest;
    pRequest = (tUSBRequest *)g_pucDataBufferIn;
    //端点 0 的最大数据包大小
    ulSize = EP0_MAX_PACKET_SIZE;
    //获取端点 0 中的数据
    USBEndpointDataGet(USB0_BASE,
                    USB_EP_0,
                    g_pucDataBufferIn,
                    &ulSize);
    //判断是否有数据读出
    if(! ulSize)
    {
        return;
    }
    //判断是否是标准请求
    if((pRequest->bmRequestType & USB_RTYPE_TYPE_M) != USB_RTYPE_STANDARD)
```

```c
{
    //如果不是标准请求,通过callback函数返回给用户处理
    if(g_psUSBDevice[0].psInfo->sCallbacks.pfnRequestHandler)
    {
        g_psUSBDevice[0].psInfo->sCallbacks.pfnRequestHandler(
            g_psUSBDevice[0].pvInstance, pRequest);
    }
    else
    {
        USBDCDStallEP0(0);
    }
}
else
{
    //标准请求,通过g_psUSBDStdRequests调用标准请求处理函数
    if((pRequest->bRequest <
        (sizeof(g_psUSBDStdRequests) / sizeof(tStdRequest))) &&
        (g_psUSBDStdRequests[pRequest->bRequest] != 0))
    {
        g_psUSBDStdRequests[pRequest->bRequest](&g_psUSBDevice[0],
                                                pRequest);
    }
    else
    {
        USBDCDStallEP0(0);
    }
}
}
```

标准请求中 USBDGetDescriptor 函数调用最多,用于描述符获取,使主机充分了解设备特性,是枚举的重要组成部分,下面是 USBDGetDescriptor 函数:

```c
static void USBDGetDescriptor(void * pvInstance, tUSBRequest * pUSBRequest)
{
    tBoolean bConfig;
    tDeviceInstance * psUSBControl;
    tDeviceInfo * psDevice;
    psUSBControl = (tDeviceInstance *)pvInstance;
    psDevice = psUSBControl->psInfo;
    USBDevEndpointDataAck(USB0_BASE, USB_EP_0, false);
    bConfig = false;
    //通过 Value 判断获取什么描述符
    switch(pUSBRequest->wValue >> 8)
```

```c
        {
            //获取设备描述符
            case USB_DTYPE_DEVICE:
            {
                //把设备描述放入 PE0 中,等待发送
                psUSBControl->pEP0Data =
                    (unsigned char *)psDevice->pDeviceDescriptor;
                psUSBControl->ulEP0DataRemain = psDevice->pDeviceDescriptor[0];
                break;
            }
            //获取配置描述符
            case USB_DTYPE_CONFIGURATION:
            {
                const tConfigHeader * psConfig;
                const tDeviceDescriptor * psDeviceDesc;
                unsigned char ucIndex;
                ucIndex = (unsigned char)(pUSBRequest->wValue & 0xFF);
                psDeviceDesc =
                    (const tDeviceDescriptor *)psDevice->pDeviceDescriptor;
                if(ucIndex >= psDeviceDesc->bNumConfigurations)
                {
                    USBDCDStallEP0(0);
                    psUSBControl->pEP0Data = 0;
                    psUSBControl->ulEP0DataRemain = 0;
                }
                else
                {
                    psConfig = psDevice->ppConfigDescriptors[ucIndex];
                    psUSBControl->ucConfigSection = 0;
                    psUSBControl->ucSectionOffset = 0;
                    psUSBControl->pEP0Data = (unsigned char *)
                                    psConfig->psSections[0]->pucData;
                    psUSBControl->ulEP0DataRemain =
                                    USBDCDConfigDescGetSize(psConfig);
                    psUSBControl->ucConfigIndex = ucIndex;
                    bConfig = true;
                }
                break;
            }
            //字符串描述符
            case USB_DTYPE_STRING:
            {
```

```c
        long lIndex;
        lIndex = USBDStringIndexFromRequest(pUSBRequest->wIndex,
                                            pUSBRequest->wValue & 0xFF);
        if(lIndex == -1)
        {
            USBDCDStallEP0(0);
            break;
        }
        psUSBControl->pEP0Data =
            (unsigned char *)psDevice->ppStringDescriptors[lIndex];
        psUSBControl->ulEP0DataRemain =
            psDevice->ppStringDescriptors[lIndex][0];
        break;
    }
    default:
    {
        if(psDevice->sCallbacks.pfnGetDescriptor)
        {
            psDevice->sCallbacks.pfnGetDescriptor(psUSBControl->
                                            pvInstance,pUSBRequest);
            return;
        }
        else
        {
            USBDCDStallEP0(0);
        }
        break;
    }
}
//判断是否有数据要发送
if(psUSBControl->pEP0Data)
{
    if(psUSBControl->ulEP0DataRemain > pUSBRequest->wLength)
    {
        psUSBControl->ulEP0DataRemain = pUSBRequest->wLength;
    }
    if(! bConfig)
    {
        //发送上面的配置信息
        USBDEP0StateTx(0);
    }
    else
```

```
            {
                //发送端点0的状态信息
                USBDEP0StateTxConfig(0);
            }
        }
    }
```

下面是一个枚举过程：

```
Start Demo：
Init Hardware................
LED && KEY   Ok!.............
USB_STATE_IDLE...............            (端点0处于USB_STATE_IDLE状态)
0x0008  0x0080  0x0006  0x0100  0x0000  0x0040   (主机请求数据:数据长度+数据包
                                                  内容)
USBDGetDescriptor...............         (解析为获取描述符)
USBDGetDescriptor USB_DTYPE_DEVICE............  (获取设备描述符,并发送设备描述
                                                  符)
USB_STATE_STATUS..............           (端点0处于USB_STATE_STATUS状态)
USB_STATE_IDLE................           (端点0处于USB_STATE_IDLE状态)
0x0008  0x0000  0x0005  0x0002  0x0000  0x0000   (主机请求数据:数据长度+数据包内
                                                  容)
USBDSetAddress................           (设置设备地址请求)
USB_STATE_STATUS..............
DEV is 0x0002.................           (设置设备地址为2)
USB_STATE_IDLE................
0x0008  0x0080  0x0006  0x0100  0x0000  0x0012
USBDGetDescriptor..............
USBDGetDescriptor USB_DTYPE_DEVICE..............(设置地址后,再获取设备描述符)
USB_STATE_STATUS..............
USB_STATE_IDLE................
0x0008  0x0080  0x0006  0x0200  0x0000  0x0009
USBDGetDescriptor..............
USBDGetDescriptor USB_DTYPE_CONFIGURATION..............(获取配置描述符)
USB_STATE_STATUS..............
USB_STATE_IDLE................
0x0008  0x0080  0x0006  0x0300  0x0000  0x00ff
USBDGetDescriptor..............
USBDGetDescriptor USB_DTYPE_STRING..............   (获取字符串描述符)
USB_STATE_STATUS..............
USB_STATE_IDLE................
0x0008  0x0080  0x0006  0x0303  0x0409  0x00ff
USBDGetDescriptor..............
```

```
USBDGetDescriptor USB_DTYPE_STRING..............
String is 0x0003,Vaule is 0x0409,Index is 0x0003
USB_STATE_STATUS..............
USB_STATE_IDLE..............
0x0008  0x0080  0x0006  0x0200  0x0000  0x00ff
USBDGetDescriptor..............
USBDGetDescriptor USB_DTYPE_CONFIGURATION..............
USB_STATE_STATUS..............
USB_STATE_IDLE..............
0x0008  0x0080  0x0006  0x0300  0x0000  0x00ff
USBDGetDescriptor..............
USBDGetDescriptor USB_DTYPE_STRING..............
USB_STATE_STATUS..............
USB_STATE_IDLE..............
0x0008  0x0080  0x0006  0x0302  0x0409  0x00ff
USBDGetDescriptor..............
USBDGetDescriptor USB_DTYPE_STRING..............
String is 0x0002,Vaule is 0x0409,Index is 0x0002
USB_STATE_STATUS..............
USB_STATE_IDLE..............
0x0008  0x0080  0x0006  0x0300  0x0000  0x00ff
USBDGetDescriptor..............
USBDGetDescriptor USB_DTYPE_STRING..............
USB_STATE_STATUS..............
USB_STATE_IDLE..............
0x0008  0x0080  0x0006  0x0302  0x0409  0x00ff
USBDGetDescriptor..............
USBDGetDescriptor USB_DTYPE_STRING..............
String is 0x0002,Vaule is 0x0409,Index is 0x0002
USB_STATE_STATUS..............
USB_STATE_IDLE..............
USB_STATE_IDLE..............
```

以上是枚举的实际过程,从上面的枚举过程中可以看出枚举的一般过程:上电检测→获取设备描述符→设置设备地址→再次获取设备描述符→获取配置描述符→获取字符串描述符→枚举成功。对于不同系统,枚举过程是不一样的,以上是在 Windows xp 下的枚举过程,在 Linux 下,枚举过程有很大不同。

1.5 USB 主机开发流程

如图 1-4 所示,为 USB 主机开发流程图。USB 主机开发流程相对复杂,①首

第1章 USB 基础

先确认 USB 系统开发类型,是 USB 主机、USB 设备还是 OTG;②如果确定是 USB 主机,必须明确控制 USB 设备的类型:HID、UDIO、CDC、HUB、IMAGE 等,编写驱动;③查找相关设备手册,控制 USB 设备枚举;④编写应用程序,在枚举成功后,主要是进行数据处理,控制 USB 设备。以上就是 USB 主机开发的主要流程,其中最主要的是驱动编写,它确定了能否与 USB 设备进行正常通信,如果驱动不了 USB 设备,一切空谈。

图 1-4 USB 主机开发流程图

1.6 USB OTG 介绍

USB OTG 是最近几年才发展起来的一种新的 USB 技术,主要解决移动、手持设备等设备的数据处理、设备控制问题。在之前,某一个 USB 系统确定为主机或者设备后,不能进行变换,是 USB 主机,将永远是 USB 主机,想变成 USB 设备是很困难的,就像普通 PC 机一样,它具有 USB 功能,但只是 USB 主机,不能用作 USB 设备。为了解决这个问题,让一个设备既有主机功能,又有设备功能,便出现了 USB OTG。现在所有手持仪器、便携式设备都具有 USB 主机功能,也有设备功能,比如平板电脑,它可以作为 U 盘,存储器存储文件,同时也是 USB 主机,可以读取 U 盘信息。在 USB OTG 中,主机称作为 A 设备,设备称为 B 设备,但是该系统任意时刻只能为 A 设备或者 B 设备,在 USB OTG 中没有 USB 主机与 USB 设备的名称,取而代之叫 A 设备、B 设备,只是具有主机与设备的功能而已。

USB OTG 使用 mini USB 接头,此接头除了具有普通 USB 头的 4 条线之外,还有一条控制线——INDEX 线,此线确定该设备为 A 设备还是 B 设备,同时也是 A、B 设备切换的物理控制线路。USB OTG 开发流程相对简单,可以看成是 USB 主机与设备的综合开发,其难点是控制 A、B 设备切换的过程。

1.7 小 结

本章介绍了 USB 的基本概念,初步讲解了 USB 的相关术语、USB 主机、USB 设备、USB OTG 开发流程及注意事项,为后面 USB 系统开发打下基础。通过本章的学习,可以让读者初步了解 USB 的概念、USB 的优点,可让读者初步建立 USB 工作模式与开发流程。本章跳过了 USB 协议,如果读者想了解更详细的 USB 协议,请参考 USB 官方数据手册。同时,本书也会在后面的章节陆续提及相关协议。

第 2 章

Stellaris 的 USB 处理器

Stellaris 的 USB 处理器支持 USB 2.0 全速 USB 主机、设备、OTG,此次推出的产品在 ARM Cortex - M3 内核的基础上集成了 USB 功能。此外,LM3S3000 系列的每个 MCU 都提供了多个已预先编入节省内存的 ROM 的 StellarisWare 软件特性。

第 1 章介绍了 USB 基础知识,本章主要介绍 Luminary 的 Stellaris 所提供 USB 处理器、USB 主机、USB 设备、USB OTG、USB 寄存器、USB 处理器等一系列与 USB 开发相关的基础知识。通过本章的学习,可以加深对使用 Stellaris 所提供的 USB 处理器进行简单 USB 系统开发的了解。

2.1 Stellaris 处理器简介

Luminary Stellaris 系列微控制器包含运行在 50 MHz 频率下的 ARM Cortex - M3 MCU 内核、嵌入式 Flash 和 SRAM、一个低压降的稳压器(LDO)、集成掉电复位(BOR)和上电复位功能(POR)、模拟比较器、10 位 ADC、SSI、GPIO、看门狗、通用定时器(GPTM)、UART、I^2C、运动控制 PWM 以及正交编码器(quadrature encoder)输入。提供的外设直接通向引脚,不需要特性复用,这个丰富的特性集非常适合楼宇和家庭自动化、工厂自动化和控制、工控电源设备、步进电机、有刷和无刷 DC 电机和 AC 感应电动机等应用。

LM3S 系统微控制的分类如下:

(1) 第 1 代:Sandstorm Class(沙暴系列)——包括 LM3S100 系列、LM3S300 系列、LM3S600 系列、LM3S800 系列。

➢ 唾手可取的首款 ARM Cortex - M3 内核控制器。

➢ 起价仅 $1。

➢ 最高工作频率达 50 MHz,单周期存储器最高到 64 KB Flash/8 KB SRAM。

➢ 高度的 IP 融合带来精细的运动控制性能。

(2) 第 2 代:Fury Class(狂暴系列)——包括 LM3S1000 系列、LM3S2000 系列、LM3S6000 系列以及 LM3S8000 系列。

➢ 扩展了 Sandstorm 系列,整合了以太网 MAC 物理层和 CAN 控制器。

第 2 章 Stellaris 的 USB 处理器

- 将存储空间增大到 256 KB Flash/64KB SRAM。
- 进一步优化了电池供电的应用场合。
- 增加了更多的外设诸如 UART、I^2C、SSI 和 QEI。

(3) 第 3 代:Dust Devil Class(旋风系列)。
- 包括 LM3S1000 系列,LM3S3000 系列以及 LM3S5000 系列。
- 改进的 Stellaris 系列提供了整合的 USB OTG、主机、从机可供选择。
- 增加了 DMA,提高了 GPIO 的驱动能力,增加了额外的 PWM 输出。
- 为高级运动控制增加了错误输入检测。
- BootLoader 和外设驱动库被固化在了 ROM 中。
- 提供了小型的 LQFP 封装的选择(64 pin LQFP)。

(4) 第 4 代:Tempest Class(飓风系列)。
- 包括 LM3S2000 系列,LM3S5000 系列以及 LM3S9000 系列。
- 在较低的功耗下获得了更高的性能(80 MHz 和 100 MHz,1.2 V internal supplies)。
- 提供了强大的外部高速芯片级总线互连。
- 增加了增强的子系统:双 ADCs,扩展 ROM 固化软件,精密振荡器,I^2S 接口。
- 扩展了以太网的网络连通性,CAN 和 USB 的选择和互连。

下面以 LM3S8962 为参照讲解其特性,其他型号处理器请参阅对应数据手册。

1. 32 位 RISC 性能

- 采用为小封装应用方案而优化的 32 位 ARM Cortex–M3 v7M 架构。
- 提供系统时钟、包括一个简单的 24 位写清零、递减、自装载计数器,同时具有灵活的控制机制。
- 仅采用与 Thumb 兼容的 Thumb–2 指令集以获取更高的代码密度。
- 工作频率为 50 MHz。
- 硬件除法和单周期乘法。
- 集成嵌套向量中断控制器(NVIC),使中断的处理更为简捷。
- 36 中断具有 8 个优先等级。
- 带存储器保护单元(MPU),提供特权模式来保护操作系统的功能。
- 非对齐式数据访问,使数据能够更为有效地安置到存储器中。
- 精确的位操作(bit–banding),不仅最大限度地利用了存储器空间而且还改良了对外设的控制。

2. 内部存储器

- 256 KB 单周期 Flash。
- 可由用户管理 对 Flash 块的保护,以 2 KB 为单位。
- 可由用户管理对 Flash 的编程。

- 可由用户定义和管理的 Flash 保护块。
- 64 KB 单周期访问的 SRAM。

3. 通用定时器

① 4 个通用定时器模块(GPTM),每个模块都能提供 2 个 16 位的定时器/计数器。每个通用定时器模块都能够被设置为独立运作的定时器或事件计数器(总共有 8 个);可用作单个 32 位的定时器(最多 4 个)或者用作单个 32 位的实时时钟(RTC)以捕获事件,用作脉宽调制输出(PWM),或者用作模数转换的触发器。

② 32 位定时器模式。
- 可编程单次触发定时器。
- 可编程周期定时器。
- 当接入 32.768 kHz 外部时钟输入时可作为实时时钟使用。
- 在调试时,当控制器发出 CPU 暂停标志时,用户可以设定暂停定时器的周期或单次触发模式。
- ADC 事件触发器。

③ 16 位定时器模式。
- 通用定时器功能,并带一个 8 位的预分频器。
- 可编程单次触发定时器。
- 可编程周期定时器。
- 在调试时,当控制器发出 CPU 暂停标志时,用户可设定暂停周期或者单次模式下的计数。
- ADC 事件触发器。

④ 16 位输入捕获模式。
- 提供输入边沿计数捕获功能。
- 提供输入边沿时间捕获功能。

⑤ 16 位 PWM 模式。
- 简单的 PWM 模式,对 PWM 信号输出的取反可由软件编程决定。

4. 兼容 ARM FiRM 的看门狗定时器

- 32 位向下计数器,带可编程的装载寄存器。
- 带使能功能的独立看门狗时钟。
- 带中断屏蔽功能的可编程中断产生逻辑。
- 软件跑飞时可锁定寄存器以提供保护。
- 带使能/禁能的复位产生逻辑。
- 在调试时,当控制器发出 CPU 暂停标志时,用户可以设定暂停定时器的周期。

5. CAN

- 支持 CAN 协议版本 2.0 PartA/B。

- 传输位速率可达 1 Mbps。
- 32 个消息对象,每个都带有独立的标识符屏蔽。
- 可屏蔽的中断。
- 可禁止 TTCAN 的自动重发模式。
- 可编程设定的自循检操作。

6. 10/100 以太网控制器

- 符合 IEEE 802.3-2002 规范。
- 在 100 Mbps 和 10 Mbps 速率运作下支持全双工和半双工的运作方式。
- 集成 10/100 Mbps 收发器(PHY 物理层)。
- 自动 MDI/MDI-X 交叉校验。
- 可编程 MAC 地址。
- 节能和断电模式。

7. 同步串行接口(SSI)

- 主机或者从机方式运作。
- 可编程控制的时钟位速率和预分频。
- 独立的发送和接收 FIFO,8×16 位宽的深度。
- 可编程控制的接口,可与 Freescale 的 SPI 接口,MICROWIRE 或者 TI 器件的同步串行接口相连。
- 可编程决定数据帧的大小,范围为 4～16 位。
- 内部循环自检模式可用于诊断/调试。

8. UART

- 3 个完全可编程控制的 16C550 型 UART,支持 IrDA。
- 带有独立的 16×8 发送(TX)以及 16×12 接收(RX)FIFO,可减轻 CPU 中断服务的负担。
- 可编程的波特率产生器,并带有分频器。
- 可编程设置 FIFO 长度,包括 1 字节深度的操作,以提供传统的双缓冲接口。
- FIFO 触发水平可设为 1/8、1/4、1/2、3/4、和 7/8。
- 标准异步通信位:开始位、停止位、奇偶位。
- 无效起始位检测。
- 行中止的产生和检测。

9. ADC

- 独立和差分输入配置。
- 用作单端输入时有 3 个 10 位的通道(输入)。
- 采样速率为 500 000 次/秒。
- 灵活、可配置的模数转换。

第 2 章　Stellaris 的 USB 处理器

- 4 个可编程的采样转换序列，1~8 个入口长，每个序列均带有相应的转换结果 FIFO。
- 每个序列都可以由软件或者内部事件(定时器、模拟比较器、PWM 或 GPIO)触发。
- 片上温度传感器。

10. 模拟比较器

- 3 个独立集成的模拟比较器。
- 可以把输出配置为：驱动输出引脚、产生中断或启动 ADC 采样序列。
- 比较两个外部引脚输入或者将外部引脚输入与内部可编程参考电压相比较。

11. I²C

- 在标准模式下主机和从机接收和发送操作的速度可达 100 kbps，在快速模式下可达 400 kbps。
- 中断的产生。
- 主机带有仲裁和时钟同步功能、支持多个主机以及 7 位寻址模式。

12. PWM

① 2 个 PWM 信号发生模块，每个模块都带有 1 个 16 位的计数器、2 个比较器、1 个 PWM 信号发生器以及 1 个死区发生器。

② 1 个 16 位的计数器。

- 运行在递减或递增/递减模式。
- 输出频率由一个 16 位的装载值控制。
- 可同步更新装载值。
- 当计数器的值到达零或者装载值的时候产成输出信号。

③ 2 个 PWM 比较器。

- 比较器值的更新可以同步。
- 在匹配的时候产生输出信号。

④ PWM 信号发生器。

- 根据计数器和 PWM 比较器的输出信号来产生 PWM 输出信号。
- 可产生两个独立的 PWM 信号。

⑤ 死区发生器。

- 产生 2 个带有可编程死区延时的 PWM 信号，适合驱动半 H 桥(half - H bridge)。
- 可以被旁路，不修改输入 PWM 信号。

⑥ 灵活的输出控制模块，每个 PWM 信号都具有 PWM 输出使能。

- 每个 PWM 信号都具有 PWM 输出使能。
- 每个 PWM 信号都可以选择将输出反相(极性控制)。

- 每个 PWM 信号都可以选择进行故障处理。
- PWM 发生器模块的定时器同步。
- PWM 发生器模块的定时器/比较器更新同步。
- PWM 发生器模块中断状态被汇总。

⑦ 可启动一个 ADC 采样序列。

13. QEI

- 硬件位置积分器追踪编码器的位置。
- 使用内置的定时器进行速率捕获。
- 在出现索引脉冲、速度定时器时间到、方向改变以及检测到正交错误时产生中断。

14. GPIO

① 高达 5~43 个 GPIO，具体数目取决于配置。
② 输入/输出可承受 5 V。
③ 中断产生可编程为边沿触发或电平检测。
④ 在读和写操作中通过地址线进行位屏蔽。
⑤ 可启动一个 ADC 采样序列。
⑥ GPIO 端口配置的可编程控制。

- 弱上拉或下拉电阻。
- 2 mA、4 mA 和 8 mA 端口驱动。
- 8 mA 驱动的斜率控制。
- 开漏使能。
- 数字输入使能。

15. 功　率

- 片内低压差（LDO）稳压器，具有可编程的输出电压，用户可调节的范围为 2.25~2.75 V。
- 休眠模块处理 3.3 V 通电/断电序列，并控制内核的数字逻辑和模拟电路。
- 控制器的低功耗模式：睡眠模式和深度睡眠模式。
- 外设的低功耗模式：软件控制单个外设的关断。
- LDO 带有检测不可调整电压和自动复位的功能，可由用户控制使能。
- 3.3 V 电源掉电检测，可通过中断或复位来报告。

16. 灵活的复位源

- 上电复位。
- 复位引脚有效。
- 掉电（BOR）检测器向系统发出电源下降的警报。
- 软件复位。

第2章 Stellaris 的 USB 处理器

- 看门狗定时器复位。
- 内部低压差(LDO)稳压器输出变为不可调整。

17. 其他特性

- 6 个复位源。
- 可编程的时钟源控制。
- 可对单个外设的时钟进行选通以节省功耗。
- 遵循 IEEE 1149.1-1990 标准的测试访问端口(TAP)控制器。
- 通过 JTAG 和串行线接口进行调试访问。
- 完整的 JTAG 边界扫描。

工业范围内遵循 RoHS 标准的 100 脚 LQFP 封装。

Luminary 的 Stellaris 提供的一系列微控制器是首款基于 Cortex-M3 的控制器,它们为对成本尤其敏感的嵌入式微控制器应用方案带来了高性能的 32 位运算能力。这些具备领先技术的芯片使用户能够以传统的 8 位和 16 位器件的价位来享受 32 位的性能,而且所有型号都以小占位面积的封装形式提供。因为 Luminary 被 TI 公司收购,所以 LM3S 系列产品都归 TI 公司所有,并在芯片上印有 TI 公司的 logo。图 2-1 为 LM3S9B96 封装图,100 引脚 LQFP 封装。

图 2-1　LM3S9B96 封装图

图 2-2 为 LM3S5749 引脚图,100 引脚 LQFP 封装,此类 ARM Cortex-M3 处理器提供了一个高性能、低成本的平台,可满足系统对降低存储需求、简化引脚数以及降低功耗 3 方面的要求,与此同时,它还提供出色的计算性能和优越的系统中断响应能力。

Stellaris 系列芯片能够提供高效的性能、广泛的集成功能以及按照要求定位的选择,适用于各种关注成本并明确要求具有过程控制以及连接能力的应用方案。LM3S3000、LM3S5000、LM9000 系列,支持最大主频为 80 MHz,128 KB Flash,32~64 KB SRAM,LQFP-64/LQFP-100 封装,集成 CAN 控制器、睡眠模块、正交编码器、ADC、带死区 PWM、温度传感器、模拟比较器、UART、SSI、通用定时器、I^2C、CCP、DMA 控制器等片内外设,芯片内部固化驱动库,支持 USB Host/Device/OTG 功能,可运行在全速和低速模式,它符合 USB 2.0 标准,包含挂起和唤醒信号。它包含 32 个端点,其中包含 2 个用于控制传输的专用连接端点(一个用于输入,一个用于输出),其他 30 个端点带有可软件动态定义大小的 FIFO 以支持多包队列。FIFO 支持 μDMA,可有效降低系统资源的占用。USB Device 启动方式灵活,可软件控制是否在启动时连接。USB 控制器遵从 OTG 标准的会话请求协议(SRP)和主机协商协议(HNP)。

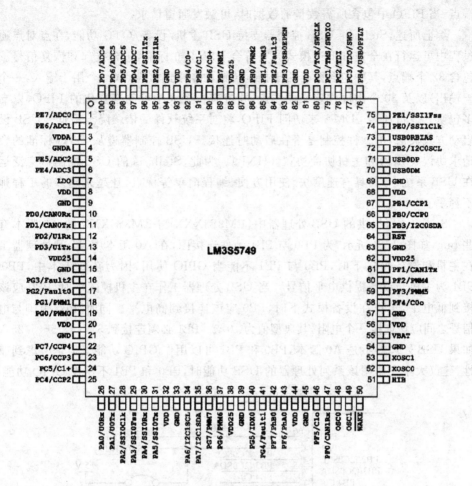

图 2-2 LM3S5749 引脚图

Stellaris USB 模块特性如下：

a. 符合 USB IF 认证标准。

b. 支持 USB 2.0 全速模式(12 Mbps)和低速模式(1.5 Mbps)。

c. 集成 PHY。

d. 传输类型：控制传输(control)，中断传输(interrupt)，批次传输(bulk)，等时传输(isochronous)。

e. 32 端点：1 个专用的输入控制端点和 1 个专用的输出控制端点；15 个可配置的输入端点和 15 个可配置的输出端点。不同型号处理器，其端点个数不一样，应用时注意请查看相关数据手册。

f. 4 KB 专用端点内存空间：一个端点可定义为双缓存 1 023 字节的等时传输。

g. 支持 VBUS 电压浮动(droop)和有效 ID 检测，并产生中断信号。

h. 高效传输 μDMA：用于发送和接收的独立通道多达 3 个输入端点和 3 个输出

第 2 章 Stellaris 的 USB 处理器

端点,当 FIFO 中包含所需数量的数据时,可触发通道请求。

综上所述,Stellaris USB 控制器支持 USB 主机/设备/OTG 功能,在点对点通信过程中可运行在全速和低速模式。它符合 USB 2.0 标准,包含挂起和恢复信号。它包含 32 个端点,其中包含两个用于控制传输的专用连接端点(一个用于输入,一个用于输出)以及 30 个由固件定义的端点,并带有一个大小可动态变化的 FIFO,以支持多包队列。可通过 μDMA 来访问 FIFO,将对系统软件的依赖降至最低。USB 设备启动方式灵活,可软件控制是否在启动时连接。USB 控制器遵从 OTG 标准的会话请求协议(SRP)和主机协商协议(HNP)。为此,Stellaris 的 USB 控制器广泛运用在 USB 系统中,其具有速度快、使用方便、编程简单等优点,让越来越多的工程师爱不释手。

Stellaris 所提供的 USB 处理器中,LM3S3XXX 与 LM3S5XXX 的 A0 版本有一些 bug,如图 2-3 所示,为 PB0 与 PB1 硬件连接图,在 A0 版本中,USB 处理器工作在主机和设备模式下时,PB0 与 PB1 不能当 GPIO 使用,因为在此版本中,PB0 与 PB1 为主机与设备提供电平信号。当 USB 处理器工作在主机模式下时,PB0 应该连接到低电平,工作在设备模式下时,PB0 应该连接到高电平。同时,PB0 引脚与电压信号之间应该连接一个电阻,其典型值为 10 Ω。PB1 必须连接到 5 V(4.75~5.25 V)。如果 USB 处理器不是 A0 版本,PB0 和 PB1 可以用作 GPIO 功能。但是考虑到兼容性,建议大家在使用该系列处理器的 USB 功能时,PB0 和 PB1 不用作 GPIO 功能。

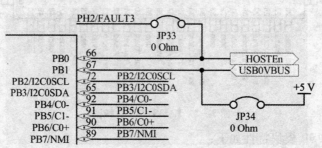

图 2-3 PB0 与 PB1 硬件连接图

如图 2-4 所示,为 LM3S3XXX 与 LM3S5XXX 系列的 USB 功能引脚连接图,USB 处理器工作在主机模式下时,USB 功能引脚连接与设备模式下一样。① USB0RBIAS,连接方式固定,为 USB 模拟电路内部必须要有的 9.1 kΩ 电阻(1% 精度),普通贴片即可满足其需求;② USB0DP 和 USB0DM,USB0 的双向差分数据引脚,连接方式固定,分别连接到 USB 规范中的 D+ 和 D− 中,在使用时,请特别注意 D+ 和 D− 的连接方式。USB0EPEN 是 USB 主机电源输出使能引脚,主机输出电源使能信号,高电平有效,用于使能外部电源。如图 2-5 所示,通过 USB 处理器的 USB0EPEN 引脚输出高电平使能 VBUS 电源。USB0PFLT 是主机模式下的外部电源异常输入引脚,指示外部电源的错误状态,低电平有效。当 TPS2051 的 VBUS 电源输出电流小于 1 mA 或者大于 500 mA 时,从 OCn 引脚输出低电平,开漏输出,通过 USB 处理器的 USB0PFLT 上拉,读取 OCn 的电平变化,并产生中断。注意,USB0EPEN 和 USB0PFLT 主要工作在主机模式下,为设备提供 VBUS 电源,当然可以不使用这两个引脚,直接通过主板 5 V 提供 VBUS 电源,这会缺少 USB 电源欠压过流保护,但电路简单。开发人员可以根据自己设计需要,自行剪裁硬件,但是 USB0RBIAS、USB0DP、USB0DM 必不可少,如是 A0 版本处理器,工作在主机和设备模式下时,PB0 与 PB1 不能当 GPIO 使用,PB1 接 5 V;主机模式下,PB0 接地;设备模式下,PB0 接 5 V。

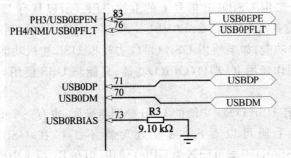

图 2-4 USB 功能引脚连接图

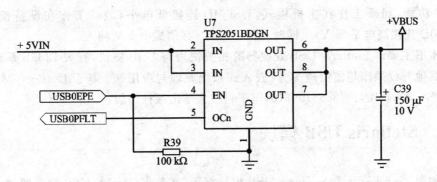

图 2-5 Vbus 控制电路图

对 LM3S3XXX、LM3S5XXX、LM3S9XXX 系列处理器,有一部分具有 USB OTG 功能,与传统不具有 USB OTG 功能的处理器比较,PB0 作为 USB0ID 引脚,此引脚用于检测 USB ID 信号的状态。此时 USBPHY 将在内部启用一个上拉电阻,通过外部元件(USB 连接器)检测 USB 控制器的初始状态(即电缆的 A 侧设置下拉电阻,B 侧设置上拉电阻)。该类 Stellaris 支持 OTG 标准的会话请求协议(SRP)和主机协商协议(HNP),提供完整的 OTG 协商。回话请求协议(SRP)允许连接在 USB 线缆 B 端的 B 类设备向连接在 A 端的 A 类设备发出请求,通知 A 类设备打开 VBUS 电源。主机协商协议(HNP)用于在初始会话请求协议提供供电后,决定 USB 线缆哪端的设备作为 USB Host 主控制器。

当该设备连接到非 OTG 外设或设备时,控制器可以检测出线缆终端所使用的设备类型,并提供一个寄存器来指示该控制器是用作主机控制器还是用作设备控制器。之上所提供的动作都是由 USB 控制器自动处理的。基于这种自动探测机制,系统使用 A 类/B 类连接器取代 A/B 类连接器,可支持完整的会话请求协议和主机协商协议。另外,USB 控制器还提供对连接到非 OTG 的外设或主机控制器的支持。可将 USB 控制器设置为专用的主机或设备模式,此时,USB0VBUS 和 USB0ID 引脚可被设置作为 GPIO 使用。但当 USB 控制器用作自供电设备时,必须将 GPIO 输入引脚或模拟比较器输入引脚连接到 VBUS 引脚上,并配置为在 VBUS 掉电时会产生中断。该中断用于禁用 USB0DP 信号上的上拉电阻。所以具有 USB OTG 功能的处理器工作在 A 设备模式下时,USB0VBUS 和 USB0ID 引脚可通过软件设置为通用 IO 端口(GPIO)功能;如果使用 USB OTG 去模拟 USB 主机功能时,USB0VBUS 和 USB0ID 不能用作通用 IO 端口(GPIO)功能,只能归 USB 使用,并且 USB0ID 必须接地。

注意:对于具有 USB OTG 功能的 USB 处理器来说,也有一个 bug 要回避,根据官方提供的数据手册可以看出,只要处理器不工作在 USB OTG 模式下,USB0VBUS 和 USB0ID 引脚可配置为 GPIO 使用,但在 B1 版本中(其他版本也有此类 bug)无论 USB 处理器工作在什么模式,USB0ID 必须归 USB 使用,不能用于 GPIO 功能,如果工作在主机模式,USB0ID 连接低电平(地);工作在设备模式,USB0ID 连接高电平(5 V)。同时 USB0VBUS 必须提供 5 V 输入。

本书主要讲 Stellaris USB 处理器的相关使用与 USB 协议,有关 LM3S 系列处理器其他外设操作与使用请参考《嵌入式系统原理与应用——基于 Cortex - M3 和 μC/OS - II》,ISBN:978564709310,电子科技大学出版社出版。

2.2　Stellaris USB 模块

如图 2 - 6 所示,为 Stellaris USB 模块框图,在数据输入时,USB 数据通过 D+(DP)和 DM(D-)进行物理传输至 USB 模块中,在 USB PHY 和 UTM 同步模块的作用下,把模拟量的 D+ 和 D- 差分电信号处理成数字量的字节数据,并用于数据包编\解码模块解码数据,通过传入的数据,确定该数据传至哪一个端点,并放入 FIFO RAM 中的 RX 缓存里,产生中断,提交给处理器处理;在数据输出时,数据放在 FIFO 中,通过处理器调度,把数据传输到数据包编\解码模块进数据编码处理,并通过 USB PHY 与 UTM 同步模块,把数据传至 D+、D- 总线上。

一些 USB 控制器的信号引脚是由 GPIO 端口复用实现的,这些引脚在复位时默认设置为 GPIO 引脚。当需要使用 USB 功能时,应将相关 GPIO 备选功能选择寄存器(GPIOAFSEL)中的 AFSEL 位置位,表示启用 GPIO 的备选功能;同时还应通过 GPIO 端口控制寄存器(GPIOCTRL)的 PMCn 位域进行控制,表示将 USB 信号分配

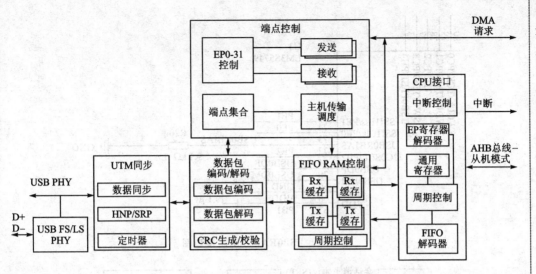

图 2-6 Stellaris USB 模块框图

给指定的 GPIO 引脚。USB0VBUS 和 USB0ID 信号通过清除 GPIO 数字使能寄存器（GPIODEN）中相应的 DEN 位来配置。

对于 LM3S9000 系列的处理器，当用于 OTG 模式时，由于 USB0VBUS 和 USB0ID 是 USB 专用的引脚，不需要配置，直接连接到 USB 连接器的 VBUS 和 ID 信号。如果 USB 控制器专用于主机或设备，USB 通用控制和状态寄存器（USBGPCS）中的 DEVMODOTG 和 DEVMOD 位用于连接 USB0VBUS 和 USB0ID 到内部固定电平，释放 PB0 和 PB1 引脚用于通用 GPIO。当用作自供电的设备时，需要监测 VBUS 值，来确定主机是否断开 VBUS，从而禁止自供电设备 D+/D- 上的上拉电阻。此功能可通过将一个标准 GPIO 连接到 VBUS 实现，在使用 USB 模块时，其主频应在 20 MHz 以上。LM3S3000 与 LM3S5000 系列通过 PB0 来判断 USB 工作在主机还是设备模式。在使用 USB 模块时，其主频应在 30 MHz 以上。

2.2.1 功能描述

如图 2-7 所示，为 USB0RAIAS 电阻接口图，在引脚 USB0RBIAS 和地之间需要接一个 1% 精度的 9.1 kΩ 电阻，且该电阻离 USB0RBIAS 引脚越近越好。由于损耗在该电阻上的功率很小，可以采用贴片电阻，目前大部分的 9.1 kΩ 贴片电阻都满足其要求。

Stellaris USB 支持 OTG 标准的会话请求协议（SRP）和主机协商协议（HNP），提供完整的 OTG 协商。如图 2-8 所示，为 SRP 会话请求协议图，允许连接在 USB 线缆 B 端的 B 类设备向连接在 A 端的 A 类设备发出请求，通知 A 类设备打开 VBUS 电源。

第 2 章　Stellaris 的 USB 处理器

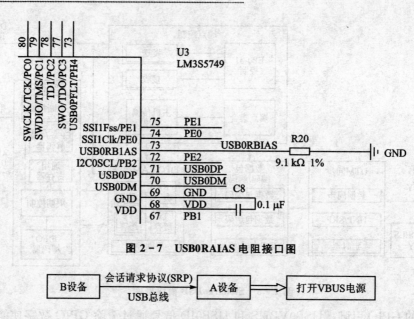

图 2-7　USB0RAIAS 电阻接口图

图 2-8　SRP 回话请求协议图

主机协商协议(HNP)用于在初始会话请求协议提供供电后,决定 USB 线缆哪端的设备作为 USB Host 主控制器。当连接到非 OTG 外设或设备时,OTG 控制器可以探测出线缆的另一端接入的是 USB Host 主机还是 Device 设备,并通过一个寄存器指示 OTG 运行在 Host 主机还是 Device 设备上,上述过程是 USB 控制器自动完成的。基于这种自动探测机制,系统使用 A 类/B 类连接器取代 AB 类连接器,可支持与另外的 OTG 设备完整的 OTG 协商。

另外,USB 控制器支持接入非 OTG 外设或 Host 主控制器。它可以被设置为专用于 Host 或 Device 功能,此时 USB0VBUS 和 USB0ID 引脚可被设置作为 GPIO 口使用。当 USB 控制器被用作自供电的 Device 时,必须将 VBUS 接到 GPIO 或模拟比较器的输入,在 VBUS 掉电时产生中断,用于关闭 USB0DP 的上拉电阻。

2.2.2　USB 控制器作为 USB 设备

USB 控制器作为 USB 设备(Device)操作时,如图 2-9 所示,USB 设备输入/输出事务图,输入事务通过使用端点的发送端点寄存器,由端点的发送接口进行控制;输出事务通过使用端点的接收端点寄存器,由端点的接收接口进行控制,这是因为输入/输出事务是定义为相对于 USB 主机的数据传输事务。

当配置端点的 FIFO 大小时,需要考虑最大数据包大小:

① 批量传输:批量端点的 FIFO 可配置为最大包长(最大 64 B),如果使用双包缓存,则需要配置为 2 倍于最大包长大小。

② 中断传输:中断端点的 FIFO 可配置为最大包长(最大 64 B),如果使用双包

第 2 章 Stellaris 的 USB 处理器

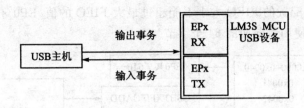

图 2-9 USB 设备输入/输出事务图

缓存,则需要配置为 2 倍于最大包长大小。

③ 等时传输:等时端点的 FIFO 比较灵活,最大支持 1 023 B。

④ 控制传输:USB 设备可能指定一个独立的控制端点,但在大多数情况,USB 设备使用 USB 控制器的端点 0 作为专用的控制端点。

1. 端 点

USB 控制器作为设备运行时,最多可提供两个专用的控制端点(输入和输出) EP0 和用于与主机通信的 30 个可配置的端点(15 个输入和 15 个输出),具体端点数量请参考相应处理器的数据手册。LM3S3000 与 LM3S5000 除两个专用控制端点(输入和输出)EP0 外,用于与主机通信可配置的端点数量少至 6 个(3 个输入和 3 个输出)。端点的端点号和方向与对应的相关寄存器有直接联系。比如,当主机发送数据到端点 1,所有的配置和数据存于端点 1 发送到寄存器接口中。端点 0 是专用的控制端点,用于枚举期间的端点 0 的所有控制传输或其他端点 0 的控制请求。端点 0 使用 USB 控制器 FIFO 内存(RAM)的前 64 个字节,此内存对于输入事务和输出事务是共享的。其余 30 个端点可配置为控制端点、批量端点、中断端点或等时端点。它们应被作为 15 个可配置的输入端点和 15 个可配置的输出端点来对待。这些成对的端点的输入和输出端点不需要配置为相同类型,比如端点对(endpoint pairs)的输出部分可以设置为批量端点,而输入部分可以设置为中断端点。每个端点的 FIFO 的地址和大小可以根据应用需求来修改。

2. 输入事务

输入事务是相对主机而言的,输入事务的数据通过 EPx - OUT 的 FIFO 来处理。如图 2-10 所示,为 USB FIFO 配置图,15 个可配置的输入端点的 FIFO 大小由 USB 发送 FIFO 起始地址寄存器(USBTXFIFOADD)决定,传输时发送端点 FIFO 中的最大数据包大小可编程配置,该大小由写入该端点的 USB 端点 n 的最大发送数据寄存器(USBTXMAXPn)中的值决定,USBTXMAXPn 值不能大于 FIFO 的大小。端点的 FIFO 可配置为双包缓存或单包缓存,当双包缓存使能时,FIFO 中可缓冲两个数据包,这需要 FIFO 至少为两个数据包大小。当不使用双包缓存时,即使数据包的大小小于 FIFO 大小的一半,也只能缓冲一个数据包。可以使用 USBTXFIFOADD 和 USBTXFIFOSZ 配置动态 FIFO 的大小,但是在使用时请注意,这两个寄存器的配置需要 USBEPIDX(USB 索引寄存器)来确定对哪一个端点进行动态 FIFO 配置,其配置的 FIFO 大小不能超过最大 FIFO 的值;EP0 不可配置为动态 FIFO 大小,固定使用最前面的 64 个字节。

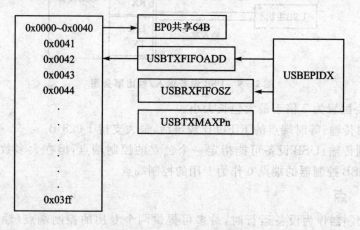

图 2-10 USB FIFO 配置图

注意理解以下概念:

(1) USBTXMAXPn:USB 端点 n 的最大发送数据寄存器,是指 USB 端点 n 一次最大发送数据包的字节大小,比如,可以通过 USBTXMAXP1 定义一次最大发送 8 个字节,那么每次最多发送 8 个字节。

(2) USBTXFIFOADD:USB 发送 FIFO 起始地址寄存器,与 USBEPIDX 一起确定某个端点的 FIFO 缓存的首地址,用于存放被发送数据 FIFO 空间的首地址。

(3) USBTXFIFOSZ:配置 USB 发送数据时,其数据缓存空间大小,与 USBEPIDX 一起使用,与 USBTXFIFOADD 操作方式一样。

注意 USBTXFIFOSZ 和 USBTXMAXPn 的区别:如图 2-11 所示,为 USB FIFO 大小与最大包区别图,可通过 USBTXFIFOSZ 和 USBTXFIFOADD 分别定义 FIFO 的空间大小与首地址分别为 64 B 与 0x0050,USBTXMAXPn 定义其发送数据时最大包数据大小为 32 B,从图中可以看出,TXFIFO 为发送数据提供缓存空间,而发送数据的最大包由 USBTXMAXPn 确定。TXFIFO 的大小与发送数据最大包的大小是两个概念。

图 2-11 USB FIFO 大小与最大包区别

(4) 单包缓冲：如果发送端点 FIFO 的大小小于该端点最大包长的两倍时，只能使用单包缓冲，在 FIFO 中缓冲一个数据包。当数据包已装载到 TXFIFO 中时，USB 端点 n 发送控制和状态低字节寄存器 USBTXCSRLn 中的 TXRDY 位必须被置位，如果 USB 端点 n 发送控制和状态高字节寄存器 USBTXCSRHn 中的 AUTOSET 位被置 1，TXRDY 位将在最大包长的包装载到 FIFO 中时自动置位；如果数据包小于最大包长，TXRDY 位必须手动置位。当 TXRDY 位被手动或自动置 1 时，表明要发送的数据包已准备好。如果数据包成功发送，TXRDY 位和 FIFONE 位将被清 0，同时产生相应的中断信号，此时下一包数据可装载到 FIFO 中。

(5) 双包缓存：如果发送端点 FIFO 的大小至少两倍于该端点最大包长时，允许使用双包缓存，FIFO 中可以缓冲两个数据包。当数据包已装载到 TXFIFO 中时，USBTXCSRLn 中的 TXRDY 位必须被置位，如果寄存器 USBTXCSRHn 中的 AUTOSET 位被置 1，TXRDY 位将在最大包长的包装载到 FIFO 中时自动置位；如果数据包小于最大包长，TXRDY 位必须手动置位。当 TXRDY 位被手动或自动置 1 时，表明要发送的数据包已准备好。在装载完第一个包后，TXRDY 位立即清除，同时产生中断信号；此时第二个数据包可装载到 TXFIFO 中，TXRDY 位重新置位（手动或自动），此时，两个要发送的包都已准备好，如果任一数据包成功发送，TXRDY 位和 FIFONE 位将被清 0，同时产生相应的发送端点中断信号，此时下一包数据可装载到 TXFIFO 中。寄存器 USBTXCSRLn 中的 FIFONE 位的状态表明此时可以装载几个包，如果 FIFONE 位置 1，表明 FIFO 中还有一个包未发送，只能装载一个数据包；如果 FIFONE 位为 0，表明 FIFO 中没有未发送的包，可以装载两个数据包。

如果 USB 发送双包缓存禁止寄存器 USBTXPKTBUFDIS 中的 EPn 位置位，相应的端点禁止双包缓存。此位默认为置 1，需要使能双包缓存时必须清 0 该位。

3. 输出事务

如图 2-9 所示，USB 设备输入/输出事务图，输出事务是相对主机而言的，输出事务的数据通过接收端点的 FIFO 来处理。15 个可配置的输出端点的 FIFO 大小由 USB 接收 FIFO 起始地址寄存器（USBRXFIFOADD）决定，传输时接收端点 FIFO 中的最大数据包大小可编程配置，其大小由写入该端点的 USB 端点 n 最大接收数据寄存器（USBRXMAXPn）中的值决定。端点的 FIFO 可配置为双包缓存或单包缓存，当双包缓存使能时 FIFO 中可缓冲两个数据包。当不使用双包缓存时，即使数据包的大小小于 FIFO 大小的一半，也只能缓冲一个数据包。

(1) 单包缓存：如果接收端点 FIFO 的大小小于该端点最大包长的两倍时，只能使用单包缓冲，在 FIFO 中缓冲一个数据包。当数据包已接收到 RXFIFO 中时，USB 端点 n 接收控制和状态低字节寄存器 USBRXCSRLn 中的 RXRDY 和 FULL 位置位，同时发出相应的中断信号，表明接收 FIFO 中有一个数据包需要读出。当数据包从 FIFO 中读出时，RXRDY 位必须被清 0 以允许接收后面的数据包，同时向 USB 主机发送确认信号。如果 USB 端点 n 接收控制和状态高字节寄存器 USBRXCSRHn

中的 AUTOCl 位被置 1,RXRDY 和 FULL 位将在最大包长的包从 FIFO 中读出时自动清 0;如果数据包小于最大包长,RXRDY 位必须被手动清 0。

(2) 双包缓存:如果接收端点的 FIFO 大小不小于该端点的最大包长的 2 倍时,可以使用双缓冲机制缓存两个数据包。当第一个数据包被接收缓存到 RXFIFO,寄存器 USBRXCSRLn 中的 RXRDY 位置位,同时产生相应的接收端点中断信号,指示有一个数据包需要从 RXFIFO 中读出。当第一个数据包被接收时,寄存器 USBRX-CSRLn 的 FULL 位不置位,该位只有在第二个数据包被接收缓存到 FIFO 时才置位。当从 FIFO 从读出一个包时,RXRDY 位必须清 0 以允许接收后面的包。如果 USB 端点 n 接收控制和状态高字节寄存器 USBRXCSRHn 中的 AUTOCl 位置位 1,RXRDY 位将在最大包长的包从 FIFO 中读出时自动清 0;如果数据包小于最大包长,RXRDY 位必须被手动清 0。当 RXRDY 位清 0 时,FULL 位为 1,USB 控制器先清除 FULL 位,然后再置位 RXRDY 位,表明 FIFO 中的另一个数据包等待被读出。

如果 USB 接收双包缓存禁止寄存器 USBRXDPKTBUFDIS 中的 EPn 位置位,相应的端点禁止双包缓存。此位默认置 1,需要使能双包缓存时必须清 0 该位。

4. 调 度

传输事务由 Host 主机控制器调度决定,Device 设备无法控制事务调度。设备等待 Host 主控制器发出请求,随时可建立传输事务。当传输事务完成或由于某些原因被终止时,会产生中断信号。当 Host 主控制器发出请求,而 Device 设备还没有准备好时,设备会返回一个 NAK 忙信号。

5. 设备挂起(SUSPEND)

USB 总线空闲达 3 ms 时,USB 控制器自动进入挂起(SUSPEND)模式。如果 USB 中断使能寄存器 USBIE 中使能挂起(SUSPEND)中断,会发出一个中断信号。当 USB 控制器进入挂起模式,USB PHY 也将进入挂起模式。当检测到唤醒(RE-SUME)信号时,USB 控制器退出挂起模式,同时使 USB PHY 退出挂起模式,此时如果唤醒中断使能,将产生中断信号。设置 USB 电源寄存器 USBPOWER 中的 RE-SUME 位同样可以强制 USB 控制器退出挂起模式。当此位置位,USB 控制器退出挂起模式,同时在总线上发出唤醒信号。RESUME 位必须在 10 ms(最大 15 ms)后清 0 来结束唤醒信号。为满足电源功耗需求,LM3S USB 控制器可进入深睡眠模式。

6. 帧起始

当 USB 控制器运行在设备模式,它每 1 ms 收到一次主机发出的帧起始包(SOF)。当收到 SOF 包时,包中包含的 11 位帧号写入 USB 帧值寄存器 USB-FRAME 中,同时发出 SOF 中断信号,由应用程序处理。一旦 USB 控制器开始收到 SOF 包,它将预期每 1 ms 收到 1 次。如果超过 1.003 58 ms 没有收到 SOF 包,将假定此包丢失,寄存器 USBFRAME 也将不更新。当 SOF 包重新成功接收时,USB 控制器继续,并重新同步这些脉冲。

使用 SOF 中断可以每 1 ms 获得一个基本时钟,可当 tick 基本定时时钟使用,并且使用方便,在移植文件系统时,需要间隔 10 ms 周期调用 disk_timerproc 函数,此处的 10 ms 就可以使用 SOF 中断产生,并调用 disk_timerproc 函数。

7. USB 复位

当 USB 控制器处于设备模式,如果检测到 USB 总线上的复位信号,USB 控制器将自动:清除地址寄存器 USBFADDR,清除 USB 端点索引寄存器(USBEPIDX),清空所有端点 FIFO 数据,清除所有控制/状态寄存器,使能所有端点中断,产生复位中断信号。对于刚接入到主机的 USB 设备,USB 复位意味着要进行枚举了。

2.2.3 USB 控制器作为主机

如图 2-12 所示,为 USB 主机与设备连接图,当 Stellaris USB 控制器运行在主机模式时,可通过①号连接方式与其他 USB 设备进行点对点通信,也可通过②号连接方式连接到集线器(HUB),使用集线器与多个设备进行通信。USB 控制器支持全速和低速设备。它自动执行必要的事务传输,允许 USB 2.0 集线器使用低速设备和全速设备。支持控制传输、批量传输、等时传输和中断传输。输入事务由端点的接收接口进行控制;输出事务使用端点的发送端点寄存器。

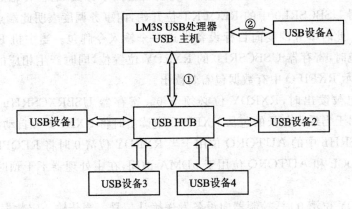

图 2-12 USB 主机与设备连接图

当配置端点的 FIFO 大小时,需要考虑最大数据包的大小。

1. 端 点

端点寄存器用于控制 USB 端点接口,通过接口可与设备进行通信。主机端点由 1 个专用控制输入端点、1 个专用控制输出端点、15 个可配置的输出端点和 15 个可配置的输入端点组成。控制端点只能与设备的端点 0 进行控制传输,用于设备枚举或其他使用设备端点 0 的控制功能。控制端点输入输出事务共享一个 FIFO 存储空间,并在 FIFO 的前 64 B。其余输入和输出端点可配置为:控制传输端点、批量传输端点、中断传输端点或等时传输端点。输入和输出控制有成对的 3 组寄存器,它们可以与不同类型的端点以及不同设备的不同端点进行通信。例如,如图 2-13 所示,为 USB 主机端点与设备端点通信图,USB 主机 EP2 端点可分开控制,USB 主机 EP2-OUT 与 USB 设备 1 的 EP1-OUT 进行批量数据通信,同时 USB 主机 EP2-IN 与 USB 设备 2 的 EP2-IN 进行中断数据传输。FIFO 的地址和大小可以利用软件设置,并且可以指定用于某一个端点输入或者输出传输。

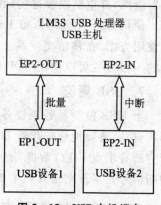

图 2-13 USB 主机端点与设备端点通信图

无论点对点通信还是集线器通信,在访问设备之前,必须设置端点 n 的接收功能地址寄存器 USBRXFUNCADDRn 和端点 n 的发送功能地址寄存器 USBTXFUNCADDRn。LM3S 系列 USB 控制器支持通过集线器连接设备,一个寄存器就可以实现,该寄存器说明集线器地址和每个传输的端口。在本书中的所有例程都未使用集线器访问设备。

2. 输入事务

输入事务,相对主机而言是数据输入,与设备输出事务类似,但传输数据必须通过设置寄存器 USBCSRL0 中的 REQPKT 位开始,向事务调度表明此端点存在一个活动的传输。此时事务调度向目标设备发送一个输入令牌包。当主机 RXFIFO 中接收到数据包时,寄存器 USBCSRL0 的 RXRDY 位置位,同时产生相应的接收端点中断信号,指示 RXFIFO 中有数据包需要读出。

当数据包被读出时,RXRDY 位必须清 0。寄存器 USBRXCSRHn 中的 AUTOCL 位可用于当最大包长的包从 RXFIFO 中读出时将 RXRDY 位自动清 0。寄存器 USBRXCSRHn 中的 AUTORQ 位用于当 RXRDY 位清 0 时将 REQPKT 位自动置位。AUTOCL 和 AUTORQ 位用于 μDMA 访问,在主处理器不干预时完成批量传输。

当 RXRDY 位清 0 时,控制器向设备发送确认信号。当传输一定数据包时,需要将端点 n 的 USBRQPKTCOUNTn 寄存器配置为要传输包的数量,每一次传输,USBRQPKTCOUNTn 寄存器中的值减 1,当减到 0 时,AUTORQ 位清 0 来阻止后面试图进行的数据传输。如传输数量未知,USBRQPKTCOUNTn 寄存器必须清 0,AUTORQ 位保持置位,直到收到结束包,AUTORQ 位才清 0(小于 USBRXMAXPn 寄存器中的 MAXLOAD 值)。

3. 输出事务

当数据包装载到 TXFIFO 中时,USBTXCSRLn 寄存器中的 TXRDY 位必须置位。如果置位了 USBTXCSRHn 寄存器中的 AUTOSET 位,当最大包长的数据包装载到 TXFIFO 中时，TXRDY 位自动置位。此外，AUTOSET 位与 μDMA 控制器配合使用，可以在不需要软件干预的情况下完成批量传输。

4. 调　度

调度由 USB 主机控制器自动处理。中断传输可以是每 1 帧进行一次，也可以每 255 帧进行一次，可以在 1 帧到 255 之间以 1 帧增量调度。批量端点不允许调度参数，但在设备的端点不响应时，允许 NAK 超时。等时端点可以在每 1 帧到每 2^{16} 帧之间调度（2 的幂）。

USB 控制器维持帧计数，并发送 SOF 包，SOF 包发送后，USB 主机控制器检查所有配置好的端点，寻找激活的传输事务。REQPKT 位置位的接收端点、TXRDY 或 FIFONE 位置位的发送端点，被视为传输事务已经激活，等待主机查询。

如果传输建立在一帧的第一个调度周期，而且端点的间隔计数器减到 0，则等时传输和中断传输开始。所以每个端点的中断传输和等时传输每 N 帧才发生一次，N 是通过 USB 主机端点 n 的 USBTXINTERVALn 寄存器或 USB 主机端点 n 的 USBRXINTERVALn 寄存器设置的间隔。

如果在帧中下一个 SOF 包之前提供足够的时间完成传输，则激活的批量传输立即开始。如果传输需要重发时（例如，收到 NAK 或设备未响应），需要在调度器先检查完其他所有端点是否有激活的传输之后，传输才能重传。这保证了一个发送大量 NAK 响应的端点不阻塞总线上的其他传输正常进行。

5. USB 集线器

以下过程只适用于 USB 2.0 集线器的主机。当低速设备或全速设备通过 USB 2.0 集线器连接到 USB 主机时，集线器地址和端口信息必须记录在相应的 USB 端点 n 的 USBRXHUBADDRn 寄存器和 USB 端点 n 的 USBRXHUBPORTn 寄存器或者 USB 端点 n 的 USBTXHUBADDRn 寄存器和 USB 端点 n 的 USBTXHUBPORTn 寄存器。

此外，设备的运行速度（全速或低速）必须记录在 USB 端点 0 的 USBTYPE0 寄存器，和设备访问主机 USB 端点 n 的 USBTXTYPEn 寄存器，或者主机 USB 端点 n 的 USBRXTYPEn 寄存器。

对于集线器通信，这些寄存器的设置记录了 USB 设备当前相应端点的配置。为了支持更多数量的设备，USB 主机控制器允许通过更新这些寄存器配置来实现。

2.2.4　OTG 模式

OTG 就是 On The Go，正在进行中的意思，可以进行"主机与设备"模式切换。USB OTG 允许使用时才给 VBUS 上电，不使用 USB 总线时，则关断 VBUS。VBUS 由总线上的 A 设备提供电源。OTG 控制器通过 PHY 采样 ID 输入信号分辨 A 设备和 B 设备。ID 信号拉低时，检测到插入 A 设备（表示 OTG 控制器作为 A 设备角色);ID 信号为高时,检测到插入 B 设备(表示 OTG 控制器作为 B 设备角色)。注意当在 OTG A 和 OTG B 之间切换时,控制器保留所有的寄存器内容。关于 OTG 的使用不在本书重点介绍。

2.3 寄存器描述

使用 USB 处理器进行 USB 主机、设备、OTG 开发离不开相关寄存器操作,USB 本身就相当复杂,寄存器相当多,下面就几个常用的寄存器进行介绍。本节主要讲主机与设备模式下的寄存器使用,关于 GTO 模式下寄存器使用不具体讲解,因为 GTO A 设备相当于主机,GTO B 设备相当于设备。如表 2-1 所列,为 USB 寄存器表。

表 2-1 USB 寄存器

寄存器名称	类型	复位值	寄存器描述
USBFADDR	R/W	0x00	USB 地址
USBPOWER	R/W	0x20	USB 电源
USBTXIS	RO	0x0000	发送中断状态
USBRXIS	RO	0x0000	接收中断状态
USBTXIE	R/W	0xFFFF	发送中断使能
USBRXIE	R/W	0xFFFE	接收中断使能
USBIS	RO	0x00	USB 处理模块中断状态
USBIE	R/W	0x06	USB 处理模块中断使能
USBFRAME	RO	0x0000	USB 帧值
USBEPIDX	R/W	0x00	端点索引
USBTEST	R/W	0x00	测试模式
USBFIFOn	R/W	0x0000.0000	端点 n FIFO
USBDEVCTL	R/W	0x80	设备控制
USBTXFIFOSZ	R/W	0x00	动态 TXFIFO 大小
USBRXFIFOSZ	R/W	0x00	动态 RXFIFO 大小
USBTXFIFOADD	R/W	0x0000	动态 TXFIFO 起始地址
USBRXFIFOADD	R/W	0x0000	动态 RXFIFO 起始地址
USBCONTIM	R/W	0x5C	USB 连接时序
USBVPLEN	R/W	0x3C	USB OTG VBUS 脉冲时序
USBFSEOF	R/W	0x77	USB 全速模式下最后的传输与帧结束时序
USBLSEOF	R/W	0x72	USB 低速模式下最后的传输与帧结束时序
USBTXFUNCADDRn	R/W	0x00	发送端点 n 功能地址

续表 2-1

寄存器名称	类型	复位值	寄存器描述
USBTXHUBADDRn	R/W	0x00	发送端点 n 集线器(Hub)地址
USBTXHUBPORTn	R/W	0x00	发送端点 n 集线器(Hub)端口
USBRXFUNCADDRn	R/W	0x00	接收端点 n 功能地址(n 不为 0)
USBRXHUBADDRn	R/W	0x00	接收端点 n 集线器(Hub)地址(n 不为 0)
USBRXHUBPORTn	R/W	0x00	接收端点 n 集线器(Hub)端口(n 不为 0)
USBCSRL0	W1C	0x00	端点 0 控制和状态低字节
USBCSRH0	W1C	0x00	端点 0 控制和状态高字节
USBCOUNT0	RO	0x00	端点 0 接收字节数量
USBTYPE0	R/W	0x00	端点 0 类型
USBNAKLMT	R/W	0x00	USB NAK 限制
USBTXMAXPn	R/W	0x0000	发送端点 n 最大传输数据
USBTXCSRLn	R/W	0x00	发送端点 n 控制和状态低字节
USBTXCSRHn	R/W	0x00	发送端点 n 控制和状态高字节
USBRXMAXPn	R/W	0x0000	接收端点 n 最大传输数据
USBRXCSRLn	R/W	0x00	接收端点 n 控制和状态低字节
USBRXCSRHn	R/W	0x00	接收端点 n 控制和状态高字节
USBRXCOUNTn	RO	0x00	端点 n 接收字节数量
USBTXTYPEn	R/W	0x00	端点 n 主机发送配置类型
USBTXINTERVALn	R/W	0x00	端点 n 主机发送间隔
USBRXTYPEn	R/W	0x00	端点 n 主机配置接收类型
USBRXINTERVALn	R/W	0x00	端点 n 主机接收巡检间隔
USBRQPKTCOUNTn	R/W	0x0000	端点 n 块传输中请求包数量
USBRXDPKTBUFDIS	R/W	0x0000	接收双包缓存禁止
USBTXDPKTBUFDIS	R/W	0x0000	发送双包缓存禁止
USBEPC	R/W	0x0000.0000	USB 外部电源控制
USBEPCRIS	RO	0x0000.0000	USB 外部电源控制原始中断状态
USBEPCIM	R/W	0x0000.0000	USB 外部电源控制中断屏蔽
USBEPCISC	R/W	0x0000.0000	USB 外部电源控制中断状态和清除
USBDRRIS	RO	0x0000.0000	USB 设备唤醒(RESUME)原始中断状态
USBDRIM	R/W	0x0000.0000	USB 设备唤醒(RESUME)中断屏蔽
USBDRISC	W1C	0x0000.0000	USB 设备唤醒(RESUME)中断状态和清除
USBGPCS	R/W	0x0000.0000	USB 通用控制和状态
USBVDC	R/W	0x0000.0000	VBUS 浮动控制

续表 2-1

寄存器名称	类型	复位值	寄存器描述
USBVDCRIS	RO	0x0000.0000	VBUS 浮动控制原始中断状态
USBVDCIM	R/W	0x0000.0000	VBUS 浮动控制中断屏蔽
USBVDCISC	R/W	0x0000.0000	VBUS 浮动控制中断状态\清除
USBIDVRIS	RO	0x0000.0000	ID 有效检测原始中断状态
USBIDVIM	R/W	0x0000.0000	ID 有效检测中断屏蔽
USBIDVISC	R/W1C	0x0000.0000	ID 有效检测中断状态与清除
USBDMASEL	R/W	0x0033.2211	DMA 选择

2.3.1 控制状态寄存器

(1) USBFADDR,设备地址寄存器。包含 7 位设备地址,将通过 SET AD-DRESS(USBREQ_SET_ADDRESS)请求收到的地址(pUSBRequest→wValue 值)写入该寄存器。如表 2-2 所列,为 USBFADDR 寄存器描述表。

表 2-2 USBFADDR 寄存器

位	名称	类型	复位	描述
8	保留	RO	0	保持不变
[7:0]	FUNCADDR	R/W	0	设备地址

例如,在枚举时会用到设置 USB 设备地址:

```
void USBDevAddrSet(unsigned long ulBase, unsigned long ulAddress)
{
    HWREGB(ulBase + USB_O_FADDR) = (unsigned char)ulAddress;
}
```

在设备模式下,设定地址。其中 ulBase 为设备基地址,LM3S USB 处理器一般只有一个 USB 模块,所以只有 USB0_BASE(0x40050000);ulAddress 为主机 USBREQ_SET_ADDRESS 请求命令时 tUSBRequest.wValue 的值。

(2) USBPOWER,USB 电源控制寄存器,有效数据为低 8 位。主要负责 USB 总线电源控制、总线复位、挂起、掉电等操作。如表 2-3 所列,为主机模式下 USBPOWER 寄存器描述表,USBPOWER 在主机模式下,用于控制挂起和恢复信号,以及一些基本操作,如表 2-4 所列,为设备模式下 USBPOWER 寄存器描述表,在设备模式下,负责控制挂起、恢复信号、等时传输控制、软件断开连接 DP、DM 引脚。

表 2-3 主机模式下 USBPOWER 寄存器

位	名称	类型	复位	描述
[7:4]	保留	RO	0x02	保持不变
3	RESET	R/W	0	总线复位
2	RESUME	R/W	0	总线唤醒,置位后 20 ms 后软件清除
1	SUSPEND	R/W1S	0	总线挂起
0	PWRDNPHY	R/W	0	PHY 掉电

表 2-4 设备模式下 USBPOWER 寄存器

位	名称	类型	复位	描述
[5:4]	保留	RO	0x02	保持不变
7	ISOUP	R/W	0	ISO 传输时,TXRDY 置位时,收到输入令牌包,将发送 0 长度的数据包
6	SOFTCONN	R/W	0	软件连接/断开 USB
3	RESET	RO	0	总线复位
2	RESUME	R/W	0	总线唤醒,置位后 10 ms 后软件清除
1	SUSPEND	RO	0	总线挂起
0	PWRDNPHY	R/W	0	PHY 掉电

为了使用方便,宏定义了每个位模块功能,如下所示。

```
#define USB_POWER_ISOUP       0x00000080    // Isochronous Update
#define USB_POWER_SOFTCONN    0x00000040    // Soft Connect/Disconnect
#define USB_POWER_RESET       0x00000008    // RESET Signaling
#define USB_POWER_RESUME      0x00000004    // RESUME Signaling
#define USB_POWER_SUSPEND     0x00000002    // SUSPEND Mode
#define USB_POWER_PWRDNPHY    0x00000001    // Power Down PHY
```

例如:在主机模式下,把 USB 总线挂起。

```
void USBHostSuspend(unsigned long ulBase)
{
    //ulBase 为设备基地址
    HWREGB(ulBase + USB_O_POWER) |= USB_POWER_SUSPEND;
}
```

当 LM3S USB 处理器作为主机模式时,这个函数将会让 USB 总线进入挂起状态,并且这个函数只能用于主机模式下。

例如:在主机模式下,复位 USB 总线。

```c
void USBHostReset(unsigned long ulBase, tBoolean bStart)
{
    //ulBase 为设备基地址,bStart 确定复位开始/结束
    if(bStart)
    {
        HWREGB(ulBase + USB_O_POWER) | = USB_POWER_RESET;
    }
    else
    {
        HWREGB(ulBase + USB_O_POWER) & = ~USB_POWER_RESET;
    }
}
```

当这个函数被调用时,如果 bStart 参数设定为 true,则将会在 USB 总线上启动一个复位,然后至少延时 20 ms,才能再次调用这个函数,此时的 bStart 参数设定为 false,完成 USB 总线复位。只能用于主机模式下。

例如:在主机模式下,唤醒 USB 总线,唤醒开始 20 ms 后必须手动结束。

```c
void USBHostResume(unsigned long ulBase, tBoolean bStart)
{
    //ulBase 为设备基地址,bStart 确定唤醒开始/结束
    if(bStart)
    {
        HWREGB(ulBase + USB_O_POWER) | = USB_POWER_RESUME;
    }
    else
    {
        HWREGB(ulBase + USB_O_POWER) & = ~USB_POWER_RESUME;
    }
}
```

当在设备模式时,这个函数将 USB 控制器带出挂起状态。如果 bStart 参数为 true,则这个函数将会启动恢复信号。然后再至少 10 ms,至多 15 ms 的延时之后再调用这个函数,此时的 bStart 参数应该设定为 false。当在主机模式时,调用这个函数将会令从设备离开挂起状态。这个调用首先让 bStart 参数为 true 来启动恢复信号。主机应用程序需要在延时至少 20 ms 之后再度调用这个函数,但是此时的 bStart 参数应该设定为 false。这样就会引起控制器完成一个 USB 总线上的恢复信号。

例如:在设备模式下,设备连接/断开主机。

```c
void USBDevConnect(unsigned long ulBase)
{
```

```
    //ulBase 为设备基地址
    HWREGB(ulBase + USB_O_POWER) |= USB_POWER_SOFTCONN;
}
void USBDevDisconnect(unsigned long ulBase)
{
    //ulBase 为设备基地址
    HWREGB(ulBase + USB_O_POWER) &= (~USB_POWER_SOFTCONN);
}
```

在设备模式下,USBDevConnect 函数让 USB 设备连接 USB 控制器到 USB 总线上,USBDevDisconnect 函数让 USB 设备从 USB 总线上移除 USB 控制器,并让 USB D+/D-处于三态。

(3) USBFRAME,16 位只读寄存器,保存最新收到的帧编号。如表 2-5 所列,为 USBFRAME 寄存器描述表。

表 2-5 USBFRAME 寄存器

位	名 称	类 型	复 位	描 述
[15:11]	保留	RO	0	保持不变
[10:0]	FRAME	RO	0	帧编号

例如:获取最新帧编号。

```
unsigned long USBFrameNumberGet(unsigned long ulBase)
{
    //ulBase 为设备基地址
    return(HWREGH(ulBase + USB_O_FRAME));
}
```

(4) USBTEST,测试模式,回应 USBREQ_SET_FEATURE 的 USB_FEATURE_TEST_MODE 命令,使 USB 控制器进入测试模式,任何时候,只能有一位被设置,但不用于正常操作。如表 2-6 所列,为主机模式下 USBTEST 寄存器描述表。FORCEH 为 1 时,无论 USB 控制器连接到什么外设,当将 SESSION 位置位时,强制 USB 控制器进入主机模式。此时忽略 USB0DP 和 USB0DM 信号的状态。如果不将 SESSION 位清零,即使在与设备断开连接时,该 USB 控制器仍保持主机模式。如果 FORCEH 位保持为 1,则 SESSION 位下次置位时,USB 控制器重新进入主机模式。FIFOACC 为 1 时,将端点 0 发送 FIFO 中的包传送到端点 0 的接收 FIFO 中。FORCEFS 为 1 时,接收到 USB 复位信号后,强制 USB 控制器运行在全速模式。否则,USB 控制器运行在低速模式;如表 2-7 所列,为设备模式下 USBTEST 寄存器描述表。FIFOACC 为 1 时,将端点 0 发送 FIFO 中的包传送到端点 0 的接收 FIFO 中。FORCEFS 为 1 时,接收到 USB 复位信号后,强制 USB 控制器运行在全速模式。否则,USB 控制器运行在低速模式。

表 2-6 主机模式下的 USBTEST 寄存器

位	名 称	类 型	复 位	描 述
[4:0]	保留	RO	0	保持不变
7	FORCEH	R/W	0	强制为主机模式
6	FIFOACC	R/W1S	0	将端点 0 的 TXFIFO 数据传送到 RXFIFO
5	FORCEFS	R/W	0	强制全速模式

在设备模式下,USBTEST 寄存器描述如表 2-7 所列。

表 2-7 设备模式下的 USBTEST 寄存器

位	名 称	类 型	复 位	描 述
[4:0]、7	保留	RO	0	保持不变
6	FIFOACC	R/W1S	0	将端点 0 的 TXFIFO 数据传送到 RXFIFO
5	FORCEFS	R/W	0	强制全速模式

(5) USBDEVCTL,USB 设备控制器,提供 USB 控制器当前状态信息,可指示接入设备是全速还是低速,只能在主机模式下访问。如表 2-8 所列,为 USBDEVCTL 寄存器描述表。

表 2-8 USBDEVCTL 寄存器

位	名 称	类 型	复 位	描 述
7	DEV	RO	1	设备模式,OTG A 端
6	FSDEV	RO	0	全速设备
5	LSDEV	RO	0	低速设备
[4:3]	VBUS	RO	0	VBUS 电平
2	HOST	RO	0	Host 主机模式
1	HOSTREQ	R/W	0	主机请求
0	SESSION	R/W	0	会话开始/结束

为了使用方便,宏定义了每个位模块功能,如下所示。

```
#define USB_DEVCTL_DEV         0x00000080  // Device Mode
#define USB_DEVCTL_FSDEV       0x00000040  // Full-Speed Device Detected
#define USB_DEVCTL_LSDEV       0x00000020  // Low-Speed Device Detected
#define USB_DEVCTL_VBUS_M      0x00000018  // VBUS Level
#define USB_DEVCTL_VBUS_NONE   0x00000000  // Below SessionEnd
#define USB_DEVCTL_VBUS_SEND   0x00000008  // Above SessionEnd, below AValid
```

```
#define USB_DEVCTL_VBUS_AVALID    0x00000010    // Above AValid, below VBUSValid
#define USB_DEVCTL_VBUS_VALID     0x00000018    // Above VBUSValid
#define USB_DEVCTL_HOST           0x00000004    // Host Mode
#define USB_DEVCTL_HOSTREQ        0x00000002    // Host Request
#define USB_DEVCTL_SESSION        0x00000001    // Session Start/End
```

例如:获取当前 USB 工作模式。

```
unsigned long USBModeGet(unsigned long ulBase)
{
    return(HWREGB(ulBase + USB_O_DEVCTL) &
          (USB_DEVCTL_DEV | USB_DEVCTL_HOST | USB_DEVCTL_SESSION |
          USB_DEVCTL_VBUS_M));
}
```

返回下面参数:

```
#define USB_DUAL_MODE_HOST          0x00000001
#define USB_DUAL_MODE_DEVICE        0x00000081
#define USB_DUAL_MODE_NONE          0x00000080
#define USB_OTG_MODE_ASIDE_HOST     0x0000001d
#define USB_OTG_MODE_ASIDE_NPWR     0x00000001
#define USB_OTG_MODE_ASIDE_SESS     0x00000009
#define USB_OTG_MODE_ASIDE_AVAL     0x00000011
#define USB_OTG_MODE_ASIDE_DEV      0x00000019
#define USB_OTG_MODE_BSIDE_HOST     0x0000009d
#define USB_OTG_MODE_BSIDE_DEV      0x00000099
#define USB_OTG_MODE_BSIDE_NPWR     0x00000081
#define USB_OTG_MODE_NONE           0x00000080
```

在 OTG 模式下返回以下值:

USB_OTG_MODE_ASIDE_HOST:指示控制器在主机模式下,连接 A 端。

USB_OTG_MODE_ASIDE_DEV:指示控制器在从机模式下,连接 A 端。

USB_OTG_MODE_BSIDE_HOST:指示控制器在主机模式下,连接 B 端。

USB_OTG_MODE_BSIDE_DEV:指示控制器在从机模式下,连接 B 端。当没有线连接的时候启动检测,则这个模式是默认的模式。

USB_OTG_MODE_NONE:指示为无模式。

双模式控制时:

USB_DUAL_MODE_HOST:指示控制器作为主机。

USB_DUAL_MODE_DEVICE:指示控制器作为从机。

USB_DUAL_MODE_NONE:指示控制器没有激活,为无模式。

第 2 章　Stellaris 的 USB 处理器

(6) USBCONTIM，连接时序控制寄存器，此 8 位寄存器用于配置连接和协商的延时。如表 2-9 所列，为 USBCONTIM 寄存器描述表。

表 2-9　USBCONTIM 寄存器

位	名 称	类 型	复 位	描 述
[7:4]	WTCON	R/W	0x5	连接等待
[3:0]	WTID	R/W	0xC	ID 等待

WTCON，此位域可按需求配置等待延时，满足连接/断开的滤波需求；单位为 533.3 ns，复位后为 0x5，默认为 2.667 μs。

WTID，此位域配置 ID 等待延时，等待 ID 值有效时才使能 OTG 的 ID 检测。单位为 4.369 ms，复位后为 0xC，默认为 52.43 ms。

(7) USBVPLEN，VBUS 脉冲时序配置寄存器，此 8 位寄存器用于配置 VBUS 脉冲充电的持续时间。如表 2-10 所列，为 USBVPLEN 寄存器描述表。

表 2-10　USBVPLEN 寄存器

位	名 称	类 型	复 位	描 述
[7:0]	VPLEN	R/W	0x3c	VBUS 脉冲宽度

VPLEN，用于配置 VBUS 脉冲充电的持续时间，单位 546.1 μs，默认为 32.77 ms。本寄存器只能在 OTG 模式下访问。

(8) USBFSEOF，USB 全速模式下最后的传输与帧结束时序寄存器。此 8 位寄存器用于配置全速模式下允许最后的传输开始与帧结束(EOF)之间的最小时间间隔。如表 2-11 所列，为 USBFSEOF 寄存器描述表。

表 2-11　USBFSEOF 寄存器

位	名 称	类 型	复 位	描 述
[7:0]	FSEOFG	R/W	0x77	全速模式帧间隙

FSEOFG，在全速传输中用于配置最后的传输与帧结束之间的最小时间间隙，单位为 533.3 ns，默认为 63.46 μs。

(9) USBLSEOF，低速模式下最后的传输与帧结束时序寄存器，此 8 位寄存器用于配置低速模式下允许最后的传输开始与帧结束(EOF)之间的最小时间间隔。如表 2-12 所列，为 USBLSEOF 寄存器描述表。

表 2-12　USBLSEOF 寄存器

位	名 称	类 型	复 位	描 述
[7:0]	LSEOFG	R/W	0x72	低速模式帧间隙

LSEOFG，在低速传输中用于配置最后的传输与帧结束之间的最小时间间隙，单位为 1.067 μs，默认为 121.6 μs。

注意：USBCONTIM、USBVPLEN、USBFSEOF、USBLSEOF 一般不用重新配置，保持默认状态就能满足 USB 通信要求。

（10）USBGPCS，USB 通用控制状态寄存器，提供内部 ID 信号的状态。如表 2-13 所列，为 USBGPCS 寄存器描述表。

表 2-13 USBGPCS 寄存器

位	名称	类型	复位	描述
[31:2]	保留	RO	0	保持不变
1	DEVMODOTG	R/W	0	使能 ID 控制模式
0	DEVMOD	R/W	0	模式控制

在 OTG 模式，由于 USB0VBUS 和 USB0ID 是 USB 控制器专用的引脚，不需要被配置，可直接连接到 USB 连接器的 VBUS 和 ID 信号。如果 USB 控制器专用于主机或设备，USB 通用控制和状态（USBGPCS）寄存器中的 DEVMODOTG 和 DEVMOD 位用于将 USB0VBUS 和 USB0ID 连接到内部固定电平，从而释放 PB0 和 PB1 引脚，以用于 GPIO。当用作自供电的设备时，需要监测 VBUS 值，来确定主机是否断开 VBUS，从而禁止自供电设备 D+/D- 上的上拉电阻。此功能可通过将一个标准 GPIO 连接到 VBUS 实现。

为了使用方便，宏定义了每个位模块功能，如下所示。

```
#define USB_GPCS_DEVMODOTG      0x00000002    // Enable Device Mode
#define USB_GPCS_DEVMOD         0x00000001    // Device Mode
```

例如：设置主机模式。

```
void USBHostMode(unsigned long ulBase)
{
    //ulBase 为设备基地址
    HWREGB(ulBase + USB_O_GPCS) &= ~(USB_GPCS_DEVMOD);
}
```

参数 ulBase 指定 USB 模块寄存器基址。这个函数改变 USB 控制器模式到主机模式。只用于有主机和从机模式的 MCU，不是具有 OTG 功能的。

2.3.2　中断控制

所谓中断，就是指 CPU 在执行程序的过程中，由于某种外部或内部事件的作用（如外部设备请求与 CPU 传送数据或 CPU 在执行程序的过程中出现了异常），强迫 CPU 停止当前正在执行的程序而转去为该事件服务，待事件服务结束后，又能自动返回到被中断了的程序中继续执行。由于 CPU 正在执行的原程序被暂停执行,所以称为中断。LM3S 的 USB 系列处理器提供一组使用方便的 USB 中断控制寄存器,完成 USB 中断设置、中断标志、中断标志清除等功能。

(1) USBTXIS,发送中断状态寄存器,是一个 16 位的只读寄存器,用于指示端点 0 和发送端点 1~15 的哪个中断是有效的。如表 2-14 所列,为 USBTXIS 寄存器描述表。

表 2-14 USBTXIS 寄存器

位	名称	类型	复位	描述
[15:0]	EPn	RO	0	端点发送中断标志

寄存器中 EPn 位域的含意取决于设备的模式。EP1 到 EP15 位指示 USB 控制器正在发送数据;在主机模式,这些位适用于输出端点;而在设备模式,这些位适用于输入端点。EP0 位比较特殊,在主机和设备模式中,该位指示一个控制输入或控制输出端点产生了中断。注意:与这些端点相关的位如果没有配置,其返回值一直为 0。对该寄存器的读操作,会清除所有已激活的中断。

(2) USBRXIS,接收中断状态寄存器。是一个 16 位的只读寄存器,用于指示接收端点 1~15 中哪个中断是有效的。如表 2-15 所列,为 USBRXIS 寄存器描述表。

表 2-15 USBRXIS 寄存器

位	名称	类型	复位	描述
[15:1]	EPn	R/W	0	端点接收中断标志

(3) USBTXIE,发送中断启用寄存器,是 16 位寄存器,用于为 USBTXIS 寄存器的中断提供中断启用位。如表 2-16 所列,为 USBTXIE 寄存器描述表。

表 2-16 USBTXIE 寄存器

位	名称	类型	复位	描述
[15:0]	EPn	RO	0	端点发送中断标志

如果 USBTXIE 寄存器的某位置位,相应的 USBTXIS 寄存器中的中断位也置 1,则 USB 中断对中断控制器有效。如果某位清零,尽管相应的 USBTXIS 寄存器中的中断位置 1,但 USB 中断对中断控制器仍是无效的。复位时,所有的中断均是启用的。

(4) USBRXIE,接收中断使能寄存器,是 16 位寄存器,用于为 USBRXIS 寄存器的中断提供中断启用位。如表 2-17 所列,为 USBRXIE 寄存器描述表。

表 2-17 USBRXIE 寄存器

位	名称	类型	复位	描述
[15:1]	EPn	R/W	0	端点接收中断标志

如果 USBRXIE 寄存器的某位置位，相应的 USBTXIS 寄存器中的中断位也置1，则 USB 中断对中断控制器有效。如果某位清零，尽管相应的 USBTXIS 寄存器中的中断位置位，但 USB 中断对中断控制器仍是无效的。复位时，所有的中断启用。

（5）USBIS，通用中断状态寄存器，是一个 8 位只读寄存器，用来指示哪个 USB 中断有效。读此寄存器，所有的中断将被清除。如表 2-18 所列，为 USBIS 寄存器描述表。

表 2-18 USBIS 寄存器

位	名称	类型	复位	描述
7	VBUSERR	RO	0	VBUS 中断（只有主机模式使用）
6	SESREQ	RO	0	会话请求中断（只有主机模式使用）
5	DISCON	RO	0	连接断开中断
4	CONN	RO	0	连接中断
3	SOF	RO	0	帧起始中断
2	BABBLE	RO	0	Babble 中断
1	RESUME	RO	0	唤醒中断
0	SUSPEND	RO	0	挂起中断（只有设备模式使用）

（6）USBIE，通用中断使能寄存器，是 8 位寄存器，为 USBIS 寄存器中的每个中断提供中断启用位。设备模式中，复位时，中断 1 和中断 2 是启用的。如表 2-19 所列，为 USBIE 寄存器描述表。

表 2-19 USBIE 寄存器

位	名称	类型	复位	描述
7	VBUSERR	RO	0	VBUS 中断使能（只有主机模式使用）
6	SESREQ	RO	0	会话请求中断使能（只有主机模式使用）
5	DISCON	RO	0	连接断开中断使能
4	CONN	RO	0	连接中断使能
3	SOF	RO	0	帧起始中断使能
2	BABBLE	RO	0	Babble 中断使能
1	RESUME	RO	0	唤醒中断使能
0	SUSPEND	RO	0	挂起中断使能（只有设备模式使用）

第 2 章 Stellaris 的 USB 处理器

此外还有 USBEPC、USBEPCRIS、USBEPC、USBEPCRIS、USBEPCIM、USBEPCISC 电源管理中断；USBIDVISC、USBIDVIM、USBIDVRIS 等 ID 检测中断控制；USBDRRIS、USBDRIM、USBDRISC 唤醒中断控制；USBVDC、USBVDCRIS、USBVDCIM、USBVDCISC 的 VBUS Droop 控制中断。这些中断使用较少，不在此详细描述。

例如，写一组函数，管理 USB 中断。USBIntEnable(unsigned long ulBase, unsigned long ulFlags) 使能 USB 中断、USBIntDisable(unsigned long ulBase, unsigned long ulFlags) 禁止 USB 中断，unsigned long USBIntStatus(unsigned long ulBase) 获取中断标志。

USBIntEnable()、USBIntDisable() 的 ulFlags 参数和 USBIntStatus() 返回参数，如下所示。

```
#define USB_INTCTRL_ALL           0x000003FF     // All control interrupt sources
#define USB_INTCTRL_STATUS        0x000000FF     // Status Interrupts
#define USB_INTCTRL_VBUS_ERR      0x00000080     // VBUS Error
#define USB_INTCTRL_SESSION       0x00000040     // Session Start Detected
#define USB_INTCTRL_SESSION_END   0x00000040     // Session End Detected
#define USB_INTCTRL_DISCONNECT    0x00000020     // Disconnect Detected
#define USB_INTCTRL_CONNECT       0x00000010     // Device Connect Detected
#define USB_INTCTRL_SOF           0x00000008     // Start of Frame Detected
#define USB_INTCTRL_BABBLE        0x00000004     // Babble signaled
#define USB_INTCTRL_RESET         0x00000004     // Reset signaled
#define USB_INTCTRL_RESUME        0x00000002     // Resume detected
#define USB_INTCTRL_SUSPEND       0x00000001     // Suspend detected
#define USB_INTCTRL_MODE_DETECT   0x00000200     // Mode value valid
#define USB_INTCTRL_POWER_FAULT   0x00000100     // Power Fault detected
//端点中断控制
#define USB_INTEP_ALL             0xFFFFFFFF     // Host IN Interrupts
#define USB_INTEP_HOST_IN         0xFFFE0000     // Host IN Interrupts
#define USB_INTEP_HOST_IN_15      0x80000000     // Endpoint 15 Host IN Interrupt
#define USB_INTEP_HOST_IN_14      0x40000000     // Endpoint 14 Host IN Interrupt
#define USB_INTEP_HOST_IN_13      0x20000000     // Endpoint 13 Host IN Interrupt
#define USB_INTEP_HOST_IN_12      0x10000000     // Endpoint 12 Host IN Interrupt
#define USB_INTEP_HOST_IN_11      0x08000000     // Endpoint 11 Host IN Interrupt
#define USB_INTEP_HOST_IN_10      0x04000000     // Endpoint 10 Host IN Interrupt
#define USB_INTEP_HOST_IN_9       0x02000000     // Endpoint 9 Host IN Interrupt
#define USB_INTEP_HOST_IN_8       0x01000000     // Endpoint 8 Host IN Interrupt
#define USB_INTEP_HOST_IN_7       0x00800000     // Endpoint 7 Host IN Interrupt
#define USB_INTEP_HOST_IN_6       0x00400000     // Endpoint 6 Host IN Interrupt
#define USB_INTEP_HOST_IN_5       0x00200000     // Endpoint 5 Host IN Interrupt
#define USB_INTEP_HOST_IN_4       0x00100000     // Endpoint 4 Host IN Interrupt
```

```
#define USB_INTEP_HOST_IN_3       0x00080000    // Endpoint 3 Host IN Interrupt
#define USB_INTEP_HOST_IN_2       0x00040000    // Endpoint 2 Host IN Interrupt
#define USB_INTEP_HOST_IN_1       0x00020000    // Endpoint 1 Host IN Interrupt
#define USB_INTEP_DEV_OUT         0xFFFE0000    // Device OUT Interrupts
#define USB_INTEP_DEV_OUT_15      0x80000000    // Endpoint 15 Device OUT Interrupt
#define USB_INTEP_DEV_OUT_14      0x40000000    // Endpoint 14 Device OUT Interrupt
#define USB_INTEP_DEV_OUT_13      0x20000000    // Endpoint 13 Device OUT Interrupt
#define USB_INTEP_DEV_OUT_12      0x10000000    // Endpoint 12 Device OUT Interrupt
#define USB_INTEP_DEV_OUT_11      0x08000000    // Endpoint 11 Device OUT Interrupt
#define USB_INTEP_DEV_OUT_10      0x04000000    // Endpoint 10 Device OUT Interrupt
#define USB_INTEP_DEV_OUT_9       0x02000000    // Endpoint 9 Device OUT Interrupt
#define USB_INTEP_DEV_OUT_8       0x01000000    // Endpoint 8 Device OUT Interrupt
#define USB_INTEP_DEV_OUT_7       0x00800000    // Endpoint 7 Device OUT Interrupt
#define USB_INTEP_DEV_OUT_6       0x00400000    // Endpoint 6 Device OUT Interrupt
#define USB_INTEP_DEV_OUT_5       0x00200000    // Endpoint 5 Device OUT Interrupt
#define USB_INTEP_DEV_OUT_4       0x00100000    // Endpoint 4 Device OUT Interrupt
#define USB_INTEP_DEV_OUT_3       0x00080000    // Endpoint 3 Device OUT Interrupt
#define USB_INTEP_DEV_OUT_2       0x00040000    // Endpoint 2 Device OUT Interrupt
#define USB_INTEP_DEV_OUT_1       0x00020000    // Endpoint 1 Device OUT Interrupt
#define USB_INTEP_HOST_OUT        0x0000FFFE    // Host OUT Interrupts
#define USB_INTEP_HOST_OUT_15     0x00008000    // Endpoint 15 Host OUT Interrupt
#define USB_INTEP_HOST_OUT_14     0x00004000    // Endpoint 14 Host OUT Interrupt
#define USB_INTEP_HOST_OUT_13     0x00002000    // Endpoint 13 Host OUT Interrupt
#define USB_INTEP_HOST_OUT_12     0x00001000    // Endpoint 12 Host OUT Interrupt
#define USB_INTEP_HOST_OUT_11     0x00000800    // Endpoint 11 Host OUT Interrupt
#define USB_INTEP_HOST_OUT_10     0x00000400    // Endpoint 10 Host OUT Interrupt
#define USB_INTEP_HOST_OUT_9      0x00000200    // Endpoint 9 Host OUT Interrupt
#define USB_INTEP_HOST_OUT_8      0x00000100    // Endpoint 8 Host OUT Interrupt
#define USB_INTEP_HOST_OUT_7      0x00000080    // Endpoint 7 Host OUT Interrupt
#define USB_INTEP_HOST_OUT_6      0x00000040    // Endpoint 6 Host OUT Interrupt
#define USB_INTEP_HOST_OUT_5      0x00000020    // Endpoint 5 Host OUT Interrupt
#define USB_INTEP_HOST_OUT_4      0x00000010    // Endpoint 4 Host OUT Interrupt
#define USB_INTEP_HOST_OUT_3      0x00000008    // Endpoint 3 Host OUT Interrupt
#define USB_INTEP_HOST_OUT_2      0x00000004    // Endpoint 2 Host OUT Interrupt
#define USB_INTEP_HOST_OUT_1      0x00000002    // Endpoint 1 Host OUT Interrupt
#define USB_INTEP_DEV_IN          0x0000FFFE    // Device IN Interrupts
#define USB_INTEP_DEV_IN_15       0x00008000    // Endpoint 15 Device IN Interrupt
#define USB_INTEP_DEV_IN_14       0x00004000    // Endpoint 14 Device IN Interrupt
#define USB_INTEP_DEV_IN_13       0x00002000    // Endpoint 13 Device IN Interrupt
#define USB_INTEP_DEV_IN_12       0x00001000    // Endpoint 12 Device IN Interrupt
#define USB_INTEP_DEV_IN_11       0x00000800    // Endpoint 11 Device IN Interrupt
#define USB_INTEP_DEV_IN_10       0x00000400    // Endpoint 10 Device IN Interrupt
```

第 2 章　Stellaris 的 USB 处理器

```c
#define USB_INTEP_DEV_IN_9      0x00000200   // Endpoint 9 Device IN Interrupt
#define USB_INTEP_DEV_IN_8      0x00000100   // Endpoint 8 Device IN Interrupt
#define USB_INTEP_DEV_IN_7      0x00000080   // Endpoint 7 Device IN Interrupt
#define USB_INTEP_DEV_IN_6      0x00000040   // Endpoint 6 Device IN Interrupt
#define USB_INTEP_DEV_IN_5      0x00000020   // Endpoint 5 Device IN Interrupt
#define USB_INTEP_DEV_IN_4      0x00000010   // Endpoint 4 Device IN Interrupt
#define USB_INTEP_DEV_IN_3      0x00000008   // Endpoint 3 Device IN Interrupt
#define USB_INTEP_DEV_IN_2      0x00000004   // Endpoint 2 Device IN Interrupt
#define USB_INTEP_DEV_IN_1      0x00000002   // Endpoint 1 Device IN Interrupt
#define USB_INTEP_0             0x00000001   // Endpoint 0 Interrupt
```

中断使能函数如下所示。

```c
void USBIntEnable(unsigned long ulBase, unsigned long ulFlags)
{
    //发送中断控制
    if(ulFlags & (USB_INT_HOST_OUT | USB_INT_DEV_IN | USB_INT_EP0))
    {
        HWREGH(ulBase + USB_O_TXIE) |=
            ulFlags & (USB_INT_HOST_OUT | USB_INT_DEV_IN | USB_INT_EP0);
    }
    //接收中断控制
    if(ulFlags & (USB_INT_HOST_IN | USB_INT_DEV_OUT))
    {
        HWREGH(ulBase + USB_O_RXIE) |=
            ((ulFlags & (USB_INT_HOST_IN | USB_INT_DEV_OUT)) >>
             USB_INT_RX_SHIFT);
    }
    //通用中断控制
    if(ulFlags & USB_INT_STATUS)
    {
        HWREGB(ulBase + USB_O_IE) |=
            (ulFlags & USB_INT_STATUS) >> USB_INT_STATUS_SHIFT;
    }
    //电源中断控制
    if(ulFlags & USB_INT_POWER_FAULT)
    {
        HWREG(ulBase + USB_O_EPCIM) = USB_EPCIM_PF;
    }
    //ID 中断控制
    if(ulFlags & USB_INT_MODE_DETECT)
    {
        HWREG(USB0_BASE + USB_O_IDVIM) = USB_IDVIM_ID;
```

中断禁止函数如下所示。

```c
void USBIntDisable(unsigned long ulBase, unsigned long ulFlags)
{
    //发送中断控制
    if(ulFlags & (USB_INT_HOST_OUT | USB_INT_DEV_IN | USB_INT_EP0))
    {
        HWREGH(ulBase + USB_O_TXIE) &=
            ~(ulFlags & (USB_INT_HOST_OUT | USB_INT_DEV_IN | USB_INT_EP0));
    }
    //接收中断控制
    if(ulFlags & (USB_INT_HOST_IN | USB_INT_DEV_OUT))
    {
        HWREGH(ulBase + USB_O_RXIE) &=
            ~((ulFlags & (USB_INT_HOST_IN | USB_INT_DEV_OUT)) >>
            USB_INT_RX_SHIFT);
    }
    //通用中断控制
    if(ulFlags & USB_INT_STATUS)
    {
        HWREGB(ulBase + USB_O_IE) &=
            ~((ulFlags & USB_INT_STATUS) >> USB_INT_STATUS_SHIFT);
    }
    //电源中断控制
    if(ulFlags & USB_INT_POWER_FAULT)
    {
        HWREG(ulBase + USB_O_EPCIM) = 0;
    }
    //ID 中断控制
    if(ulFlags & USB_INT_MODE_DETECT)
    {
        HWREG(USB0_BASE + USB_O_IDVIM) = 0;
    }
}
```

获取中断标志并清除函数如下所示。

```c
unsigned long USBIntStatus(unsigned long ulBase)
{
    unsigned long ulStatus;
    ulStatus = (HWREGB(ulBase + USB_O_TXIS));
```

```c
    ulStatus |= (HWREGB(ulBase + USB_O_RXIS) << USB_INT_RX_SHIFT);
    ulStatus |= (HWREGB(ulBase + USB_O_IS) << USB_INT_STATUS_SHIFT);
    if(HWREG(ulBase + USB_O_EPCISC) & USB_EPCISC_PF)
    {
        ulStatus |= USB_INT_POWER_FAULT;
        HWREGB(ulBase + USB_O_EPCISC) |= USB_EPCISC_PF;
    }

    if(HWREG(USB0_BASE + USB_O_IDVISC) & USB_IDVRIS_ID)
    {
        ulStatus |= USB_INT_MODE_DETECT;
        HWREG(USB0_BASE + USB_O_IDVISC) |= USB_IDVRIS_ID;
    }
    return(ulStatus);
}
```

端点中断使用较为频繁,可单独控制。

```c
void USBIntDisableEndpoint(unsigned long ulBase, unsigned long ulFlags)
{
    HWREGH(ulBase + USB_O_TXIE) &=
        ~(ulFlags & (USB_INTEP_HOST_OUT | USB_INTEP_DEV_IN | USB_INTEP_0));
    HWREGH(ulBase + USB_O_RXIE) &=
        ~((ulFlags & (USB_INTEP_HOST_IN | USB_INTEP_DEV_OUT)) >>
          USB_INTEP_RX_SHIFT);
}
void USBIntEnableEndpoint(unsigned long ulBase, unsigned long ulFlags)
{
    HWREGH(ulBase + USB_O_TXIE) |=
            ulFlags & (USB_INTEP_HOST_OUT | USB_INTEP_DEV_IN | USB_INTEP_0);
    HWREGH(ulBase + USB_O_RXIE) |=
        ((ulFlags & (USB_INTEP_HOST_IN | USB_INTEP_DEV_OUT)) >>
          USB_INTEP_RX_SHIFT);
}
unsigned long USBIntStatusEndpoint(unsigned long ulBase)
{
    unsigned long ulStatus;
    ulStatus = HWREGH(ulBase + USB_O_TXIS);
    ulStatus |= (HWREGH(ulBase + USB_O_RXIS) << USB_INTEP_RX_SHIFT);
    return(ulStatus);
}
```

2.3.3 端点寄存器

端点是主机与设备之间通信的目的或来源,本节介绍 LM3S 系列 USB 处理器中对端点控制的端点寄存器的功能和使用。

(1) USBEPIDX,USB 端点索引寄存器。8 位寄存器 USBEPIDX 与寄存器 USBTXFIFOSZ、USBRXFIFOSZ、USBTXFIFOADD 以及 USBRXFIFOADD 联合使用,配置端点缓冲区的 FIFO 大小和起始地址。如表 2-20 所列,为 USBEPIDX 寄存器描述表。

表 2-20 USBEPIDX 寄存器

位	名称	类型	复位	描述
[7:4]	保留	RO	0	保留
[3:0]	EPIDX	R/W	0	端点索引

(2) USBRXFIFOSZ 和 USBTXFIFOSZ,USB 接收/发送 FIFO 大小寄存器。USBRXFIFOSZ 和 USBTXFIFOSZ 寄存器描述如表 2-21 所列。

表 2-21 USBRXFIFOSZ 和 USBTXFIFOSZ 寄存器

位	名称	类型	复位	描述
[7:5]	保留	RO	0	保留
4	DPB	R/W	0	双包缓存
[3:0]	SIZE	R/W	0	最大包 $8 \times (2^{SIZE})$

(3) USBRXFIFOADD 和 USBTXFIFOADD,USB 接收/发送 FIFO 首地址寄存器。此 8 位寄存器与 USBEPIDX 联合使用,设置被发送/接收端点的动态大小。如表 2-22 所列,为 USBRXFIFOADD 和 USBTXFIFOADD 寄存器描述表。

表 2-22 USBRXFIFOADD 和 USBTXFIFOADD 寄存器

位	名称	类型	复位	描述
[15:9]	保留	RO	0	保留
[8:0]	ADDR	R/W	0	最大包 $8 \times SIZE$

注意:对端点 FIFO 进行访问时,与 USBEPIDX 联合使用的寄存器只有 USBRXFIFOSZ、USBTXFIFOSZ、USBRXFIFOADD、USBTXFIFOADD 共 4 个。

例如:写一个函数,控制端点的 FIFO,程序如下。

```
static void  USBIndexWrite(unsigned long ulBase, unsigned long ulEndpoint,
                unsigned long ulIndexedReg, unsigned long ulValue,
                unsigned long ulSize)
```

```
{
    unsigned long ulIndex;
    // 保存当前索引寄存器值
    ulIndex = HWREGB(ulBase + USB_O_EPIDX);
    // 写新值到端点索引寄存器
    HWREGB(ulBase + USB_O_EPIDX) = ulEndpoint;
    //根据寄存器大小写入新值到FIFO寄存器
    if(ulSize = = 1)
    {
        HWREGB(ulBase + ulIndexedReg) = ulValue;
    }
    else
    {
        HWREGH(ulBase + ulIndexedReg) = ulValue;
    }
    //恢复以前索引寄存器的值
    HWREGB(ulBase + USB_O_EPIDX) = ulIndex;
}
```

注意,USBRXFIFOSZ 和 USBTXFIFOSZ 用于定义 FIFO 大小,主要功能是缓存发送或者接收数据,不定义最大传输数据包的字节大小。

(4) USBTXFUNCADDRn,发送端点功能地址寄存器。在主机模式下,用于设置端点 n 访问的目标地址。如表 2-23 所列,为 USBTXFUNCADDRn 寄存器描述表。

表 2-23 USBTXFUNCADDRn 寄存器

位	名称	类型	复位	描述
7	保留	RO	0	保留
[6:0]	ADDR	R/W	0	设备总线地址

(5) USBTXHUBPORTn,发送端点 n 集线器端口号寄存器。如表 2-24 所列,为 USBTXHUBPORTn 寄存器描述表。

表 2-24 USBTXHUBPORTn 寄存器

位	名称	类型	复位	描述
7	保留	RO	0	保留
[6:0]	ADDR	R/W	0	集线器端口

(6) USBTXHUBADDRn,发送端点 n 集线器地址寄存器。如表 2-25 所列,为 USBTXHUBADDRn 寄存器描述表。

表 2-25 USBTXHUBADDRn 寄存器

位	名称	类型	复位	描述
7	MULTTRAN	RO	0	多路开关
[6:0]	ADDR	R/W	0	集线器地址

(7) USBRXFUNCADDRn,接收端点功能地址寄存器。在主机模式下,用于设置端点 n 访问的目标地址。如表 2-26 所列,为 USBRXFUNCADDRn 寄存器描述表。

表 2-26 USBRXFUNCADDRn 寄存器

位	名称	类型	复位	描述
7	保留	RO	0	保留
[6:0]	ADDR	R/W	0	设备总线地址

(8) USBRXHUBPORTn,接收端点 n 集线器端口号寄存器。如表 2-27 所列,为 USBRXHUBPORTn 寄存器描述表。

表 2-27 USBRXHUBPORTn 寄存器

位	名称	类型	复位	描述
7	保留	RO	0	保留
[6:0]	ADDR	R/W	0	集线器端口

(9) USBRXHUBADDRn,接收端点 n 集线器地址寄存器,如表 2-28 所列,为 USBRXHUBADDRn 寄存器描述表。

表 2-28 USBRXHUBADDRn 寄存器

位	名称	类型	复位	描述
7	MULTTRAN	RO	0	多路开关
[6:0]	ADDR	R/W	0	集线器地址

注意:USBTXFUNCADDR0、USBTXHUBPORT0、USBTXHUBADDR0 同时用于端点 0 的接收和发送。

例如:写一个函数,主机的端点访问某个设备地址,程序如下。

USBHostAddrSet() 的 ulEndpoint 参数:

```
#define USB_EP_0        0x00000000    // Endpoint 0
#define USB_EP_1        0x00000010    // Endpoint 1
#define USB_EP_2        0x00000020    // Endpoint 2
```

```
#define USB_EP_3                    0x00000030    // Endpoint 3
#define USB_EP_4                    0x00000040    // Endpoint 4
#define USB_EP_5                    0x00000050    // Endpoint 5
#define USB_EP_6                    0x00000060    // Endpoint 6
#define USB_EP_7                    0x00000070    // Endpoint 7
#define USB_EP_8                    0x00000080    // Endpoint 8
#define USB_EP_9                    0x00000090    // Endpoint 9
#define USB_EP_10                   0x000000A0    // Endpoint 10
#define USB_EP_11                   0x000000B0    // Endpoint 11
#define USB_EP_12                   0x000000C0    // Endpoint 12
#define USB_EP_13                   0x000000D0    // Endpoint 13
#define USB_EP_14                   0x000000E0    // Endpoint 14
#define USB_EP_15                   0x000000F0    // Endpoint 15
#define NUM_USB_EP                  16            // Number of supported endpoints
void USBHostAddrSet(unsigned long ulBase, unsigned long ulEndpoint,
                    unsigned long ulAddr, unsigned long ulFlags)
{
    //根据 ulFlags 设置发送还是接收地址
    if(ulFlags & USB_EP_HOST_OUT)
    {
        HWREGB(ulBase + USB_O_TXFUNCADDR0 + (ulEndpoint >> 1)) = ulAddr;
    }
    else
    {
        HWREGB(ulBase + USB_O_TXFUNCADDR0 + 4 + (ulEndpoint >> 1)) = ulAddr;
    }
}
```

在主机模式时,连接到端点上的设备设定一个地址。ulBase,指定 USB 模块寄存器基址;ulEndpoint,指定端点;ulAddr 参数为目标设备的地址,这个地址是以后要通信的地址;ulFlags 参数指定是否为 IN 或 OUT 端点。

例如:写一个函数,主机端点通过集线器访问某个设备地址,程序如下。

```
void USBHostHubAddrSet(unsigned long ulBase, unsigned long ulEndpoint,
                       unsigned long ulAddr, unsigned long ulFlags)
{
    //根据 ulFlags 设置发送还是接收地址
    if(ulFlags & USB_EP_HOST_OUT)
    {
        HWREGB(ulBase + USB_O_TXHUBADDR0 + (ulEndpoint >> 1)) = ulAddr;
    }
    else
```

```
    {
        HWREGB(ulBase + USB_O_TXHUBADDR0 + 4 + (ulEndpoint >> 1)) = ulAddr;
    }
}
```

这个函数设定连接到一个端点上的设备的 Hub 地址。ulBase,指定 USB 模块寄存器基址;ulEndpoint,指定端点;ulAddr,指定要设定的地址;ulFlags,指定是否为 IN 或 OUT 端点。并且函数只能在主机模式下调用。

(10) USBTXMAXPn,发送端点 n 最大传输数据寄存器,定义发送端点单次可传输的最大数据长度。如表 2-29 所列,为 USBTXMAXPn 寄存器描述表。

表 2-29 USBTXMAXPn 寄存器

位	名 称	类 型	复 位	描 述
[15:11]	保留	RO	0	保留
[10:0]	MAXLOAD	R/W	0	单次传输数据最大字节数

(11) USBRXCOUNTn,USB 端点 n 接收字节数寄存器,从 RXFIFO 中读出数据的字节数。如表 2-30 所列,为 USBRXCOUNTn 寄存器描述表。

表 2-30 USBRXCOUNTn 寄存器

位	名 称	类 型	复 位	描 述
[15:13]	保留	RO	0	保留
[12:0]	COUNT	R/W	0	接收字节数

(12) USBTXTYPEn,主机发送类型寄存器,配置发送端点的目标端点号,传输协议,以及其运行速度。如表 2-31 所列,为 USBTXTYPEn 寄存器描述表。

表 2-31 USBTXTYPEn 寄存器

位	名 称	类 型	复 位	描 述
[7:6]	SPEED	R/W	0	运行速度
[5:4]	PROTO	R/W	0	协议
[3:0]	TEP	R/W	0	目标端点号

(13) USBTXTYPEn,主机发送类型寄存器,配置发送端点的目标端点号,传输协议,以及其运行速度。如表 2-32 所列,为 USBTXTYPEn 寄存器描述表。

表 2-32　USBTXTYPEn 寄存器

位	名称	类型	复位	描述
[7:6]	SPEED	R/W	0	运行速度
[5:4]	PROTO	R/W	0	协议
[3:0]	TEP	R/W	0	目标端点号

（14）USBTXINTERVALn，主机发送间隔寄存器。如表 2-33 所列，为 USBTXINTERVALn 寄存器描述表。

表 2-33　USBTXINTERVALn 寄存器

位	名称	类型	复位	描述
[7:0]	TXPOLL / NAKLMT	R/W	0	NAK 超时时间间隔

（15）USBRXTYPEn，主机接收类型寄存器，配置接收端点的目标端点号，传输协议，以及其运行速度。如表 2-34 所列，为 USBRXTYPEn 寄存器描述表。

表 2-34　USBRXTYPEn 寄存器

位	名称	类型	复位	描述
[7:6]	SPEED	R/W	0	运行速度
[5:4]	PROTO	R/W	0	协议
[3:0]	TEP	R/W	0	目标端点号

（16）USBRXINTERVALn，主机接收间隔寄存器。如表 2-35 所列，为 USBRXINTERVALn 寄存器描述表。

表 2-35　USBRXINTERVALn 寄存器

位	名称	类型	复位	描述
[7:0]	TXPOLL / NAKLMT	R/W	0	NAK 超时时间间隔

（17）USBRQPKTCOUNTn，块传输请求包数量寄存器。如表 2-36 所列，为 USBRQPKTCOUNTn 寄存器描述表。

表 2-36　USBRQPKTCOUNTn 寄存器

位	名称	类型	复位	描述
[15:0]	COUNT	R/W	0	块传输包数量

（18）USBRXDPKTBUFDIS，接收双包缓存禁止寄存器，如表 2-37 所列，为 USBRXDPKTBUFDIS 寄存器描述表。

表 2-37 USBRXDPKTBUFDIS 寄存器

位	名称	类型	复位	描述
[15:0]	EPn	R/W	0	接收双包缓存禁止

(19) USBTXDPKTBUFDIS,发送双包缓存禁止寄存器,如表 2-38 所列,为 USBTXDPKTBUFDIS 寄存器描述表。

表 2-38 USBTXDPKTBUFDIS 寄存器

位	名称	类型	复位	描述
[15:0]	EPn	R/W	0	发送双包缓存禁止

(20) USBFIFOn,FIFO 端点寄存器,主要进行 FIFO 访问。写操作,将向 TXFIFO 中写入数据;读操作,将从 RXFIFO 中读出数据。如表 2-39 所列,为 USBFIFOn 寄存器描述表。

表 2-39 USBFIFOn 寄存器

位	名称	类型	复位	描述
[31:0]	EPDATA	R/W	0	端点数据

(21) USBCSRL0,端点 0 控制和状态低字节寄存器,为端点 0 提供控制和状态位。如表 2-40 所列,为 USBCSRL0 寄存器描述表。

表 2-40 USBCSRL0 寄存器

位	名称	类型	复位	描述 主机/设备
7	NAKTO/SETENDC	R/W	0	NAK 超时/SETEND 清 0
6	STATUS/RXRDYC	R/W	0	状态包/清 RXRDY 位
5	REQPKT/STALL	R/W	0	请求包/发送 STALL 握手
4	ERROR/SETEND	R/W	0	错误/Setup End
3	SETUP/DATAEND	R/W	0	建立令牌包/数据结束
2	STALLED	R/W	0	端点挂起
1	TXRDY	R/W	0	发送包准备好
0	RXRDY	R/W	0	接收包准备好

(22) USBCSRH0,端点 0 控制和状态高字节寄存器,为端点 0 提供控制和状态位。如表 2-41 所列,为 USBCSRH0 寄存器描述表。

表 2-41 USBCSRH0 寄存器

位	名 称	类 型	复 位	描 述
[7:3]	保留	RO	0	保留
2	DTWE	R/W	0	数据切换写使能(只有主机模式)
1	DT	R/W	0	数据切换(只有主机模式)
0	FLUSH	R/W	0	清空 FIFO

(23) USBTXCSRLn,发送端点 n(非端点 0)控制和状态低字节寄存器,为端点 n 提供控制和状态位。如表 2-42 所列,为 USBTXCSRLn 寄存器描述表。

表 2-42 USBTXCSRLn 寄存器

位	名 称	类 型	复 位	描 述 主机/设备
7	NAKTO	R/W	0	NAK 超时(只有主机)
6	CLRDT	R/W	0	清除数据转换
5	STALLED	R/W	0	端点挂起
4	SETUP/STALL	R/W	0	建立令牌包/发送 STALL
3	FLUSH	R/W	0	清空 FIFO
2	ERROR/UNDRN	R/W	0	错误/欠运转
1	FIFONE	R/W	0	FIFO 不空
0	TXRDY	R/W	0	发送包准备好

(24) USBTXCSRHn,发送端点 n(非端点 0)控制和状态高字节寄存器,为端点 n 提供控制和状态位。如表 2-43 所列,为 USBTXCSRHn 寄存器描述表。

表 2-43 USBTXCSRHn 寄存器

位	名 称	类 型	复 位	描 述
7	AUTOSET	R/W	0	自动置位
6	ISO	R/W	0	ISO 传输(只有设备模式)
5	MODE	R/W	0	模式
4	DMAEN	R/W	0	DMA 请求使能
3	FDT	R/W	0	强制数据切换
2	DMAMOD	R/W	0	DMA 请求模式
1	DTWE	R/W	0	数据切换写使能(只有主机模式)
0	DT	R/W	0	数据切换(只有主机模式)

(25) USBRXCSRLn,接收端点 n 控制和状态低字节寄存器,如表 2-44 所列,为 USBRXCSRLn 寄存器描述表。

表 2-44 USBRXCSRLn 寄存器

位	名称	类型	复位	描述 主机/设备
7	CLRDT	R/W	0	清除数据转换
6	STALLED	R/W	0	端点挂起
5	REQPKT/ STALL	R/W	0	请求包/发送 STALL 握手
4	FLUSH	R/W	0	清空 FIFO
3	DATAERR\NAKTO/ DATAERR	R/W	0	数据错误\NAK 超时 /数据错误
2	ERROR/OVER	R/W	0	错误/Overrun
1	FULL	R/W	0	错误
0	RXRDY	R/W	0	接收包准备好

(26) USBRXCSRHn，接收端点 n 控制和状态高字节寄存器，为接收端点提供额外的控制和状态位。如表 2-45 所列，为 USBRXCSRHn 寄存器描述表。

表 2-45 USBRXCSRHn 寄存器

位	名称	类型	复位	描述
7	AUTOCL	R/W	0	自动清除
6	AUTORQ	R/W	0	自动请求
5	DMAEN	R/W	0	DMA 请求使能
4	PIDERR	R/W	0	PID 错误
3	DMAMOD	R/W	0	DMA 请求模式
2	DTWE	R/W	0	数据切换写使能（只有主机模式）
1	DT	R/W	0	数据切换（只有主机模式）
0	保留	RO	0	保留

例如：写一个函数，配置端点，程序如下。
USBHostEndpointConfig() 和 USBDevEndpointConfigSet() 的 ulFlags 参数如下所示。

```
#define USB_EP_AUTO_SET        0x00000001  // Auto set feature enabled
#define USB_EP_AUTO_REQUEST    0x00000002  // Auto request feature enabled
#define USB_EP_AUTO_CLEAR      0x00000004  // Auto clear feature enabled
#define USB_EP_DMA_MODE_0      0x00000008  // Enable DMA access using mode 0
#define USB_EP_DMA_MODE_1      0x00000010  // Enable DMA access using mode 1
#define USB_EP_MODE_ISOC       0x00000000  // Isochronous endpoint
#define USB_EP_MODE_BULK       0x00000100  // Bulk endpoint
#define USB_EP_MODE_INT        0x00000200  // Interrupt endpoint
#define USB_EP_MODE_CTRL       0x00000300  // Control endpoint
```

第 2 章 Stellaris 的 USB 处理器

```
#define USB_EP_MODE_MASK        0x00000300    // Mode Mask
#define USB_EP_SPEED_LOW        0x00000000    // Low Speed
#define USB_EP_SPEED_FULL       0x00001000    // Full Speed
#define USB_EP_HOST_IN          0x00000000    // Host IN endpoint
#define USB_EP_HOST_OUT         0x00002000    // Host OUT endpoint
#define USB_EP_DEV_IN           0x00002000    // Device IN endpoint
#define USB_EP_DEV_OUT          0x00000000    // Device OUT endpoint
```

ulFIFOSize 参数如下所示。

```
#define USB_FIFO_SZ_8           0x00000000    // 8 byte FIFO
#define USB_FIFO_SZ_16          0x00000001    // 16 byte FIFO
#define USB_FIFO_SZ_32          0x00000002    // 32 byte FIFO
#define USB_FIFO_SZ_64          0x00000003    // 64 byte FIFO
#define USB_FIFO_SZ_128         0x00000004    // 128 byte FIFO
#define USB_FIFO_SZ_256         0x00000005    // 256 byte FIFO
#define USB_FIFO_SZ_512         0x00000006    // 512 byte FIFO
#define USB_FIFO_SZ_1024        0x00000007    // 1 024 byte FIFO
#define USB_FIFO_SZ_2048        0x00000008    // 2 048 byte FIFO
#define USB_FIFO_SZ_4096        0x00000009    // 4 096 byte FIFO
#define USB_FIFO_SZ_8_DB        0x00000010    // 8 byte double buffered FIFO
#define USB_FIFO_SZ_16_DB       0x00000011    // 16 byte double buffered FIFO
#define USB_FIFO_SZ_32_DB       0x00000012    // 32 byte double buffered FIFO
#define USB_FIFO_SZ_64_DB       0x00000013    // 64 byte double buffered FIFO
#define USB_FIFO_SZ_128_DB      0x00000014    // 128 byte double buffered FIFO
#define USB_FIFO_SZ_256_DB      0x00000015    // 256 byte double buffered FIFO
#define USB_FIFO_SZ_512_DB      0x00000016    // 512 byte double buffered FIFO
#define USB_FIFO_SZ_1024_DB     0x00000017    // 1 024 byte double buffered FIFO
#define USB_FIFO_SZ_2048_DB     0x00000018    // 2 048 byte double buffered FIFO
```

宏定义端点转化到其状态控制寄存器地址：

```
#define EP_OFFSET(Endpoint)     (Endpoint - 0x10)
```

主机端点配置函数如下所示。

```
void USBHostEndpointConfig(unsigned long ulBase, unsigned long ulEndpoint,
                           unsigned long ulMaxPayload,
                           unsigned long ulNAKPollInterval,
                           unsigned long ulTargetEndpoint, unsigned long ulFlags)
{
    unsigned long ulRegister;
    //判断是否是端点 0,端点 0 的发送与接收配置使用同一寄存器
    if(ulEndpoint == USB_EP_0)
    {
```

```c
            HWREGB(ulBase + USB_O_NAKLMT) = ulNAKPollInterval;
            HWREGB(ulBase + EP_OFFSET(ulEndpoint) + USB_O_TYPE0) =
                ((ulFlags & USB_EP_SPEED_FULL) ? USB_TYPE0_SPEED_FULL :
                 USB_TYPE0_SPEED_LOW);
    }
    //端点 1~15 配置
    else
    {
        ulRegister = ulTargetEndpoint;
        if(ulFlags & USB_EP_SPEED_FULL)
        {
            ulRegister |= USB_TXTYPE1_SPEED_FULL;
        }
        else
        {
            ulRegister |= USB_TXTYPE1_SPEED_LOW;
        }
        switch(ulFlags & USB_EP_MODE_MASK)
        {
            case USB_EP_MODE_BULK:
            {
                ulRegister |= USB_TXTYPE1_PROTO_BULK;
                break;
            }
            case USB_EP_MODE_ISOC:
            {
                ulRegister |= USB_TXTYPE1_PROTO_ISOC;
                break;
            }
            case USB_EP_MODE_INT:
            {
                ulRegister |= USB_TXTYPE1_PROTO_INT;
                break;
            }
            case USB_EP_MODE_CTRL:
            {
                ulRegister |= USB_TXTYPE1_PROTO_CTRL;
                break;
            }
        }
        //发送/接收端点配置
        if(ulFlags & USB_EP_HOST_OUT)
```

```c
    {
        HWREGB(ulBase + EP_OFFSET(ulEndpoint) + USB_O_TXTYPE1) =
            ulRegister;
        HWREGB(ulBase + EP_OFFSET(ulEndpoint) + USB_O_TXINTERVAL1) =
            ulNAKPollInterval;
        HWREGB(ulBase + EP_OFFSET(ulEndpoint) + USB_O_TXMAXP1) =
            ulMaxPayload;
        ulRegister = 0;
        if(ulFlags & USB_EP_AUTO_SET)
        {
            ulRegister |= USB_TXCSRH1_AUTOSET;
        }
        if(ulFlags & USB_EP_DMA_MODE_1)
        {
            ulRegister |= USB_TXCSRH1_DMAEN | USB_TXCSRH1_DMAMOD;
        }
        else if(ulFlags & USB_EP_DMA_MODE_0)
        {
            ulRegister |= USB_TXCSRH1_DMAEN;
        }
        HWREGB(ulBase + EP_OFFSET(ulEndpoint) + USB_O_TXCSRH1) =
            (unsigned char)ulRegister;
    }
    else
    {
        HWREGB(ulBase + EP_OFFSET(ulEndpoint) + USB_O_RXTYPE1) =
            ulRegister;
        HWREGB(ulBase + EP_OFFSET(ulEndpoint) + USB_O_RXINTERVAL1) =
            ulNAKPollInterval;
        ulRegister = 0;
        if(ulFlags & USB_EP_AUTO_CLEAR)
        {
            ulRegister |= USB_RXCSRH1_AUTOCL;
        }
        //DMA 控制
        if(ulFlags & USB_EP_DMA_MODE_1)
        {
            ulRegister |= USB_RXCSRH1_DMAEN | USB_RXCSRH1_DMAMOD;
        }
        else if(ulFlags & USB_EP_DMA_MODE_0)
        {
            ulRegister |= USB_RXCSRH1_DMAEN;
```

```c
        }
        HWREGB(ulBase + EP_OFFSET(ulEndpoint) + USB_O_RXCSRH1) =
            (unsigned char)ulRegister;
    }
}
```

设备端点配置函数如下所示。

```c
void USBDevEndpointConfigSet(unsigned long ulBase, unsigned long ulEndpoint,
                             unsigned long ulMaxPacketSize, unsigned long ulFlags)
{
    unsigned long ulRegister;
    if(ulFlags & USB_EP_DEV_IN)
    {
        HWREGB(ulBase + EP_OFFSET(ulEndpoint) + USB_O_TXMAXP1) =
            ulMaxPacketSize;
        ulRegister = 0;
        if(ulFlags & USB_EP_AUTO_SET)
        {
            ulRegister |= USB_TXCSRH1_AUTOSET;
        }
        if(ulFlags & USB_EP_DMA_MODE_1)
        {
            ulRegister |= USB_TXCSRH1_DMAEN | USB_TXCSRH1_DMAMOD;
        }
        else if(ulFlags & USB_EP_DMA_MODE_0)
        {
            ulRegister |= USB_TXCSRH1_DMAEN;
        }
        if((ulFlags & USB_EP_MODE_MASK) == USB_EP_MODE_ISOC)
        {
            ulRegister |= USB_TXCSRH1_ISO;
        }
        HWREGB(ulBase + EP_OFFSET(ulEndpoint) + USB_O_TXCSRH1) =
            (unsigned char)ulRegister;
        HWREGB(ulBase + EP_OFFSET(ulEndpoint) + USB_O_TXCSRL1) =
            USB_TXCSRL1_CLRDT;
    }
    else
    {
        HWREGB(ulBase + EP_OFFSET(ulEndpoint) + USB_O_RXMAXP1) =
            ulMaxPacketSize;
```

```c
        ulRegister = 0;
        if(ulFlags & USB_EP_AUTO_CLEAR)
        {
            ulRegister = USB_RXCSRH1_AUTOCL;
        }
        if(ulFlags & USB_EP_DMA_MODE_1)
        {
            ulRegister |= USB_RXCSRH1_DMAEN | USB_RXCSRH1_DMAMOD;
        }
        else if(ulFlags & USB_EP_DMA_MODE_0)
        {
            ulRegister |= USB_RXCSRH1_DMAEN;
        }
        if((ulFlags & USB_EP_MODE_MASK) == USB_EP_MODE_ISOC)
        {
            ulRegister |= USB_RXCSRH1_ISO;
        }
        HWREGB(ulBase + EP_OFFSET(ulEndpoint) + USB_O_RXCSRH1) =
            (unsigned char)ulRegister;
        HWREGB(ulBase + EP_OFFSET(ulEndpoint) + USB_O_RXCSRL1) =
            USB_RXCSRL1_CLRDT;
    }
}
```

从 FIFO 中读取数据函数如下所示。

```c
long USBEndpointDataGet(unsigned long ulBase, unsigned long ulEndpoint,
                        unsigned char * pucData, unsigned long * pulSize)
{
    unsigned long ulRegister, ulByteCount, ulFIFO;
    if(ulEndpoint == USB_EP_0)
    {
        ulRegister = USB_O_CSRL0;
    }
    else
    {
        ulRegister = USB_O_RXCSRL1 + EP_OFFSET(ulEndpoint);
    }
    //判断数据是否准备好
    if((HWREGH(ulBase + ulRegister) & USB_CSRL0_RXRDY) == 0)
    {
        * pulSize = 0;
        return(-1);
```

```c
    }
    //获取要读取的数据个数
    ulByteCount = HWREGH(ulBase + USB_O_COUNT0 + ulEndpoint);
    ulByteCount = (ulByteCount < *pulSize) ? ulByteCount : *pulSize;
    *pulSize = ulByteCount;
    ulFIFO = ulBase + USB_O_FIFO0 + (ulEndpoint >> 2);
    //从 FIFO 中读取
    for(; ulByteCount > 0; ulByteCount--)
    {
        *pucData++ = HWREGB(ulFIFO);
    }
    return(0);
}
```

写数据到 FIFO 中的函数如下所示。

```c
long USBEndpointDataPut(unsigned long ulBase, unsigned long ulEndpoint,
                        unsigned char *pucData, unsigned long ulSize)
{
    unsigned long ulFIFO;
    unsigned char ucTxPktRdy;
    if(ulEndpoint == USB_EP_0)
    {
        ucTxPktRdy = USB_CSRL0_TXRDY;
    }
    else
    {
        ucTxPktRdy = USB_TXCSRL1_TXRDY;
    }
    //判断是否可以写入数据
    if(HWREGB(ulBase + USB_O_CSRL0 + ulEndpoint) & ucTxPktRdy)
    {
        return(-1);
    }
    //计算 FIFO 地址
    ulFIFO = ulBase + USB_O_FIFO0 + (ulEndpoint >> 2);
    //写入 FIFO
    for(; ulSize > 0; ulSize--)
    {
        HWREGB(ulFIFO) = *pucData++;
    }
    return(0);
}
```

发送刚写入 FIFO 的数据函数如下所示。

```
long USBEndpointDataSend(unsigned long ulBase, unsigned long ulEndpoint,
                         unsigned long ulTransType)
{
    unsigned long ulTxPktRdy;
    if(ulEndpoint == USB_EP_0)
    {
        ulTxPktRdy = ulTransType & 0xff;
    }
    else
    {
        ulTxPktRdy = (ulTransType >> 8) & 0xff;
    }
    // 判断发送数据包是否准备好
    if(HWREGB(ulBase + USB_O_CSRL0 + ulEndpoint) & USB_CSRL0_TXRDY)
    {
        return(-1);
    }
    //发送刚写入的数据
    HWREGB(ulBase + USB_O_CSRL0 + ulEndpoint) = ulTxPktRdy;
    return(0);
}
```

以上介绍了常用 USB 寄存器原理与操作方式，不常用的寄存器请读者参考相关数据手册。通过以上对寄存器的操作可以开发 USB 主机与设备模块，可以完成全部 USB 2.0 支持的全速和低速系统开发。一般情况下，不使用寄存器级编程，为了方便开发人员编程、缩短开发周期，LM3S 处理器官方已经编写好 C 语言 driverlib.lib 库，并免费为开发人员提供，在 LM3S5000 系列 MCU 中已将 driverlib.lib 库预先放入 ROM 中，节省程序存储空间，并且调用速度更快。第 3 章将重点介绍 Stellaris 处理器的 USB 底层驱动函数，这些函数已经编译到 driverlib.lib 中，只要加入 usb.h 头文件就可以进行程序开发。第 4 章及以后的章节会介绍使用 usblib.lib 开发 USB 主机与设备模块。

2.4 USB 处理器配置使用

与其他处理器一样，在默认情况下，所有外设资源都处于停用状态，如果确定要使用某些外设资源时，必须先进行资源初始化配置。初始化配置分为 4 步：

第 2 章 Stellaris 的 USB 处理器

(1) 内核配置

使用 USB 处理器前必须配置 RCGC2 寄存器，使能其外设时钟；使能 USB 的 PLL，并为 PHY 提供正常的工作时钟；使能相应的 USB 中断。

(2) 引脚配置

配置 RCGC2 使能相应的 GPIO 模块；配置 GPIOPCTL 寄存器中的 PMCn 位，分配 USB 信号到合适的引脚上（根据官方提供的数据手册，进行芯片配置，部份芯片不需要此步骤）；作为设备时，必须禁止向 VBUS 供电，从外部主机控制器的总线上获取电源。通常使用 USB0EPEN 信号用于控制外部稳压器，禁止使能外部稳压器，避免同时驱动电源引脚 USB0VBUS。

(3) 端点配置

在主机或设备模式开始发起通信之前，必须先配置端点寄存器。在主机模式时，端点配置过程在端点寄存器和设备的端点之间建立连接。在设备模式时，设备枚举之前必须先配置端点。

(4) 建立通信

硬件配置完成后还需进行协议设置，比如描述符、类协议等，与 USB 通信相关的协议部分都需要考虑。

例如：使用 USB 处理器实现鼠标，在进行正常通信之前，要进行以下配置（即步骤(1)、(2)完成功能，步骤(3)、(4)完成端点配置、建立通信，在本书第 5.4.3 小节，USB 鼠标开发中有详细讲解。）。

```
//配置内核 CPU 时钟,作设备时不能低于 20 MHz
HWREG(SYSCTL_RCC) &= ~(SYSCTL_RCC_MOSCDIS);
SysCtlDelay(524288);
HWREG(SYSCTL_RCC) = ((HWREG(SYSCTL_RCC) &
                    ~(SYSCTL_RCC_PWRDN | SYSCTL_RCC_XTAL_M |
                    SYSCTL_RCC_OSCSRC_M)) |
                    SYSCTL_RCC_XTAL_8MHZ | SYSCTL_RCC_OSCSRC_MAIN);
SysCtlDelay(524288);
HWREG(SYSCTL_RCC) = ((HWREG(SYSCTL_RCC) & ~(SYSCTL_RCC_BYPASS |
                    SYSCTL_RCC_SYSDIV_M)) | SYSCTL_RCC_SYSDIV_4 |
                    SYSCTL_RCC_USESYSDIV);
//使能 USB 设备时钟
SysCtlPeripheralEnable(SYSCTL_PERIPH_USB0);
// 打开 USB PHY 时钟
SysCtlUSBPLLEnable();
//清除中断标志
USBIntStatusControl(USB0_BASE);
USBIntStatusEndpoint(USB0_BASE);
//使能相关中断
```

第 2 章 Stellaris 的 USB 处理器

```
USBIntEnableControl(USB0_BASE, USB_INTCTRL_RESET |
                                USB_INTCTRL_DISCONNECT |
                                USB_INTCTRL_RESUME |
                                USB_INTCTRL_SUSPEND |
                                USB_INTCTRL_SOF);
USBIntEnableEndpoint(USB0_BASE, USB_INTEP_ALL);
//开总中断
IntEnable(INT_USB0);
//引脚配置
SysCtlPeripheralEnable(SYSCTL_PERIPH_GPIOF);
GPIOPinTypeGPIOOutput(GPIO_PORTF_BASE,0xf0);
GPIOPinWrite(GPIO_PORTF_BASE,0xf0,0);
GPIOPinTypeGPIOInput(GPIO_PORTF_BASE,0x0f);
HWREG(GPIO_PORTF_BASE + GPIO_O_PUR) = 0x0f;
```

2.5 小 结

本章简单地介绍了 LM3S 处理器基本知识,并详细阐述了 LM3S USB 处理器的常用寄存器使用,举了多个实例,让读者从多方面去理解这些寄存器的配置与相关使用方法,初步对 USB 系统开发做了简单介绍。学习完本章的内容,主要掌握以下几点:①USB 通信的优点;②LM3S USB 处理器的基本编程方法;③USB 寄存器的使用方法;④USB 初始化与配置。本章是使用 LM3S USB 处理器进行 USB 系统开发的基础。

第 3 章

底层库函数

第 2 章介绍了 Stellaris 提供 USB 处理器的相关寄存器功能与操作方法。本章主要介绍使用底层驱动函数对该系列处理器编程。底层驱动函数是使用 1 个或者多个寄存器实现常用功能的函数，而开发人员只需了解这些函数的调用就可以实现相应的功能，从而减轻开发人员的负担。

3.1 底层库函数

寄存器级编程直接、效率高，但不易编写与移植。一般情况下，在使用 Cortex-M3 处理器开发系统时，不使用寄存器级编程，因为官方工程师已经把相关的寄存器操作编写成了功能函数，只需要按函数要求正确调用这些函数，就可以自动完成相关寄存器的读/写，从面实现对应系统的功能。为了让开发者在最短时间内完成产品设计，官方工程师把这些函数统一编写为外围驱动函数库，外围驱动程序库是一系列用来访问 Stellaris 系列的基于 Cortex-M3 微处理器上的外设的驱动程序。尽管从纯粹的操作系统的角度理解，它们不是驱动程序，但这些程序确实提供了一种机制，使器件的外设使用起来很容易。

对于许多应用来说，直接使用驱动程序就能满足一般应用的功能、内存或处理要求。外设驱动程序库提供了两个编程模型：直接寄存器访问模型和软件驱动程序模型。根据应用的需要或者开发者所需要的编程环境，每个模型可以独立使用或组合使用。

每个编程模型有优点也有弱点。使用直接寄存器访问模型通常能得到比使用软件驱动程序模型更少和更高效的代码。然而，要使用直接寄存器访问模型一定要求了解每个寄存器、位段、它们之间的相互作用以及任何一个外设适当操作所需的先后顺序的详细内容；而开发者使用软件驱动程序模型，则不需要知道这些详细内容，通常只需更短的时间开发应用。

驱动程序能够对外设进行完全控制，在 USB 产品开发时，可以直接使用驱动库函数编程，从而缩短开发周期。图 3-1 为驱动程序开发模型 A。同时，为了让高级编程用户体验寄存器编程的优点，在此驱动库中也提供了寄存器编程接口宏，使用此宏可以直接操作寄存器，图 3-2 为驱动程序开发模型 B。所以，此驱动函数库已经能够充分满足所有应用编程。

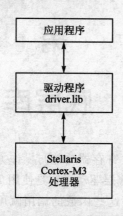

图 3-1 驱动程序开发模型 A　　　图 3-2 驱动程序开发模型 B

驱动程序包含 Stellaris 处理器的全部外设资源控制,下面就 USB 开发过程中会使用到的常用底层库函数进行分类说明。

3.2 通用库函数

通用库函数包含了内核操作、中断控制、GPIO 控制、USB 基本操作,能完成内核控制和设备调试的全部操作功能,包括系统的时钟、使能的外设、器件的配置、处理复位等;能控制嵌套向量中断控制器(NVIC),使能和禁止中断、注册中断处理程序和设置中断的优先级;提供对 8 个独立 GPIO 引脚(实际出现的引脚数取决于 GPIO 端口和器件型号)的控制;能进行寄存器级操作 USB 外设模块。

3.2.1 内核操作

在处理器使用之前要进行必要的系统配置,包括内核电压、CPU 主频、外设资源使能等;在应用程序开发中还常常用到获取系统时钟、延时等操作。这些操作都通过内核系统操作函数进行访问。内核操作函数包含在 sysctl.h 中,更多内核操作函数请参考 sysctl.h。

(1) void SysCtlLDOSet(unsigned long ulVoltage);

作用:设置 LDO 引脚的输出电压,LDO 引脚连接到处理器内核,所以也控制着处理器的内核电压。

参数:ulVoltage,LDO 电压参数,默认电压为 2.5 V。使用处理器 PLL 之前,建议开发人员设置内核电压为 2.75 V,保证能提供充足的处理器高速运行所需的电能,让系统工作更稳定。

返回:无。

例如:设置 CPU 主机为 2.75 V 程序如下。

SysCtlLDOSet(SYSCTL_LDO_2_75V);

(2) void SysCtlClockSet(unsigned long ulConfig);
作用:配置 CPU 系统主频。
参数:ulConfig,时钟配置。格式:(振荡源|晶体频率 | PLL 是否使用 |分频)。如果使用 PLL,则配置的时钟 = 200 MHz/分频。
返回:无。

本函数配置器件的时钟。输入晶体频率、使用的振荡器、PLL 的使用和系统时钟分频器全部用这个函数来配置。参数 ulConfig 是几个不同值的逻辑或,这些值中的某些值组合成组,其中只有一组值能被选用系统时钟分频,用下面的一个值来选择:SYSCTL_SYSDIV_1, SYSCTL_SYSDIV_2, SYSCTL_SYSDIV_3, …,SYSCTL_SYSDIV_64。只有 SYSCTL_SYSDIV_1 和 SYSCTL_SYSDIV_16 在 Sandstorm－class 系列器件上有效。

SYSCTL_USE_PLL 选择系统时钟来源于 PLL 输出,SYSCTL_USE_OSC 选择的系统时钟来源于主晶振,不使用 PLL 功能。

外部晶体频率用下面的一个值来选择:SYSCTL_XTAL_1MHZ, SYSCTL_XTAL_1_84MHZ, SYSCTL_XTAL_2MHZ, SYSCTL_XTAL_2_45MHZ, SYSCTL_XTAL_3_57MHZ, SYSCTL_XTAL_3_68MHZ, SYSCTL_XTAL_4MHZ, SYSCTL_XTAL_4_09MHZ, SYSCTL_XTAL_4_91MHZ, SYSCTL_XTAL_5MHZ, SYSCTL_XTAL_5_12MHZ, SYSCTL_XTAL_6MHZ, SYSCTL_XTAL_6_14MHZ, SYSCTL_XTAL_7_37MHZ, SYSCTL_XTAL_8MHZ, SYSCTL_XTAL_8_19MHZ, SYSCTL_XTAL_10MHZ, SYSCTL_XTAL_12MHZ, SYSCTL_XTAL_12_2MHZ, SYSCTL_XTAL_13_5MHZ, SYSCTL_XTAL_14_3MHZ, SYSCTL_XTAL_16MHZ,或者是 SYSCTL_XTAL_16_3MHZ。以上的值中 SYSCTL_XTAL_3_57MHZ 只在使用 PLL 时有效。在 Sandstorm 和 Fury－class 系列器件上 SYSCTL_XTAL_8_19MHZ 不可用。

振荡器源用下面的一个值来选择:SYSCTL_OSC_MAIN,SYSCTL_OSC_INT,SYSCTL_OSC_INT4,SYSCTL_OSC_INT30,或者是 SYSCTL_OSC_EXT32。在 Sandstorm - class 系列器件上 SYSCTL_OSC_INT30 和 SYSCTL_OSC_EXT32 不可用。SYSCTL_OSC_EXT32 只在有 hibernate 模式下的器件才有效,并且需要使能 hibernate 模块。

内部振荡器和主振荡器分别用:SYSCTL_INT_OSC_DIS 和 SYSCTL_MAIN_OSC_DIS 标志来禁止。为了使用外部时钟,使用 SYSCTL_USE_OSC| SYSCTL_OSC_MAIN 来选择由外部源(例如一个外部晶体振荡器)提供系统时钟。使用 SYSCTL_USE_OSC | SYSCTL_OSC_MAIN 来选择由主振荡器提供系统时钟。为了使系统时钟由 PLL 来提供,请使用 SYSCTL_USE_PLL | SYSCTL_OSC_MAIN,并根据 SYSCTL_XTAL_xxx 值选择合适的晶体。

如果选择 PLL 作为系统时钟源（即通过 SYSCTL_USE_PLL），这个函数将轮询 PLL 锁定中断来决定 PLL 是何时锁定的。如果系统控制中断的一个中断处理程序已经就绪，并且响应和清除了 PLL 锁定中断，这个函数将延迟，直至出现超时，而不是一旦 PLL 达到锁定就结束函数的执行。

例如：外接晶振 8 MHz，设置系统频率为 50 MHz。（200 MHz/4＝50 MHz），程序如下所示。

```
SysCtlClockSet(SYSCTL_XTAL_8MHZ | SYSCTL_SYSDIV_4 | SYSCTL_USE_PLL | SYSCTL_OSC_MAIN);
```

(3) unsigned long SysCtlClockGet(void);

作用：获取 CPU 主频。

参数：无。

返回：当前 CPU 主频。

(4) void SysCtlPeripheralEnable(unsigned long ulPeripheral);

作用：使能外设资源，外设使用前必须先使能，通过使能对应外设资源时钟来实现。

参数：ulPeripheral，外设资源。

返回：无。

例如：使能 GPIO 的端口 A，使能 USB0 外设资源，程序如下所示。

```
SysCtlPeripheralEnable(SYSCTL_PERIPH_GPIOA);
SysCtlPeripheralEnable(SYSCTL_PERIPH_USB0);
```

(5) void SysCtlDelay(unsigned long ulCount);

作用：延时 ulCount×3 个系统时钟周期。

参数：ulCount，延时计数。延时为 ulCount×3 个系统时钟周期。

返回：无。

例如：延时 1 s，程序如下所示。

```
SysCtlDelay(SysCtlClockGet()/ 3);
```

(6) void SysCtlUSBPLLEnable (void);

作用：使能 USB 的 PLL 模块。

返回：无。

本函数将使能 USB 控制器的 PLL 模块，即 USB 的物理层。在用 USB 连接任何外部器件时，必须被调用此函数。

例如：在使用 USB 设备时会先配置处理器，使用本节函数完成系统配置：外接晶振 6 MHz，设置内核电压为 2.75 V，配置 CPU 主频为 25 MHz，使能 USB0 模块，使能端口 F，并延时 100 ms，程序如下所示。

```
SysCtlLDOSet(SYSCTL_LDO_2_75V);
SysCtlClockSet(SYSCTL_XTAL_6MHZ | SYSCTL_SYSDIV_8| SYSCTL_USE_PLL    | SYSCTL_OSC_
            MAIN );
SysCtlPeripheralEnable(SYSCTL_PERIPH_USB0);
SysCtlPeripheralEnable(SYSCTL_PERIPH_GPIOF);
SysCtlDelay(SysCtlClockGet()/ 30);
```

3.2.2 系统中断控制

对于 Stellaris 处理器,中断类型很多、控制相当复杂。如图 3-3 所示,为 Stellaris 处理器中断控制基本的模型,有 3 层中断控制,如确定要使用中断,必须每一层中断都要配置。下面介绍两组中断函数,进行处理器中断控制。

(1) void IntEnable(unsigned long ulInterrupt);

作用:使能外设资源中断。

参数:ulInterrupt,指定要使能的外设资源中断。

返回:无。

例如:使能 USB0 中断,程序如下所示。

```
IntEnable(INT_USB0);
```

(2) void IntDisable(unsigned long ulInterrupt);

作用:禁止外设资源中断。

参数:ulInterrupt,外设资源中断。

返回:无。

例如:禁止 USB0 中断,程序如下所示。

```
IntDisable (INT_USB0);
```

(3) tBoolean IntMasterEnable(void)

作用:使能总中断。

参数:无。

返回:中断使能是否成功。

例如:使能总中断,程序如下所示。

```
IntMasterEnable();
```

(4) tBoolean IntMasterDisable(void)

作用:禁止总中断。

参数:无。

返回:中断禁止是否成功。

例如:禁止总中断,程序如下所示。

第 3 章 底层库函数

```
IntMasterDisable();
```

如图 3-3 所示，该类处理器中断控制必须进行 3 层配置；图 3-4 是中断控制实际操作模型，从图中可以看出中断控制的实际操作过程和配置顺序为从上到下依次配置。首先配置外设模块内部中断，比如使用 USBIntEnableEndpoint 使能 USB 模块的端点中断；再配置处理器外设中断，比如 IntEnable 使能 USB0 模块中断；最后配置处理器总中断，比如使用 IntMasterEnable 使能处理器总中断。如果不使能总中断，系统外设中断和外设内部中断都无法正常工作。

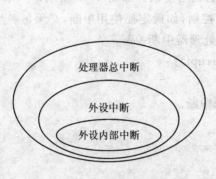 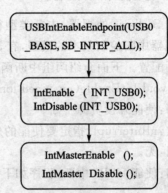

图 3-3 中断控制基本模型　　　　图 3-4 中断控制实际操作模型

3.2.3 GPIO 控制

GPIO，即通用输入输出端口。对于任意处理器来说，GPIO 占有重要的应用地位。无论是通信还是控制，都离不开 GPIO。Stellaris GPIO 模块由 7 个物理 GPIO 模块组成，分别对应端口 A、端口 B、端口 C、端口 D、端口 E、端口 F 和端口 G，各种 Stellaris Cortex-M3 的模块数量不一样，具体模块数请参考对应数据手册；每个模块支持 8 个通用输入/输出端口（GPIO），即 GPIO_PIN_0 到 GPIO_PIN_7，Stellaris Cortex-M3 可支持 5~42 个独立的 GPIO；每个 GPIO 可以根据具体使用情况独立配置成输入/输出。每种 Stellaris Cortex-M3 模块数量、模块上的 GPIO 数量不一样，但是使用方法都一样。GPIO 操作函数包含在 gpio.h 中，本书只详细讲解常用 GPIO 函数的使用。

在使用下列函数前，必须设置 SysCtlPeripheralEnable 使能对应 GPIO 模块，如下所示。

```
void GPIOPinTypeGPIOInput(unsigned long ulPort, unsigned char ucPins);
void GPIOPinTypeGPIOOutput(unsigned long ulPort, unsigned char ucPins);
void GPIOPinTypeUSBAnalog(unsigned long ulPort, unsigned char ucPins);
void GPIOPinTypeUSBDigital(unsigned long ulPort, unsigned char ucPins);
long GPIOPinRead(unsigned long ulPort, unsigned char ucPins);
```

```
void GPIOPinWrite(unsigned long ulPort, unsigned char ucPins,
                  unsigned char ucVal);
void GPIOIntTypeSet(unsigned long ulPort, unsigned char ucPins,
                    unsigned long ulIntType);
void GPIOPinIntEnable(unsigned long ulPort, unsigned char ucPins);
void GPIOPinIntDisable(unsigned long ulPort, unsigned char ucPins);
long GPIOPinIntStatus(unsigned long ulPort, tBoolean bMasked);
void GPIOPinIntClear(unsigned long ulPort, unsigned char ucPins);
```

为了使用方便,已经利用宏定义了引脚,如下所示:

```
#define GPIO_PIN_0          0x00000001    // GPIO pin 0
#define GPIO_PIN_1          0x00000002    // GPIO pin 1
#define GPIO_PIN_2          0x00000004    // GPIO pin 2
#define GPIO_PIN_3          0x00000008    // GPIO pin 3
#define GPIO_PIN_4          0x00000010    // GPIO pin 4
#define GPIO_PIN_5          0x00000020    // GPIO pin 5
#define GPIO_PIN_6          0x00000040    // GPIO pin 6
#define GPIO_PIN_7          0x00000080    // GPIO pin 7
```

(1) void GPIOPinTypeGPIOInput(unsigned long ulPort, unsigned char ucPins);

作用:把 ulPort 端口上的 ucPins 引脚设置为输入。

参数:ulPort,指定端口,GPIO_PORTn_BASE,n 取对应端口号(A~G);ucPins,指定控制引脚,GPIO_PIN_n,n 取对应引脚号(0~7)。

返回:无。

GPIO 引脚必须正确配置,以便 GPIO 输入/输出能正常工作。特别是对于 Fury-class 器件来说是很重要的,在 Furry-class 器件中,数字输入使能在默认状态下是关闭的。这个函数可为 GPIO 提供正确的配置。图 3-5 为 GPIOPinTypeGPIOInput 函数调用图。从图中可以看出 GPIOPinTypeGPIOInput 调用 GPIODirModeSet 和 GPIOPadConfigSet 来实现其配置 GPIO 为输入引脚功能,在实际编程中,也可直接使用这两个函数来实现 GPIOPinTypeGPIOInput 的功能。

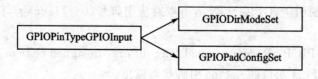

图 3-5 GPIOPinTypeGPIOInput 函数调用图

例如:设置端口 A 的 3、4 引脚为输入,程序如下所示。

```
GPIOPinTypeGPIOInput(GPIO_PORTA_BASE, (GPIO_PIN_3 | GPIO_PIN_4));
```

(2) void GPIOPinTypeGPIOOutput(unsigned long ulPort, unsigned char ucPins)

作用：把 ulPort 端口上的 ucPins 引脚设置为输出。

参数：ulPort,指定端口,GPIO_PORTn_BASE,n 取对应端口号（A～G）；ucPins,指定控制引脚,GPIO_PIN_n,n 取对应引脚号(0～7)。

返回：无。

(3) void GPIOPinTypeUSBAnalog(unsigned long ulPort, unsigned char ucPins);

作用：把 ulPort 端口上的 ucPins 引脚配置为 USB 使用的模拟信号引脚。

参数：ulPort,指定端口,GPIO_PORTn_BASE,n 取对应端口号（A～G）；ucPins,指定控制引脚,GPIO_PIN_n,n 取对应引脚号(0～7)。

返回：无。

(4) void GPIOPinTypeUSBDigital(unsigned long ulPort, unsigned char ucPins);

作用：把 ulPort 端口上的 ucPins 引脚配置为 USB 使用的数字信号引脚。

参数：ulPort,指定端口,GPIO_PORTn_BASE,n 取对应端口号（A～G）；ucPins,指定控制引脚,GPIO_PIN_n,n 取对应引脚号(0～7)。

返回：无。

GPIOPinTypeUSBAnalog 和 GPIOPinTypeUSBDigital 不能把所有引脚都设置为 USB 功能引脚，必须要相应引脚有 USB 引脚功能，才可配置。不同 USB 处理器的 USB 功能引脚不一样，具体情况请参考对应芯片的数据手册。由于以上两个函数不常用，在使用 Lm3S3XXXX 与 Lm3S5XXXX 系列时一般不使用。Lm3S9XXXX 系列或以后更高版本由于引脚复用情况较多，所以必须使用以上两个函数，确定 USB 功能引脚。

(5) void GPIOPinWrite(unsigned long ulPort, unsigned char ucPins, unsigned char ucVal);

作用：向 ulPort 端口上的 ucPins 引脚写入 ucVal 值。

参数：ulPort,指定端口,GPIO_PORTn_BASE,n 取对应端口号（A～G）；ucPins,指定控制引脚,GPIO_PIN_n,n 取对应引脚号(0～7);ucVal,待写入的数据。

返回：无。

(6) long GPIOPinRead(unsigned long ulPort, unsigned char ucPins);

作用：把 ulPort 端口的 ucPins 引脚状态读出。

参数：ulPort,指定端口,GPIO_PORTn_BASE,n 取对应端口号（A～G）；ucPins,指定控制引脚,GPIO_PIN_n,n 取对应引脚号(0～7)。

返回：ucPins 指定引脚状态组合。

例如：A 口输入数据从 B 口输出，程序如下所示。

```
GPIOPinWrite(GPIO_PORTB_BASE,0xff,GPIOPinRead(GPIO_PORTA_BASE,0xff));
//GPIO 中断类型
#define GPIO_FALLING_EDGE      0x00000000    // Interrupt on falling edge
#define GPIO_RISING_EDGE       0x00000004    // Interrupt on rising edge
#define GPIO_BOTH_EDGES        0x00000001    // Interrupt on both edges
#define GPIO_LOW_LEVEL         0x00000002    // Interrupt on low level
#define GPIO_HIGH_LEVEL        0x00000007    // Interrupt on high level
```

(7) void GPIOIntTypeSet(unsigned long ulPort, unsigned char ucPins, unsigned long ulIntType);

作用：设置 ulPort 端口的 ucPins 引脚中断类型。

参数：ulPort，指定端口，GPIO_PORTn_BASE，n 取对应端口号（A～G）；ucPins，指定控制引脚，GPIO_PIN_n，n 取对应引脚号(0～7)；ulIntType，指定中断类型。

返回：无。

(8) void GPIOPinIntEnable(unsigned long ulPort, unsigned char ucPins);

作用：使能 ulPort 端口的 ucPins 引脚中断。

参数：ulPort，指定端口，GPIO_PORTn_BASE，n 取对应端口号（A～G）；ucPins，指定控制引脚，GPIO_PIN_n，n 取对应引脚号(0～7)。

返回：无。

(9) void GPIOPinIntDisable(unsigned long ulPort, unsigned char ucPins);

作用：禁止 ulPort 端口的 ucPins 引脚中断。

参数：ulPort，指定端口，GPIO_PORTn_BASE，n 取对应端口号（A～G）；ucPins，指定控制引脚，GPIO_PIN_n，n 取对应引脚号(0～7)。

返回：无。

(10) long GPIOPinIntStatus(unsigned long ulPort, tBoolean bMasked);

作用：获取 ulPort 端口的中断状态。

参数：ulPort，指定端口，GPIO_PORTn_BASE，n 取对应端口号（A～G）；bMasked，是否获取屏蔽后的中断状态，一般使用 true。

返回：引脚中断状态。

(11) void GPIOPinIntClear(unsigned long ulPort, unsigned char ucPins);

作用：清除 ulPort 端口的 ucPins 引脚中断标志。

参数：ulPort，指定端口，GPIO_PORTn_BASE，n 取对应端口号（A～G）；ucPins，指定控制引脚，GPIO_PIN_n，n 取对应引脚号(0～7)。

返回：无。

例如：使用已学的函数，利用 GPIO 中断，完成 PC4～PC7 上按键控制 PB0、PB1、PB4、PB5 上 LED 灯的任务。KEY1 控制 LED1，KEY2 控制 LED2，KEY3 控

制LED3,KEY4控制LED4,当按键被按下,对应LED灯取当前的相反状态,如LED灯现在亮,当按下KEY时,对应的LED灯灭,再次按下KEY时,LED灯又亮,程序如下所示。

```c
#include <sysctl.h>
#include <gpio.h>
#include <hw_memmap.h>
#include <hw_ints.h>
#include <interrupt.h>
#include <lm3s8962.h>
#define   u32          unsigned long
//LED 引脚定义    PORTB
#define   LED1         GPIO_PIN_1
#define   LED2         GPIO_PIN_4
#define   LED3         GPIO_PIN_5
#define   LED4         GPIO_PIN_6
#define   LED          (LED1 | LED2 | LED3 | LED4)
//按键引脚定义    PORTC
#define   KEY1         GPIO_PIN_4
#define   KEY2         GPIO_PIN_5
#define   KEY3         GPIO_PIN_6
#define   KEY4         GPIO_PIN_7
#define   KEY          (KEY1 | KEY2 | KEY3 | KEY4)
int main(void)
{
    //设置内核电压与CPU主频
    SysCtlLDOSet(SYSCTL_LDO_2_75V);
    SysCtlClockSet(SYSCTL_XTAL_6MHZ | SYSCTL_SYSDIV_8| SYSCTL_USE_PLL   | SYSCTL_OSC_
            MAIN );
    //使能 GPIO 外设
    SysCtlPeripheralEnable(SYSCTL_PERIPH_GPIOB);
    SysCtlPeripheralEnable(SYSCTL_PERIPH_GPIOC);
    //LED IO 设置
    GPIOPinTypeGPIOOutput(GPIO_PORTB_BASE,LED);
    //KEY IO 设置
    GPIOPinTypeGPIOInput(GPIO_PORTC_BASE,KEY);
    //KEY 中断设置,配置为上升沿触发
    GPIOIntTypeSet(GPIO_PORTC_BASE,KEY,GPIO_RISING_EDGE);
    GPIOPinIntEnable(GPIO_PORTC_BASE,KEY);
    //打开端口 C 中断
    IntEnable(INT_GPIOC);
    //打开全局中断
```

```
    IntMasterEnable();
    //打开所有LED
    GPIOPinWrite(GPIO_PORTB_BASE,LED,LED);
    //等待中断
    while(1)
    {
        //add your codes!
    }
}
void GPIO_Port_C_ISR(void)
{
    u32 ulStatus = GPIOPinIntStatus(GPIO_PORTC_BASE,1);
    GPIOPinIntClear(GPIO_PORTC_BASE,KEY);
    if(ulStatus & KEY1)
    {
        GPIOPinWrite(GPIO_PORTB_BASE,LED1,~GPIOPinRead(GPIO_PORTB_BASE,LED1));
    }
    else if(ulStatus & KEY2)
        {
            GPIOPinWrite(GPIO_PORTB_BASE,LED2,~GPIOPinRead(GPIO_PORTB_BASE,
                         LED2));
        }
        else if(ulStatus & KEY3)
            {
                GPIOPinWrite(GPIO_PORTB_BASE,LED3,~GPIOPinRead(GPIO_PORTB_BASE,
                             LED3));
            }
            else if(ulStatus & KEY4)
                {
                    GPIOPinWrite(GPIO_PORTB_BASE,LED4,~GPIOPinRead(GPIO_PORTB_BASE,
                                 LED4));
                }
}
```

掌握本节介绍的 API 函数操作后，能够为外设资源控制打好基础，为学习 USB 底层驱动 API 函数做好基本功。在实际 USB 工程开发中，以上 API 函数的使用次数较多、应用范围广，在开发中占有重要地位。有关 MCU 外设的其他 API 函数请参考其他书籍和数据手册。

3.3　USB 基本操作

USB API 提供了用来访问 Stellaris USB 设备控制器或主机控制器的函数集。

第 3 章 底层库函数

API 分组如下：USBDev、USBHost、USBOTG、USBEndpoint 和 USBFIFO。USB 设备控制器只使用 USBDev 组 API 函数；USB 主机控制器只使用 USBHost 中 API 函数；OTG 接口的微控制器使用 USBOTG 组 API 函数。USB OTG 微控制器，一旦配置完，则可以使用设备或主机 API 函数。余下的 API 均可被 USB 主机和 USB 设备控制器使用，USBEndpoint 组 API 一般用来配置和访问端点，USBFIFO 组 API 则配置 FIFO 大小和位置。所以在本书中把 USB API 分为 3 类：USB 基本操作 API、设备库函数 API、主机库函数 API。"USB 基本操作 API"包含设备和主机都使用的 API：USBEndpoint 组 API、USBFIFO 组 API、其他公用 API。"设备库函数 API"只包含设备能够使用的 API 函数。"主机库函数 API"只包含主机能够使用的 API 函数。OTG 配置完成时，可以使用"设备库函数 API"和"主机库函数 API"。

本节主要介绍 USB 基本 API 函数，包含 USBEndpoint 组 API、USBFIFO 组 API 以及其他公用 API。

1. 端点控制函数

```
unsigned long USBEndpointDataAvail(unsigned long ulBase, unsigned long ulEndpoint);
long USBEndpointDataGet(unsigned long ulBase, unsigned long ulEndpoint,
                        unsigned char * pucData, unsigned long * pulSize);
long USBEndpointDataPut(unsigned long ulBase, unsigned long ulEndpoint,
                        unsigned char * pucData, unsigned long ulSize);
long USBEndpointDataSend(unsigned long ulBase, unsigned long ulEndpoint,
                         unsigned long ulTransType);
void USBEndpointDataToggleClear(unsigned long ulBase,
                                unsigned long ulEndpoint,
                                unsigned long ulFlags);
unsigned long USBEndpointStatus(unsigned long ulBase,
                                unsigned long ulEndpoint);
void USBEndpointDMAChannel(unsigned long ulBase,
                           unsigned long ulEndpoint,
                           unsigned long ulChannel);
void USBEndpointDMAEnable(unsigned long ulBase, unsigned long ulEndpoint,
                          unsigned long ulFlags);
void USBEndpointDMADisable(unsigned long ulBase,
                           unsigned long ulEndpoint,
                           unsigned long ulFlags);
```

2. FIFO 控制函数

```
unsigned long USBFIFOAddrGet(unsigned long ulBase,
                             unsigned long ulEndpoint);
void USBFIFOConfigSet(unsigned long ulBase, unsigned long ulEndpoint,
```

```
                        unsigned long ulFIFOAddress,
                        unsigned long ulFIFOSize, unsigned long ulFlags);
void USBFIFOConfigGet(unsigned long ulBase, unsigned long ulEndpoint,
                      unsigned long * pulFIFOAddress,
                      unsigned long * pulFIFOSize,
                      unsigned long ulFlags);
void USBFIFOFlush(unsigned long ulBase, unsigned long ulEndpoint,
                  unsigned long ulFlags);
```

3. 中断控制函数

```
void USBIntEnableControl(unsigned long ulBase,
                         unsigned long ulIntFlags);
void USBIntDisableControl(unsigned long ulBase,
                          unsigned long ulIntFlags);
unsigned long USBIntStatusControl(unsigned long ulBase);
void USBIntEnableEndpoint(unsigned long ulBase,
                          unsigned long ulIntFlags);
void USBIntDisableEndpoint(unsigned long ulBase,
                           unsigned long ulIntFlags);
unsigned long USBIntStatusEndpoint(unsigned long ulBase);
void USBIntEnable(unsigned long ulBase, unsigned long ulIntFlags);
void USBIntDisable(unsigned long ulBase, unsigned long ulIntFlags);
unsigned long USBIntStatus(unsigned long ulBase);
```

4. 其他控制函数

```
unsigned long USBFrameNumberGet(unsigned long ulBase);
void USBOTGSessionRequest(unsigned long ulBase, tBoolean bStart);
unsigned long USBModeGet(unsigned long ulBase);
```

端点控制函数包含所有端点控制：端点设置、端点数据发送与接收、端点 DMA 控制等，是 USB 通信中最基本的一组函数。在任何端点通信前都要进行端点控制，如果是 USB 主机，使用 USBHostEndpointConfig 配置主机端点；如果是 USB 设备，使用 USBDevEndpointConfigSet 配置设备端点。

（1）unsigned long USBEndpointDataAvail(unsigned long ulBase,
                                        unsigned long ulEndpoint)

作用：检查接收端点中有多少有效数据。

参数：ulBase，指定 USB 模块，在 Stellaris USB 处理器中只有一个 USB 模块，所以参数为 USB0_BASE。ulEndpoint，指定接收端点，USB_EP_n(n=0～15)。

返回：接收端点中的可用数据个数。

(2) long USBEndpointDataGet(unsigned long ulBase, unsigned long ulEndpoint, unsigned char * pucData, unsigned long * pulSize);

作用：从 ulEndpoint 的 RXFIFO 中读取 pulSize 指定长度的数据到 pucData 指定数据中。

参数：ulBase，指定 USB 模块，在 Stellaris USB 处理器中只有一个 USB 模块，所以参数为 USB0_BASE。ulEndpoint，指定接收端点，USB_EP_n(n=0～15)。pucData 指定接收数据的数组指针。pulSize 指定接收数据的长度。

返回：函数是否成功执行。

(3) long USBEndpointDataPut(unsigned long ulBase, unsigned long ulEndpoint, unsigned char * pucData, unsigned long ulSize);

作用：pucData 中的 ulSize 个数据放入 ulEndpoint 的 TXFIFO 中，等待发送。

参数：ulBase，指定 USB 模块，在 Stellaris USB 处理器中只有一个 USB 模块，所以参数为 USB0_BASE。ulEndpoint，指定操作的端点号，USB_EP_n(n=0～15)。pucData 指定待发送数据的数组指针。pulSize 指定发送数据的长度。

返回：函数是否成功执行。

(4) long USBEndpointDataSend(unsigned long ulBase, unsigned long ulEndpoint, unsigned long ulTransType);

作用：触发 ulEndpoint 的 TXFIFO 发送数据。

参数：ulBase，指定 USB 模块，在 Stellaris USB 处理器中只有一个 USB 模块，所以参数为 USB0_BASE。ulEndpoint，指定操作的端点号，USB_EP_n(n=0～15)。ulTransType，指定发送类型。

返回：函数是否成功执行。

ulTransType 可选参数，如下所示。

```
#define USB_TRANS_OUT       0x00000102   // Normal OUT transaction
#define USB_TRANS_IN        0x00000102   // Normal IN transaction
#define USB_TRANS_IN_LAST   0x0000010a   // Final IN transaction (for
                                         // endpoint 0 in device mode)
#define USB_TRANS_SETUP     0x0000110a   // Setup transaction (for endpoint 0)
#define USB_TRANS_STATUS    0x00000142   // Status transaction (for endpoint 0)
```

(5) void USBEndpointDataToggleClear(unsigned long ulBase, unsigned long ulEndpoint, unsigned long ulFlags);

作用：清除 Toggle。

参数：ulBase，指定 USB 模块，在 Stellaris USB 处理器中只有一个 USB 模块，所以参数为 USB0_BASE。ulEndpoint，指定操作的端点号，USB_EP_n(n=0~15)。ulFlags，指定访问 IN/OUT 端点。

返回：无。

(6) unsigned long USBEndpointStatus(unsigned long ulBase,
　　　　　　　　　　　　　　　　　　unsigned long ulEndpoint);

作用：获取 ulEndpoint 指定端点的状态。

参数：ulBase，指定 USB 模块，在 Stellaris USB 处理器中只有一个 USB 模块，所以参数为 USB0_BASE。ulEndpoint，指定操作的端点号，USB_EP_n(n=0~15)。

返回：ulEndpoint 端点的状态。

USB 工作在设备模式下时，USBEndpointStatus 返回当前设备端点状态，状态标志名字以"USB_DEV_"开头；USB 工作在主机模式下时，USBEndpointStatus 返回当前主机端点状态，状态标志名字以"USB_HOST_"开头，返回数据定义如下：

```
#define USB_HOST_IN_PID_ERROR      0x01000000  // Stall on this endpoint received
#define USB_HOST_IN_NOT_COMP       0x00100000  // Device failed to respond
#define USB_HOST_IN_STALL          0x00400000  // Stall on this endpoint received
#define USB_HOST_IN_DATA_ERROR     0x00080000  // CRC or bit-stuff error
                                               // (ISOC Mode)
#define USB_HOST_IN_NAK_TO         0x00080000  // NAK received for more than the
                                               // specified timeout period
#define USB_HOST_IN_ERROR          0x00040000  // Failed to communicate with a
                                               // device
#define USB_HOST_IN_FIFO_FULL      0x00020000  // RX FIFO full
#define USB_HOST_IN_PKTRDY         0x00010000  // Data packet ready
#define USB_HOST_OUT_NAK_TO        0x00000080  // NAK received for more than the
                                               // specified timeout period
#define USB_HOST_OUT_NOT_COMP      0x00000080  // No response from device
                                               // (ISOC mode)
#define USB_HOST_OUT_STALL         0x00000020  // Stall on this endpoint received
#define USB_HOST_OUT_ERROR         0x00000004  // Failed to communicate with a
                                               // device
#define USB_HOST_OUT_FIFO_NE       0x00000002  // TX FIFO is not empty
#define USB_HOST_OUT_PKTPEND       0x00000001  // Transmit still being transmitted
#define USB_HOST_EP0_NAK_TO        0x00000080  // NAK received for more than the
                                               // specified timeout period
#define USB_HOST_EP0_STATUS        0x00000040  // This was a status packet
#define USB_HOST_EP0_ERROR         0x00000010  // Failed to communicate with a
                                               // device
```

```
#define USB_HOST_EP0_RX_STALL    0x00000004    // Stall on this endpoint received
#define USB_HOST_EP0_RXPKTRDY    0x00000001    // Receive data packet ready
#define USB_DEV_RX_SENT_STALL    0x00400000    // Stall was sent on this endpoint
#define USB_DEV_RX_DATA_ERROR    0x00080000    // CRC error on the data
#define USB_DEV_RX_OVERRUN       0x00040000    // OUT packet was not loaded due to
                                               // a full FIFO
#define USB_DEV_RX_FIFO_FULL     0x00020000    // RX FIFO full
#define USB_DEV_RX_PKT_RDY       0x00010000    // Data packet ready
#define USB_DEV_TX_NOT_COMP      0x00000080    // Large packet split up, more data
                                               // to come
#define USB_DEV_TX_SENT_STALL    0x00000020    // Stall was sent on this endpoint
#define USB_DEV_TX_UNDERRUN      0x00000004    // IN received with no data ready
#define USB_DEV_TX_FIFO_NE       0x00000002    // The TX FIFO is not empty
#define USB_DEV_TX_TXPKTRDY      0x00000001    // Transmit still being transmitted
#define USB_DEV_EP0_SETUP_END    0x00000010    // Control transaction ended before
                                               // Data End seen
#define USB_DEV_EP0_SENT_STALL   0x00000004    // Stall was sent on this endpoint
#define USB_DEV_EP0_IN_PKTPEND   0x00000002    // Transmit data packet pending
#define USB_DEV_EP0_OUT_PKTRDY   0x00000001    // Receive data packet ready
```

(7) void USBEndpointDMAChannel(unsigned long ulBase,
　　　　　　　　　　　　　　　　unsigned long ulEndpoint,
　　　　　　　　　　　　　　　　unsigned long ulChannel);

作用:端点 DMA 控制。

参数:ulBase,指定 USB 模块,在 Stellaris USB 处理器中只有一个 USB 模块,所以参数为 USB0_BASE。ulEndpoint,指定操作的端点号,USB_EP_n(n=0～15)。ulChannel,指定 DMA 通过。

返回:无。

(8) void USBEndpointDMAEnable(unsigned long ulBase, unsigned long ulEndpoint,unsigned long ulFlags);

作用:端点 DMA 使能。

参数:ulBase,指定 USB 模块,在 Stellaris USB 处理器中只有一个 USB 模块,所以参数为 USB0_BASE。ulEndpoint,指定操作的端点号,USB_EP_n(n=0～15)。

返回:无。

(9) void USBEndpointDMADisable(unsigned long ulBase,
　　　　　　　　　　　　　　　　unsigned long ulEndpoint,
　　　　　　　　　　　　　　　　unsigned long ulFlags);

作用:端点 DMA 禁止。

参数:ulBase,指定 USB 模块,在 Stellaris USB 处理器中只有一个 USB 模块,所

以参数为 USB0_BASE。ulEndpoint,指定操作的端点号,USB_EP_n(n=0~15)。

返回:无。

在 USB 通信中,实际是端点与端点间通信,掌握端点控制函数非常必要。例如,在枚举过程中,会大量使用端点 0 与主机进行数据传输。下面的程序介绍了枚举过程中端点 0 的数据传输。

```
tDeviceInstance g_psUSBDevice[1];
static void USBDEP0StateTx(unsigned long ulIndex)
{
    unsigned long ulNumBytes;
    unsigned char * pData;
    g_psUSBDevice[0].eEP0State = USB_STATE_TX;
    //设置端点待发送
    ulNumBytes = g_psUSBDevice[0].ulEP0DataRemain;
    //端点 0 最大传输包为 64,传输数据包大于 64,就分批次传输
    if(ulNumBytes > EP0_MAX_PACKET_SIZE)
    {
        ulNumBytes = EP0_MAX_PACKET_SIZE;
    }
    //数据指针,指向待发送数组
    pData = (unsigned char *)g_psUSBDevice[0].pEP0Data;
    //分批次发送
    g_psUSBDevice[0].ulEP0DataRemain -= ulNumBytes;
    g_psUSBDevice[0].pEP0Data += ulNumBytes;
    //把待发送数据放入端点 0 的 TXFIFO 中
    USBEndpointDataPut(USB0_BASE, USB_EP_0, pData, ulNumBytes);
    // 判断 ulNumBytes 大小,如果等于端点 0 允许一次发送最大数据包的字节数,就直接
    // 发送该数据
    if(ulNumBytes == EP0_MAX_PACKET_SIZE)
    {
        //普通发送,对于主机来说,设备通过输入端点发送
        USBEndpointDataSend(USB0_BASE, USB_EP_0, USB_TRANS_IN);
    }
    else
    {
        g_psUSBDevice[0].eEP0State = USB_STATE_STATUS;
        //发送数据结束,并发送标志位
        USBEndpointDataSend(USB0_BASE, USB_EP_0, USB_TRANS_IN_LAST);
        // 判断是否有 callback 函数
        if((g_psUSBDevice[0].psInfo->sCallbacks.pfnDataSent) &&
           (g_psUSBDevice[0].ulOUTDataSize != 0))
        {
```

```
                //通过callback告诉应用程序,发送结束
                g_psUSBDevice[0].psInfo->sCallbacks.pfnDataSent(
                    g_psUSBDevice[0].pvInstance, g_psUSBDevice[0].ulOUTDataSize);
                g_psUSBDevice[0].ulOUTDataSize = 0;
            }
        }
    }
```

从上面程序中可以看出,如果发送数据大于端点最大包大小,则分批次发送,发送到最后一个数据包时,要传送包结束标志。

(10) unsigned long USBFIFOAddrGet(unsigned long ulBase,
　　　　　　　　　　　　　　　　　unsigned long ulEndpoint);

作用:通过端点获取端点对应的FIFO地址。

参数:ulBase,指定USB模块,在Stellaris USB处理器中只有一个USB模块,所以参数为USB0_BASE。ulEndpoint,指定操作的端点号,USB_EP_n(n=0～15)。

返回:FIFO地址。

(11) void USBFIFOConfigSet(unsigned long ulBase, unsigned long ulEndpoint,
　　　　　　　　　　　　　　unsigned long ulFIFOAddress,
　　　　　　　　　　　　　　unsigned long ulFIFOSize,
　　　　　　　　　　　　　　unsigned long ulFlags);

作用:FIFO配置。

参数:ulBase,指定USB模块,在Stellaris USB处理器中只有一个USB模块,所以参数为USB0_BASE。ulEndpoint,指定操作的端点号,USB_EP_n(n=0～15)。ulFIFOAddress,配置ulEndpoint端点的FIFO首地址。ulFIFOSize,FIFO大小。ulFlags,端点类型。

返回:无。

ulFIFOSize参数如下所示。

```
#define USB_FIFO_SZ_8        0x00000000    // 8 byte FIFO
#define USB_FIFO_SZ_16       0x00000001    // 16 byte FIFO
#define USB_FIFO_SZ_32       0x00000002    // 32 byte FIFO
#define USB_FIFO_SZ_64       0x00000003    // 64 byte FIFO
#define USB_FIFO_SZ_128      0x00000004    // 128 byte FIFO
#define USB_FIFO_SZ_256      0x00000005    // 256 byte FIFO
#define USB_FIFO_SZ_512      0x00000006    // 512 byte FIFO
#define USB_FIFO_SZ_1024     0x00000007    // 1 024 byte FIFO
#define USB_FIFO_SZ_2048     0x00000008    // 2 048 byte FIFO
#define USB_FIFO_SZ_4096     0x00000009    // 4 096 byte FIFO
#define USB_FIFO_SZ_8_DB     0x00000010    // 8 byte double buffered FIFO
#define USB_FIFO_SZ_16_DB    0x00000011    // 16 byte double buffered FIFO
```

```
#define USB_FIFO_SZ_32_DB        0x00000012    // 32 byte double buffered FIFO
#define USB_FIFO_SZ_64_DB        0x00000013    // 64 byte double buffered FIFO
#define USB_FIFO_SZ_128_DB       0x00000014    // 128 byte double buffered FIFO
#define USB_FIFO_SZ_256_DB       0x00000015    // 256 byte double buffered FIFO
#define USB_FIFO_SZ_512_DB       0x00000016    // 512 byte double buffered FIFO
#define USB_FIFO_SZ_1024_DB      0x00000017    // 1 024 byte double buffered FIFO
#define USB_FIFO_SZ_2048_DB      0x00000018    // 2 048 byte double buffered FIFO
```

ulFlags 参数如下所示。

```
#define USB_EP_AUTO_SET          0x00000001    // Auto set feature enabled
#define USB_EP_AUTO_REQUEST      0x00000002    // Auto request feature enabled
#define USB_EP_AUTO_CLEAR        0x00000004    // Auto clear feature enabled
#define USB_EP_DMA_MODE_0        0x00000008    // Enable DMA access using mode 0
#define USB_EP_DMA_MODE_1        0x00000010    // Enable DMA access using mode 1
#define USB_EP_MODE_ISOC         0x00000000    // Isochronous endpoint
#define USB_EP_MODE_BULK         0x00000100    // Bulk endpoint
#define USB_EP_MODE_INT          0x00000200    // Interrupt endpoint
#define USB_EP_MODE_CTRL         0x00000300    // Control endpoint
#define USB_EP_MODE_MASK         0x00000300    // Mode Mask
#define USB_EP_SPEED_LOW         0x00000000    // Low Speed
#define USB_EP_SPEED_FULL        0x00001000    // Full Speed
#define USB_EP_HOST_IN           0x00000000    // Host IN endpoint
#define USB_EP_HOST_OUT          0x00002000    // Host OUT endpoint
#define USB_EP_DEV_IN            0x00002000    // Device IN endpoint
#define USB_EP_DEV_OUT           0x00000000    // Device OUT endpoint
```

(12) void USBFIFOConfigGet(unsigned long ulBase, unsigned long ulEndpoint,
　　　　　　　　　　　　unsigned long * pulFIFOAddress,
　　　　　　　　　　　　unsigned long * pulFIFOSize,
　　　　　　　　　　　　unsigned long ulFlags);

作用：获取 FIFO 配置。

参数：ulBase，指定 USB 模块，在 Stellaris USB 处理器中只有一个 USB 模块，所以参数为 USB0_BASE。ulEndpoint，指定操作的端点号，USB_EP_n(n＝0～15)。ulFIFOAddress，获取 ulEndpoint 端点的 FIFO 首地址。ulFIFOSize，获取 FIFO 大小。ulFlags，端点类型。

返回：无。

(13) void USBFIFOFlush(unsigned long ulBase, unsigned long ulEndpoint,
　　　　　　　　　　　unsigned long ulFlags);

作用：清空 FIFO。

参数：ulBase，指定 USB 模块，在 Stellaris USB 处理器中只有一个 USB 模块，所以参数为 USB0_BASE。ulEndpoint,指定操作的端点号,USB_EP_n(n=0~15)。ulFlags,端点类型。

返回:无。

中断控制在 USB 开发中具有重要地位,USB 处理速度高达 12 MHz,每一种状态都可以触发中断,并让处理器进行数据处理;USB 中断类型多于 20 种。在这种复杂的控制中,中断管理和相应的数据处理非常重要。以下这些函数可以有效地进行USB 中断管理与控制。在使用这些函数前必须使能处理器总中断和外设模块中断。

(14) void USBIntEnableControl(unsigned long ulBase, unsigned long ulIntFlags);

作用:使能 USB 通用中断,除端点中断外的所有 USB 外设内部中断。

参数:ulBase,指定 USB 模块,在 Stellaris USB 处理器中只有一个 USB 模块,所以参数为 USB0_BASE。ulIntFlags,中断标志。

返回:无。

ulIntFlags 可用参数如下所示。

```
#define USB_INTCTRL_ALL           0x000003FF   // All control interrupt sources
#define USB_INTCTRL_STATUS        0x000000FF   // Status Interrupts
#define USB_INTCTRL_VBUS_ERR      0x00000080   // VBUS Error
#define USB_INTCTRL_SESSION       0x00000040   // Session Start Detected
#define USB_INTCTRL_SESSION_END   0x00000040   // Session End Detected
#define USB_INTCTRL_DISCONNECT    0x00000020   // Disconnect Detected
#define USB_INTCTRL_CONNECT       0x00000010   // Device Connect Detected
#define USB_INTCTRL_SOF           0x00000008   // Start of Frame Detected
#define USB_INTCTRL_BABBLE        0x00000004   // Babble signaled
#define USB_INTCTRL_RESET         0x00000004   // Reset signaled
#define USB_INTCTRL_RESUME        0x00000002   // Resume detected
#define USB_INTCTRL_SUSPEND       0x00000001   // Suspend detected
#define USB_INTCTRL_MODE_DETECT   0x00000200   // Mode value valid
#define USB_INTCTRL_POWER_FAULT   0x00000100   // Power Fault detected
```

(15) void USBIntDisableControl(unsigned long ulBase, unsigned long ulIntFlags);

作用:禁止 USB 通用中断,除端点中断外的所有 USB 外设内部中断。

参数:ulBase,指定 USB 模块,在 Stellaris USB 处理器中只有一个 USB 模块,所以参数为 USB0_BASE。ulIntFlags,中断标志。

返回:无。

(16) unsigned long USBIntStatusControl(unsigned long ulBase);

作用:获取中断标志。

参数:ulBase,指定 USB 模块,在 Stellaris USB 处理器中只有一个 USB 模块,所

以参数为 USB0_BASE。

返回：返回中断标志，与 USBIntEnableControl 的 ulIntFlags 标志相同。

(17) void USBIntEnableEndpoint(unsigned long ulBase, unsigned long ulIntFlags);

作用：使能 USB 端点中断。

参数：ulBase，指定 USB 模块，在 Stellaris USB 处理器中只有一个 USB 模块，所以参数为 USB0_BASE。ulIntFlags，中断标志。

返回：无。

ulIntFlags 可选参数如下所示。

```
#define USB_INTEP_ALL            0xFFFFFFFF    // Host IN Interrupts
#define USB_INTEP_HOST_IN        0xFFFE0000    // Host IN Interrupts
#define USB_INTEP_HOST_IN_15     0x80000000    // Endpoint 15 Host IN Interrupt
#define USB_INTEP_HOST_IN_14     0x40000000    // Endpoint 14 Host IN Interrupt
#define USB_INTEP_HOST_IN_13     0x20000000    // Endpoint 13 Host IN Interrupt
#define USB_INTEP_HOST_IN_12     0x10000000    // Endpoint 12 Host IN Interrupt
#define USB_INTEP_HOST_IN_11     0x08000000    // Endpoint 11 Host IN Interrupt
#define USB_INTEP_HOST_IN_10     0x04000000    // Endpoint 10 Host IN Interrupt
#define USB_INTEP_HOST_IN_9      0x02000000    // Endpoint 9 Host IN Interrupt
#define USB_INTEP_HOST_IN_8      0x01000000    // Endpoint 8 Host IN Interrupt
#define USB_INTEP_HOST_IN_7      0x00800000    // Endpoint 7 Host IN Interrupt
#define USB_INTEP_HOST_IN_6      0x00400000    // Endpoint 6 Host IN Interrupt
#define USB_INTEP_HOST_IN_5      0x00200000    // Endpoint 5 Host IN Interrupt
#define USB_INTEP_HOST_IN_4      0x00100000    // Endpoint 4 Host IN Interrupt
#define USB_INTEP_HOST_IN_3      0x00080000    // Endpoint 3 Host IN Interrupt
#define USB_INTEP_HOST_IN_2      0x00040000    // Endpoint 2 Host IN Interrupt
#define USB_INTEP_HOST_IN_1      0x00020000    // Endpoint 1 Host IN Interrupt
#define USB_INTEP_DEV_OUT        0xFFFE0000    // Device OUT Interrupts
#define USB_INTEP_DEV_OUT_15     0x80000000    // Endpoint 15 Device OUT Interrupt
#define USB_INTEP_DEV_OUT_14     0x40000000    // Endpoint 14 Device OUT Interrupt
#define USB_INTEP_DEV_OUT_13     0x20000000    // Endpoint 13 Device OUT Interrupt
#define USB_INTEP_DEV_OUT_12     0x10000000    // Endpoint 12 Device OUT Interrupt
#define USB_INTEP_DEV_OUT_11     0x08000000    // Endpoint 11 Device OUT Interrupt
#define USB_INTEP_DEV_OUT_10     0x04000000    // Endpoint 10 Device OUT Interrupt
#define USB_INTEP_DEV_OUT_9      0x02000000    // Endpoint 9 Device OUT Interrupt
#define USB_INTEP_DEV_OUT_8      0x01000000    // Endpoint 8 Device OUT Interrupt
#define USB_INTEP_DEV_OUT_7      0x00800000    // Endpoint 7 Device OUT Interrupt
#define USB_INTEP_DEV_OUT_6      0x00400000    // Endpoint 6 Device OUT Interrupt
#define USB_INTEP_DEV_OUT_5      0x00200000    // Endpoint 5 Device OUT Interrupt
#define USB_INTEP_DEV_OUT_4      0x00100000    // Endpoint 4 Device OUT Interrupt
#define USB_INTEP_DEV_OUT_3      0x00080000    // Endpoint 3 Device OUT Interrupt
```

```
#define USB_INTEP_DEV_OUT_2         0x00040000    // Endpoint 2 Device OUT Interrupt
#define USB_INTEP_DEV_OUT_1         0x00020000    // Endpoint 1 Device OUT Interrupt
#define USB_INTEP_HOST_OUT          0x0000FFFE    // Host OUT Interrupts
#define USB_INTEP_HOST_OUT_15       0x00008000    // Endpoint 15 Host OUT Interrupt
#define USB_INTEP_HOST_OUT_14       0x00004000    // Endpoint 14 Host OUT Interrupt
#define USB_INTEP_HOST_OUT_13       0x00002000    // Endpoint 13 Host OUT Interrupt
#define USB_INTEP_HOST_OUT_12       0x00001000    // Endpoint 12 Host OUT Interrupt
#define USB_INTEP_HOST_OUT_11       0x00000800    // Endpoint 11 Host OUT Interrupt
#define USB_INTEP_HOST_OUT_10       0x00000400    // Endpoint 10 Host OUT Interrupt
#define USB_INTEP_HOST_OUT_9        0x00000200    // Endpoint 9 Host OUT Interrupt
#define USB_INTEP_HOST_OUT_8        0x00000100    // Endpoint 8 Host OUT Interrupt
#define USB_INTEP_HOST_OUT_7        0x00000080    // Endpoint 7 Host OUT Interrupt
#define USB_INTEP_HOST_OUT_6        0x00000040    // Endpoint 6 Host OUT Interrupt
#define USB_INTEP_HOST_OUT_5        0x00000020    // Endpoint 5 Host OUT Interrupt
#define USB_INTEP_HOST_OUT_4        0x00000010    // Endpoint 4 Host OUT Interrupt
#define USB_INTEP_HOST_OUT_3        0x00000008    // Endpoint 3 Host OUT Interrupt
#define USB_INTEP_HOST_OUT_2        0x00000004    // Endpoint 2 Host OUT Interrupt
#define USB_INTEP_HOST_OUT_1        0x00000002    // Endpoint 1 Host OUT Interrupt
#define USB_INTEP_DEV_IN            0x0000FFFE    // Device IN Interrupts
#define USB_INTEP_DEV_IN_15         0x00008000    // Endpoint 15 Device IN Interrupt
#define USB_INTEP_DEV_IN_14         0x00004000    // Endpoint 14 Device IN Interrupt
#define USB_INTEP_DEV_IN_13         0x00002000    // Endpoint 13 Device IN Interrupt
#define USB_INTEP_DEV_IN_12         0x00001000    // Endpoint 12 Device IN Interrupt
#define USB_INTEP_DEV_IN_11         0x00000800    // Endpoint 11 Device IN Interrupt
#define USB_INTEP_DEV_IN_10         0x00000400    // Endpoint 10 Device IN Interrupt
#define USB_INTEP_DEV_IN_9          0x00000200    // Endpoint 9 Device IN Interrupt
#define USB_INTEP_DEV_IN_8          0x00000100    // Endpoint 8 Device IN Interrupt
#define USB_INTEP_DEV_IN_7          0x00000080    // Endpoint 7 Device IN Interrupt
#define USB_INTEP_DEV_IN_6          0x00000040    // Endpoint 6 Device IN Interrupt
#define USB_INTEP_DEV_IN_5          0x00000020    // Endpoint 5 Device IN Interrupt
#define USB_INTEP_DEV_IN_4          0x00000010    // Endpoint 4 Device IN Interrupt
#define USB_INTEP_DEV_IN_3          0x00000008    // Endpoint 3 Device OUT Interrupt
#define USB_INTEP_DEV_IN_2          0x00000004    // Endpoint 2 Device OUT Interrupt
#define USB_INTEP_DEV_IN_1          0x00000002    // Endpoint 1 Device IN Interrupt
#define USB_INTEP_0                 0x00000001    // Endpoint 0 Interrupt
```

(18) void USBIntDisableEndpoint(unsigned long ulBase, unsigned long ulInt-Flags);

作用：禁止 USB 端点中断。

参数：ulBase，指定 USB 模块，在 Stellaris USB 处理器中只有一个 USB 模块，所以参数为 USB0_BASE。ulIntFlags，中断标志。

返回:无。

(19) unsigned long USBIntStatusEndpoint(unsigned long ulBase);

作用:获取 USB 端点中断标志。

参数:ulBase,指定 USB 模块,在 Stellaris USB 处理器中只有一个 USB 模块,所以参数为 USB0_BASE。

返回:中断标志,与 USBIntEnableEndpoint 的 ulIntFlags 标志相同。

(20) void USBIntEnable(unsigned long ulBase, unsigned long ulIntFlags);

作用:使能 USB 中断,包括端点中断。

参数:ulBase,指定 USB 模块,在 Stellaris USB 处理器中只有一个 USB 模块,所以参数为 USB0_BASE。ulIntFlags,中断标志。

返回:无。

ulIntFlags 可选参数如下所示。

```
#define USB_INT_ALL            0xFF030E0F    // All Interrupt sources
#define USB_INT_STATUS         0xFF000000    // Status Interrupts
#define USB_INT_VBUS_ERR       0x80000000    // VBUS Error
#define USB_INT_SESSION_START  0x40000000    // Session Start Detected
#define USB_INT_SESSION_END    0x20000000    // Session End Detected
#define USB_INT_DISCONNECT     0x20000000    // Disconnect Detected
#define USB_INT_CONNECT        0x10000000    // Device Connect Detected
#define USB_INT_SOF            0x08000000    // Start of Frame Detected
#define USB_INT_BABBLE         0x04000000    // Babble signaled
#define USB_INT_RESET          0x04000000    // Reset signaled
#define USB_INT_RESUME         0x02000000    // Resume detected
#define USB_INT_SUSPEND        0x01000000    // Suspend detected
#define USB_INT_MODE_DETECT    0x00020000    // Mode value valid
#define USB_INT_POWER_FAULT    0x00010000    // Power Fault detected
#define USB_INT_HOST_IN        0x00000E00    // Host IN Interrupts
#define USB_INT_DEV_OUT        0x00000E00    // Device OUT Interrupts
#define USB_INT_HOST_IN_EP3    0x00000800    // Endpoint 3 Host IN Interrupt
#define USB_INT_HOST_IN_EP2    0x00000400    // Endpoint 2 Host IN Interrupt
#define USB_INT_HOST_IN_EP1    0x00000200    // Endpoint 1 Host IN Interrupt
#define USB_INT_DEV_OUT_EP3    0x00000800    // Endpoint 3 Device OUT Interrupt
#define USB_INT_DEV_OUT_EP2    0x00000400    // Endpoint 2 Device OUT Interrupt
#define USB_INT_DEV_OUT_EP1    0x00000200    // Endpoint 1 Device OUT Interrupt
#define USB_INT_HOST_OUT       0x0000000E    // Host OUT Interrupts
#define USB_INT_DEV_IN         0x0000000E    // Device IN Interrupts
#define USB_INT_HOST_OUT_EP3   0x00000008    // Endpoint 3 HOST_OUT Interrupt
#define USB_INT_HOST_OUT_EP2   0x00000004    // Endpoint 2 HOST_OUT Interrupt
#define USB_INT_HOST_OUT_EP1   0x00000002    // Endpoint 1 HOST_OUT Interrupt
```

```
#define USB_INT_DEV_IN_EP3        0x00000008    // Endpoint 3 DEV_IN Interrupt
#define USB_INT_DEV_IN_EP2        0x00000004    // Endpoint 2 DEV_IN Interrupt
#define USB_INT_DEV_IN_EP1        0x00000002    // Endpoint 1 DEV_IN Interrupt
#define USB_INT_EP0               0x00000001    // Endpoint 0 Interrupt
```

(21) void USBIntDisable(unsigned long ulBase, unsigned long ulIntFlags);

作用：禁止 USB 通用中断，包括端点中断。

参数：ulBase，指定 USB 模块，在 Stellaris USB 处理器中只有一个 USB 模块，所以参数为 USB0_BASE。ulIntFlags，中断标志。

返回：无。

(22) unsigned long USBIntStatus(unsigned long ulBase);

作用：获取中断标志。

参数：ulBase，指定 USB 模块，在 Stellaris USB 处理器中只有一个 USB 模块，所以参数为 USB0_BASE。

返回：返回中断标志，与 USBIntEnable 的 ulIntFlags 标志相同。

这些函数中 USBIntEnableControl、USBIntDisableControl、USBIntStatusControl 控制除端点外的 USB 通用中断；USBIntEnableEndpoint、USBIntDisableEndpoint、USBIntStatusEndpoint 控制 USB 端点中断，端点数量可达 16 个，包括端点 0；USBIntEnable、USBIntDisable、USBIntStatus 控制 USB 中断，包括通用中断和端点中断，但是端点中断只可以控制 4 个，端点 0 到端点 3。

(23) unsigned long USBFrameNumberGet(unsigned long ulBase);

作用：获取当前帧编号。

参数：ulBase，指定 USB 模块，在 Stellaris USB 处理器中只有一个 USB 模块，所以参数为 USB0_BASE。

返回：返回当前帧编号。

(24) void USBOTGSessionRequest(unsigned long ulBase, tBoolean bStart);

作用：OTG 启动会话。

参数：ulBase，指定 USB 模块，在 Stellaris USB 处理器中只有一个 USB 模块，所以参数为 USB0_BASE。bStart，会话启动还是停止。

返回：无。

(25) unsigned long USBModeGet(unsigned long ulBase);

作用：获取 USB 工作模式。

参数：ulBase，指定 USB 模块，在 Stellaris USB 处理器中只有一个 USB 模块，所以参数为 USB0_BASE。

返回：工作模式。

返回值如下所示。

```
#define USB_DUAL_MODE_HOST        0x00000001    // Dual mode controller is in Host
```

```
#define USB_DUAL_MODE_DEVICE      0x00000081   // Dual mode controller is in Device
#define USB_DUAL_MODE_NONE        0x00000080   // Dual mode controller mode not set
#define USB_OTG_MODE_ASIDE_HOST   0x0000001d   // OTG A side
#define USB_OTG_MODE_ASIDE_NPWR   0x00000001   // OTG A side
#define USB_OTG_MODE_ASIDE_SESS   0x00000009   // OTG A side of cable Session Valid
#define USB_OTG_MODE_ASIDE_AVAL   0x00000011   // OTG A side of the cable A valid
#define USB_OTG_MODE_ASIDE_DEV    0x00000019   // OTG A side of the cable
#define USB_OTG_MODE_BSIDE_HOST   0x0000009d   // OTG B side of the cable
#define USB_OTG_MODE_BSIDE_DEV    0x00000099   // OTG B side of the cable
#define USB_OTG_MODE_BSIDE_NPWR   0x00000081   // OTG B side of the cable
#define USB_OTG_MODE_NONE         0x00000080   // OTG controller mode is not set
```

本节主要介绍 USB 基本操作 API 函数,包含 USBEndpoint 组 API、USBFIFO 组 API、其他公用 API。通过以上函数的学习,可以对 USB 底层操作做一个大概了解,为以后的协议理解、学习打好基础。

3.4 设备库函数

设备库函数 API,只包含设备能够使用的 API 函数。要开发 USB 设备还需要与前面章节介绍的 API 函数结合,才能完成 USB 设备功能。

```
void USBDevAddrSet(unsigned long ulBase, unsigned long ulAddress);
unsigned long USBDevAddrGet(unsigned long ulBase);
void USBDevConnect(unsigned long ulBase);
void USBDevDisconnect(unsigned long ulBase);
void USBDevEndpointConfigSet(unsigned long ulBase,
                             unsigned long ulEndpoint,
                             unsigned long ulMaxPacketSize,
                             unsigned long ulFlags);
void USBDevEndpointConfigGet(unsigned long ulBase,
                             unsigned long ulEndpoint,
                             unsigned long * pulMaxPacketSize,
                             unsigned long * pulFlags);
void USBDevEndpointDataAck(unsigned long ulBase,
                           unsigned long ulEndpoint,
                           tBoolean bIsLastPacket);
void USBDevEndpointStall(unsigned long ulBase, unsigned long ulEndpoint,
                         unsigned long ulFlags);
void USBDevEndpointStallClear(unsigned long ulBase,
                              unsigned long ulEndpoint,
                              unsigned long ulFlags);
void USBDevEndpointStatusClear(unsigned long ulBase,
```

```
                          unsigned long ulEndpoint,
                          unsigned long ulFlags);
```

(1) void USBDevAddrSet(unsigned long ulBase, unsigned long ulAddress);
作用:设置设备地址。
参数:ulBase,指定 USB 模块,在 Stellaris USB 处理器中只有一个 USB 模块,所以参数为 USB0_BASE。ulAddress,为主机 USBREQ_SET_ADDRESS 请求命令时,传送给 tUSBRequest.wValue 值,用于配置设备地址。
返回:无。

(2) unsigned long USBDevAddrGet(unsigned long ulBase);
作用:获取设备地址。
参数:ulBase,指定 USB 模块,在 Stellaris USB 处理器中只有一个 USB 模块,所以参数为 USB0_BASE。
返回:设备地址。

(3) void USBDevConnect(unsigned long ulBase);
作用:软件连接到 USB 主机。
参数:ulBase,指定 USB 模块,在 Stellaris USB 处理器中只有一个 USB 模块,所以参数为 USB0_BASE。
返回:无。

(4) void USBDevDisconnect(unsigned long ulBase);
作用:软件断开设备与主机的连接。
参数:ulBase,指定 USB 模块,在 Stellaris USB 处理器中只有一个 USB 模块,所以参数为 USB0_BASE。
返回:无。

(5) void USBDevEndpointConfigSet(unsigned long ulBase, unsigned long ulEndpoint, unsigned long ulMaxPacketSize, unsigned long ulFlags);
作用:在设备模式下,配置端点的最大数据包大小,及其他配置。
参数:ulBase,指定 USB 模块,在 Stellaris USB 处理器中只有一个 USB 模块,所以参数为 USB0_BASE。ulEndpoint,指定操作的端点号,USB_EP_n(n=0~15)。ulMaxPacketSize,最大数据包大小。ulFlags,关于端点的其他配置。
返回:无。

ulFlags 可选参数如下所示。

```
#define USB_EP_AUTO_SET        0x00000001   // Auto set feature enabled
#define USB_EP_AUTO_REQUEST    0x00000002   // Auto request feature enabled
#define USB_EP_AUTO_CLEAR      0x00000004   // Auto clear feature enabled
#define USB_EP_DMA_MODE_0      0x00000008   // Enable DMA access using mode 0
```

```
#define USB_EP_DMA_MODE_1       0x00000010      // Enable DMA access using mode 1
#define USB_EP_MODE_ISOC        0x00000000      // Isochronous endpoint
#define USB_EP_MODE_BULK        0x00000100      // Bulk endpoint
#define USB_EP_MODE_INT         0x00000200      // Interrupt endpoint
#define USB_EP_MODE_CTRL        0x00000300      // Control endpoint
#define USB_EP_MODE_MASK        0x00000300      // Mode Mask
#define USB_EP_SPEED_LOW        0x00000000      // Low Speed
#define USB_EP_SPEED_FULL       0x00001000      // Full Speed
#define USB_EP_HOST_IN          0x00000000      // Host IN endpoint
#define USB_EP_HOST_OUT         0x00002000      // Host OUT endpoint
#define USB_EP_DEV_IN           0x00002000      // Device IN endpoint
#define USB_EP_DEV_OUT          0x00000000      // Device OUT endpoint
```

(6) void USBDevEndpointConfigGet(unsigned long ulBase, unsigned long ulEndpoint, unsigned long * pulMaxPacketSize, unsigned long * pulFlags);

作用:在设备模式下,获取端点配置。

参数:ulBase,指定USB模块,在Stellaris USB处理器中只有一个USB模块,所以参数为USB0_BASE。ulEndpoint,指定操作的端点号,USB_EP_n(n=0~15)。ulMaxPacketSize,返回最大数据包大小指针。ulFlags,返回端点配置指针。

返回:无。

(7) void USBDevEndpointDataAck(unsigned long ulBase, unsigned long ulEndpoint, tBoolean bIsLastPacket);

作用:在设备模式下,收到数据响应。

参数:ulBase,指定USB模块,在Stellaris USB处理器中只有一个USB模块,所以参数为USB0_BASE。ulEndpoint,指定操作的端点号,USB_EP_n(n=0~15)。bIsLastPacket,数据是否是最后一个包。

返回:无。

(8) void USBDevEndpointStall(unsigned long ulBase, unsigned long ulEndpoint, unsigned long ulFlags);

作用:在设备模式下,停止端点。

参数:ulBase,指定USB模块,在Stellaris USB处理器中只有一个USB模块,所以参数为USB0_BASE。ulEndpoint,指定操作的端点号,USB_EP_n(n=0~15)。ulFlags,指定是IN端点还是OUT端点。

返回:无。

(9) void USBDevEndpointStallClear(unsigned long ulBase, unsigned long ulEndpoint, unsigned long ulFlags);

作用:在设备模式下,清除端点的停止条件。

参数:ulBase,指定 USB 模块,在 Stellaris USB 处理器中只有一个 USB 模块,所以参数为 USB0_BASE。ulEndpoint,指定操作的端点号,USB_EP_n(n=0~15)。ulFlags,指定是 IN 端点还是 OUT 端点。

返回:无。

(10) void USBDevEndpointStatusClear(unsigned long ulBase,
　　　　　　　　　　　　　　　　　　unsigned long ulEndpoint,
　　　　　　　　　　　　　　　　　　unsigned long ulFlags);

作用:在设备模式下,清除端点的状态。

参数:ulBase,指定 USB 模块,在 Stellaris USB 处理器中只有一个 USB 模块,所以参数为 USB0_BASE。ulEndpoint,指定操作的端点号,USB_EP_n(n=0~15)。ulFlags,与 USBEndpointStatus 返回值相同,用于指定清除端点的状态。

返回:无。

以上介绍的 API 函数只能在设备模式下调用,并且在设备模式下使用频繁。如果不正确使用这些 API 函数,可能枚举不成功;枚举不成功,数据传输可能就会出现大问题,甚至使整个系统崩溃。所以开发 USB 设备,掌握这些 API 函数是非常必要的。

3.5 主机库函数

主机库函数 API,只包含主机能够使用的 API 函数。要开发 USB 主机还需要与前面章节介绍的 API 结合,才能实现 USB 主机功能。

```
void USBHostAddrSet(unsigned long ulBase, unsigned long ulEndpoint,
                    unsigned long ulAddr, unsigned long ulFlags);
unsigned long USBHostAddrGet(unsigned long ulBase,
                             unsigned long ulEndpoint,
                             unsigned long ulFlags);
void USBHostEndpointConfig(unsigned long ulBase,
                           unsigned long ulEndpoint,
                           unsigned long ulMaxPacketSize,
                           unsigned long ulNAKPollInterval,
                           unsigned long ulTargetEndpoint,
                           unsigned long ulFlags);
void USBHostEndpointDataAck(unsigned long ulBase,
                            unsigned long ulEndpoint);
void USBHostEndpointDataToggle(unsigned long ulBase,
                               unsigned long ulEndpoint,
                               tBoolean bDataToggle,
                               unsigned long ulFlags);
```

```
void USBHostEndpointStatusClear(unsigned long ulBase,
                                unsigned long ulEndpoint,
                                unsigned long ulFlags);
void USBHostHubAddrSet(unsigned long ulBase, unsigned long ulEndpoint,
                       unsigned long ulAddr, unsigned long ulFlags);
unsigned long USBHostHubAddrGet(unsigned long ulBase,
                                unsigned long ulEndpoint,
                                unsigned long ulFlags);
void USBHostPwrEnable(unsigned long ulBase);
void USBHostPwrDisable(unsigned long ulBase);
void USBHostPwrConfig(unsigned long ulBase, unsigned long ulFlags);
void USBHostPwrFaultEnable(unsigned long ulBase);
void USBHostPwrFaultDisable(unsigned long ulBase);
void USBHostRequestIN(unsigned long ulBase, unsigned long ulEndpoint);
void USBHostRequestStatus(unsigned long ulBase);
void USBHostReset(unsigned long ulBase, tBoolean bStart);
void USBHostResume(unsigned long ulBase, tBoolean bStart);
void USBHostSuspend(unsigned long ulBase);
void USBHostMode(unsigned long ulBase);
unsigned long USBHostSpeedGet(unsigned long ulBase);
```

(1) void USBHostAddrSet(unsigned long ulBase, unsigned long ulEndpoint,
　　　　　　　　　　　unsigned long ulAddr, unsigned long ulFlags);

作用：主机端点与设备通信时，设置端点访问的设备地址。

参数：ulBase，指定 USB 模块，在 Stellaris USB 处理器中只有一个 USB 模块，所以参数为 USB0_BASE。ulEndpoint，指定操作的端点号，USB_EP_n(n＝0～15)。ulFlags，指定是 IN 端点还是 OUT 端点。ulAddr，指定设备地址。

返回：无。

(2) unsigned long USBHostAddrGet(unsigned long ulBase,
　　　　　　　　　　　　　　　　unsigned long ulEndpoint,
　　　　　　　　　　　　　　　　unsigned long ulFlags);

作用：主机端点与设备通信时，获取端点访问的设备地址。

参数：ulBase，指定 USB 模块，在 Stellaris USB 处理器中只有一个 USB 模块，所以参数为 USB0_BASE。ulEndpoint，指定操作的端点号，USB_EP_n(n＝0～15)。ulFlags，指定是 IN 端点还是 OUT 端点。

返回：设备地址。

(3) void USBHostEndpointConfig(unsigned long ulBase,
　　　　　　　　　　　　　　　unsigned long ulEndpoint,
　　　　　　　　　　　　　　　unsigned long ulMaxPacketSize,

第3章　底层库函数

　　　　　　　　　　　　　　　unsigned long ulNAKPollInterval,
　　　　　　　　　　　　　　　unsigned long ulTargetEndpoint,
　　　　　　　　　　　　　　　unsigned long ulFlags);

作用：主机端点配置。

参数：ulBase，指定 USB 模块，在 Stellaris USB 处理器中只有一个 USB 模块，所以参数为 USB0_BASE。ulEndpoint，指定操作的端点号，USB_EP_n(n=0~15)。ulMaxPacketSize，端点的最大数据包大小。ulNAKPollInterval，NAK 超时限制或查询间隔，这取决于端点的类型。ulTargetEndpoint，目标端点。ulFlags，指定是 IN 端点还是 OUT 端点。

返回：无。

(4) void USBHostEndpointDataAck(unsigned long ulBase, unsigned long ulEndpoint);

作用：主机端点响应。

参数：ulBase，指定 USB 模块，在 Stellaris USB 处理器中只有一个 USB 模块，所以参数为 USB0_BASE。ulEndpoint，指定操作的端点号，USB_EP_n(n=0~15)。

返回：无。

(5) void USBHostEndpointDataToggle(unsigned long ulBase, unsigned long ulEndpoint, tBoolean bDataToggle, unsigned long ulFlags);

作用：主机端点数据转换。

参数：ulBase，指定 USB 模块，在 Stellaris USB 处理器中只有一个 USB 模块，所以参数为 USB0_BASE。ulEndpoint，指定操作的端点号，USB_EP_n(n=0~15)。bDataToggle，是否进行数据转换。ulFlags，指定是 IN 端点还是 OUT 端点。

返回：无。

(6) void USBHostEndpointStatusClear(unsigned long ulBase, unsigned long ulEndpoint, unsigned long ulFlags);

作用：清除主机端点状态。

参数：ulBase，指定 USB 模块，在 Stellaris USB 处理器中只有一个 USB 模块，所以参数为 USB0_BASE。ulEndpoint，指定操作的端点号，USB_EP_n(n=0~15)。ulFlags，指定是 IN 端点还是 OUT 端点。

返回：无。

(7) void USBHostHubAddrSet(unsigned long ulBase, unsigned long ulEndpoint, unsigned long ulAddr, unsigned long ulFlags);

作用：设置集线器地址。

参数：ulBase，指定 USB 模块，在 Stellaris USB 处理器中只有一个 USB 模块，所以参数为 USB0_BASE。ulEndpoint,指定操作的端点号,USB_EP_n(n=0～15)。ulFlags,指定是 IN 端点还是 OUT 端点。ulAddr,集线器地址。

返回:无。

(8) unsigned long USBHostHubAddrGet(unsigned long ulBase,
　　　　　　　　　　　　　　　　　　unsigned long ulEndpoint,
　　　　　　　　　　　　　　　　　　unsigned long ulFlags);

作用:获取集线器地址。

参数:ulBase,指定 USB 模块,在 Stellaris USB 处理器中只有一个 USB 模块,所以参数为 USB0_BASE。ulEndpoint,指定操作的端点号,USB_EP_n(n=0～15)。ulFlags,指定是 IN 端点还是 OUT 端点。

返回:集线器地址。

(9) void USBHostPwrEnable(unsigned long ulBase);

作用:主机使能外部电源。

参数:ulBase,指定 USB 模块,在 Stellaris USB 处理器中只有一个 USB 模块,所以参数为 USB0_BASE。

返回:无。

(10) void USBHostPwrDisable(unsigned long ulBase);

作用:主机禁止外部电源。

参数:ulBase,指定 USB 模块,在 Stellaris USB 处理器中只有一个 USB 模块,所以参数为 USB0_BASE。

返回:无。

(11) void USBHostPwrConfig(unsigned long ulBase, unsigned long ulFlags);

作用:外部电源配置。

参数:ulBase,指定 USB 模块,在 Stellaris USB 处理器中只有一个 USB 模块,所以参数为 USB0_BASE。ulFlags,电源配置。

返回:无。

(12) void USBHostPwrFaultEnable(unsigned long ulBase);

作用:电源异常使能。

参数:ulBase,指定 USB 模块,在 Stellaris USB 处理器中只有一个 USB 模块,所以参数为 USB0_BASE。

返回:无。

(13) void USBHostPwrFaultDisable(unsigned long ulBase);

作用:电源异常禁止。

参数:ulBase,指定 USB 模块,在 Stellaris USB 处理器中只有一个 USB 模块,所以参数为 USB0_BASE。

返回:无。

(14) void USBHostRequestIN(unsigned long ulBase, unsigned long ulEndpoint);

作用:发送请求。

参数:ulBase,指定 USB 模块,在 Stellaris USB 处理器中只有一个 USB 模块,所以参数为 USB0_BASE。ulEndpoint,指定操作的端点号,USB_EP_n(n=0~15)。

返回:无。

(15) void USBHostRequestStatus(unsigned long ulBase);

作用:在端点 0 上发送设备状态标准请求。

参数:ulBase,指定 USB 模块,在 Stellaris USB 处理器中只有一个 USB 模块,所以参数为 USB0_BASE。

返回:无。

(16) void USBHostReset(unsigned long ulBase, tBoolean bStart);

作用:总线复位。

参数:ulBase,指定 USB 模块,在 Stellaris USB 处理器中只有一个 USB 模块,所以参数为 USB0_BASE。bStart,复位开始还是停止。复位开始 20 ms 后软件复位停止。

返回:无。

(17) void USBHostResume(unsigned long ulBase, tBoolean bStart);

作用:总线唤醒。这个函数在设备模式下可以调用。

参数:ulBase,指定 USB 模块,在 Stellaris USB 处理器中只有一个 USB 模块,所以参数为 USB0_BASE。bStart,唤醒开始还是停止。主机模式下,唤醒开始 20 ms 后软件唤醒停止。在设备模式下,大于 10 ms 小于 15 ms。

返回:无。

(18) void USBHostSuspend(unsigned long ulBase);

作用:总线挂起。

参数:ulBase,指定 USB 模块,在 Stellaris USB 处理器中只有一个 USB 模块,所以参数为 USB0_BASE。

返回:无。

(19) void USBHostMode(unsigned long ulBase);

作用:设置 USB 处理器工作于主机模式。

参数:ulBase,指定 USB 模块,在 Stellaris USB 处理器中只有一个 USB 模块,所以参数为 USB0_BASE。

返回:无。

(20) unsigned long USBHostSpeedGet(unsigned long ulBase);

作用:获取与主机连接的设备速度。

参数:ulBase,指定 USB 模块,在 Stellaris USB 处理器中只有一个 USB 模块,所

以参数为 USB0_BASE。

返回：设备速度。

以上介绍的 API 函数（除 USBHostResume 外）只能在主机模式下调用，并且在主机模式下使用频繁。所有主机都会使用到本节介绍的 API 函数，这些函数在 USB 主机开发中占有重要地位。

例如，主机对设备进行枚举过程控制，主机标准请求过程，程序如下所示。

```
static void USBHCDEnumHandler(void)
{
    unsigned long ulEPStatus;
    unsigned long ulDataSize;
    //获取端点 0 的状态
    ulEPStatus = USBEndpointStatus(USB0_BASE, USB_EP_0);
    //判断端点 0 的状态
    if(ulEPStatus == USB_HOST_EP0_ERROR)
    {
        //如果端点 0 出现问题，清除端点 0 状态
        USBHostEndpointStatusClear(USB0_BASE, USB_EP_0, USB_HOST_EP0_ERROR);
        //清空端点 0 的 FIFO 数据
        USBFIFOFlush(USB0_BASE, USB_EP_0, 0);
        g_sUSBHEP0State.eState = EP0_STATE_ERROR;
        return;
    }
    //根据端点 0 的状态进行处理
    switch(g_sUSBHEP0State.eState)
    {
        case EP0_STATE_STATUS:
        {
            //处理接收到的数据
            if(ulEPStatus & (USB_HOST_EP0_RXPKTRDY | USB_HOST_EP0_STATUS))
            {
                //清除接收状态
                USBHostEndpointStatusClear(USB0_BASE, USB_EP_0,
                                (USB_HOST_EP0_RXPKTRDY |
                                USB_HOST_EP0_STATUS));
            }
            g_sUSBHEP0State.eState = EP0_STATE_IDLE;
            break;
        }
        case EP0_STATE_STATUS_IN:
        {
```

```c
            //主机发送状态请求
            USBHostRequestStatus(USB0_BASE);
            g_sUSBHEP0State.eState = EP0_STATE_STATUS;
            break;
        }
        case EP0_STATE_IDLE:
        {
            break;
        }
        case EP0_STATE_SETUP_OUT:
        {
            //通过端点0发送数据
            USBHCDEP0StateTx();
            break;
        }
        case EP0_STATE_SETUP_IN:
        {
            //主机发出请求
            USBHostRequestIN(USB0_BASE, USB_EP_0);
            g_sUSBHEP0State.eState = EP0_STATE_RX;
            break;
        }
        case EP0_STATE_RX:
        {
            if(ulEPStatus & USB_HOST_EP0_RX_STALL)
            {
                g_sUSBHEP0State.eState = EP0_STATE_IDLE;
                USBHostEndpointStatusClear(USB0_BASE, USB_EP_0, ulEPStatus);
                break;
            }
            if(g_sUSBHEP0State.ulBytesRemaining > MAX_PACKET_SIZE_EP0)
            {
                ulDataSize = MAX_PACKET_SIZE_EP0;
            }
            else
            {
                ulDataSize = g_sUSBHEP0State.ulBytesRemaining;
            }
            if(ulDataSize != 0)
            {
                //从端点0中获取数据
                USBEndpointDataGet(USB0_BASE, USB_EP_0, g_sUSBHEP0State.pData,
```

```c
                                    &ulDataSize);
            }
            g_sUSBHEP0State.pData += ulDataSize;
            g_sUSBHEP0State.ulBytesRemaining -= ulDataSize;
            //主机响应
            USBDevEndpointDataAck(USB0_BASE, USB_EP_0, false);
            if((ulDataSize < g_sUSBHEP0State.ulMaxPacketSize) ||
               (g_sUSBHEP0State.ulBytesRemaining == 0))
            {
                g_sUSBHEP0State.eState = EP0_STATE_STATUS;
                //数据接收完,发送 NULL 包
                USBEndpointDataSend(USB0_BASE, USB_EP_0, USB_TRANS_STATUS);
            }
            else
            {
                //主机请求
                USBHostRequestIN(USB0_BASE, USB_EP_0);
            }
            break;
        }
        case EP0_STATE_STALL:
        {
            g_sUSBHEP0State.eState = EP0_STATE_IDLE;
            break;
        }
        default:
        {
            ASSERT(0);
            break;
        }
    }
}
```

端点 0 发送数据,程序如下所示。

```c
static void USBHCDEP0StateTx(void)
{
    unsigned long ulNumBytes;
    unsigned char * pData;
    g_sUSBHEP0State.eState = EP0_STATE_SETUP_OUT;
    //设置发送数据字节数
    ulNumBytes = g_sUSBHEP0State.ulBytesRemaining;
    //判断剩下未发送数据的字节数是否大于 64
```

```
        if(ulNumBytes > 64)
        {
            ulNumBytes = 64;
        }
        // 指向待发数据
        pData = (unsigned char *)g_sUSBHEP0State.pData;
        g_sUSBHEP0State.ulBytesRemaining -= ulNumBytes;
        g_sUSBHEP0State.pData += ulNumBytes;
        // 将待发数据放入端点 0 的 FIFO 中,并通过端点 0 将其 FIFO 中的数据发送给主机
        USBEndpointDataPut(USB0_BASE, USB_EP_0, pData, ulNumBytes);
        //判断是否是最后一个包,并存入包结束标志
        if(ulNumBytes == 64)
        {
            USBEndpointDataSend(USB0_BASE, USB_EP_0, USB_TRANS_OUT);
        }
        else
        {
            USBEndpointDataSend(USB0_BASE, USB_EP_0, USB_TRANS_OUT);
            g_sUSBHEP0State.eState = EP0_STATE_STATUS_IN;
        }
    }
```

本章介绍了 USB 处理器底层 API 函数的使用,有助于加深读者对 USB 设备和主机控制的理解。对于具体的 USB 开发,使用现有的 API 函数完全可以实现。由于 USB 协议的复杂性,从第 4 章开始学习使用 Luminary Micro 公司提供的 USB 应用性库函数进行实际 USB 设备和 USB 主机系统开发。

3.6 小 结

通过本章的学习,让读者明白 Stellaris Cortex-M3 处理器使用库函数编程的理念,明白使用驱动库编程的过程,掌握 USB 系统开发时,常用的底层驱动库函数的调用方式,了解 USB 操作的基本驱动函数,在此基础上,进一步学习了 USB 设备、USB 主机等编程函数。通过本章的学习,读者可以体会到与传统 USB 系统开发相比,Cortex-M3 的 USB 处理器开发更容易,此类编程模式减轻了开发人员的编程、调试难度,降低了开发门槛,缩短了研发周期。

第 4 章

USB 库介绍

前面几章介绍了 USB 的基础知识、USB 处理器、USB 寄存器等基础性知识,利用前 3 章的知识完全可以开发 USB 系统,但是开发难度相当大。本章介绍 USB 库的基本知识,使用官方提供的 USB 库,可以进一步简化 USB 系统的开发难度。

4.1 USB 库函数简介

Luminary 为 USB 处理器提供的 USB 库函数,应用在 Stellaris 处理器上,为 USB 设备、USB 主机、OTG 开发提供 USB 协议框架和 API 函数,适用于多种开发环境:Keil、CSS、IAR、CRT、CCS 等。本书中的所有例程都在 Keil uv4 中编译。图 4-1 为 keil 工程文件组织结构。底层驱动库(DriverLib.lib)和 USB 库(USBlib.lib),这两个库还在不断更新中,进行系统开发时,请注意各个库的版本号,不同版本号的库其操作有一些不一样,请参考库的升级手册。比如现在最新的库已经包含了 Cortex-M4 的相关操作。

使用 USB 库开发时,要加入两个已经编译好的 DriverLib.lib 和 USBlib.lib。Keil 中建立 USB 开发工程文件组织结构如图 4-1 所示。

在使用 USB 库之前必须了解 USB 库的结构,有助于开发者理解与使用。USB 库分为 3 个层次:USB 设备 API、设备类驱动和设备类。图 4-2 为 USB 库架构图。

从图 4-2 中可以看出,USB 库大致可分为两类:底层驱动和 USB 协议层。最底

图 4-1 Keil 工程文件组织结构

层驱动是第 3 章讲的 USB 驱动程序,只使用 USB 驱动程序可以进行简单的 USB 开发。对于更为复杂的 USB 工程,仅仅使用驱动程序进行开发是很困难的。为了解决 USB 协议的复杂性,便引入了 USB 库,可以很方便、简单地进行复杂的 USB 工程设计。USB 库提供 3 层 API,底层为 USB 设备 API,为最基础的 USB 协议和类定义;USB 设备驱动层是在 USB 设备 API 基础上扩展的 USB 各种设备驱动,比如 HID

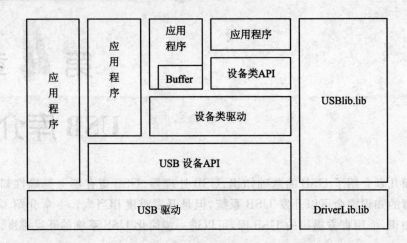

图 4-2 USB 库架构

类、CDC 类等类驱动；为了更方便程序员使用，还提供了设备类 API，扩展 USB 库的使用范围，进一步减轻开发人员的负担，可以在不用考虑更底层驱动的情况下完成 USB 工程开发。

同时开发人员在进行 USB 系统开发时有多种开发模式可供选择，USB 库提供了 4 种方案。图 4-3 为使用底层驱动开发模型。开发人员可以使用最底层的 API 驱动函数（只使用 DriverLib.lib 中的 USB 驱动函数，而不使用 USBlib.lib）进行开发，应用程序通过底层驱动与 USB 主机通信，进行数据交换。但该方案要求开发人员对 USB 协议完全了解，并熟练协议编写。

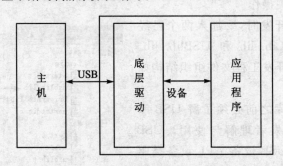

图 4-3 使用底层驱动开发模型

图 4-4 为使用底层驱动和 USB 设备驱动开发模型图。开发人员可以在最底层的 API 驱动函数基础上使用 USB 库函数提供的 USB 设备驱动函数进行 USB 系统设计，应用程序通过底层驱动和 USB 设备驱动与 USB 主机通信，进行数据交换，减轻了开发人员的负担。从图中可以看出，USB 设备驱动为 USB 底层驱动提供协议，USB 底层驱动为 USB 设备驱动提供硬件操作函数，两者紧密联系，便可以很轻松地设计出复杂的 USB 系统。该模型要求开发人员对相关子协议有所了解。

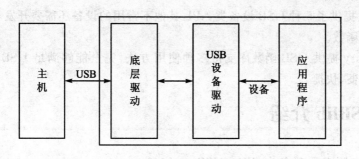

图 4-4 使用底层驱动和 USB 设备驱动开发模型

图 4-5 为使用底层驱动、USB 设备驱动、设备类驱动开发模型。开发人员可以利用最底层的 API 驱动函数、USB 设备驱动函数和设备类驱动函数,进行 USB 系统设计。设备类驱动主要提供各种 USB 设备类的驱动,比如 Audio 类驱动、HID 类驱动、Composite 类驱动、CDC 类驱动、Bulk 类驱动、Mass Storage 类驱动等 6 种基本类驱动,其他设备类驱动可以参考这 5 种驱动的源码,由开发者自己编写。使用此类开发模型,开发人员可以不用了解 USB 协议,直接进行 USB 系统编程,但是对于不属于这 5 种设备类型的 USB 系统来说,开发人员还得按照这几种驱动源码进行设备类的驱动开发。

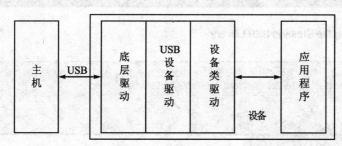

图 4-5 使用底层驱动、USB 设备驱动、设备类驱动开发模型

图 4-6 为使用 USB 库开发模型。开发人员可以利用最底层的 API 驱动函数、USB 设备驱动函数、设备类驱动函数和设备类 API,进行 USB 开发。设备类 API 主要提供各种 USB 设备类操作相关的函数,比如 HID 中的键盘、鼠标操作接口。设备

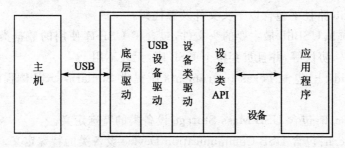

图 4-6 使用 USB 库开发模型

第 4 章　USB 库介绍

类 API 也只提供了 5 种 USB 设备类 API，其他不常用的设备还需要开发者自己编写设备类驱动函数。

Luminary 提供 USB 函数库支持多种使用方法，完全能够满足 USB 产品开发，并且使用方便、快捷。

4.2　USBlib 介绍

从 TI 官网中下载名为 SW-USBL-8049.exe 的 USB 库。图 4-7 为 USB 库安装包图，可知此 USB 库的版本号为 8 049。双击此安装包，并安装到计算机硬盘中，如图 4-8 所示，为 USB 库安装过程图。安装结束时会在安装路径下出现 USBlib 文件夹，如图 4-9 所示，为 USBlib 文件夹。

图 4-7　USB 库安装包图

在此文件夹下包含 LM3S 处理器的 USB 库（USBlib）的所有源代码。图 4-10 为 USBlib 包含的文件及文件夹。

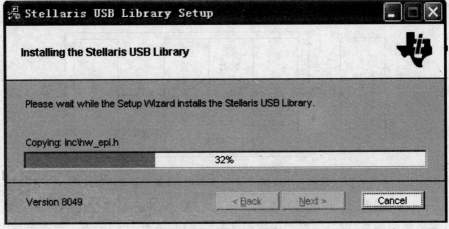

图 4-8　USB 库安装过程图

在 USBlib 路径下包含以下头文件及源代码：
- usblib.h：USBlib 最主要的头文件，包含了 USB 库使用的数据类型、结构体、宏定义、协议等，并且贯穿整个 USB 库，无处不用。
- usb-ids.h：包含 Texas Instruments 所提供 Stellaris USB 例程的 USB VID 和 PID。
- usbmsc.h：包含 USB Mass Storage 设备类的特殊定义。
- usbcdc.h：包含 USB Communication Device 设备类的特殊定义。
- usbhid.h：包含 USB Human Interface Device 人机接口类的特殊定义。

第 4 章 USB 库介绍

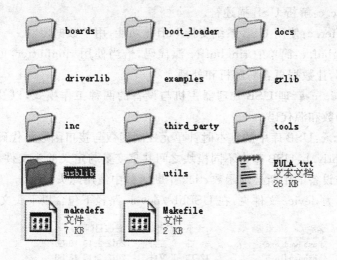

图 4-9 USBlib 文件夹

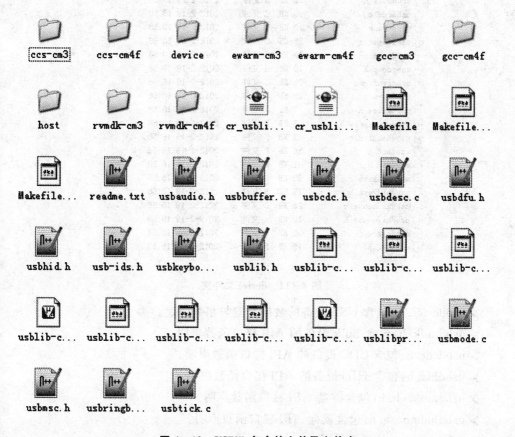

图 4-10 USBlib 包含的文件及文件夹

- usbdesc.c:解析 USB 描述符。
- usbbuffer.c:使用 USB 系统支持 buffer 功能,用于数据传输。
- usbringbuf.c:简单的 ringbuffer 源代码,恰当使用 ringbuffer,可提高内存的利用率,让数据更容易解析和处理。
- usbmode.c:管理 USB 处理器主机与设备的两种工作模式,以及两种模式切换的函数和源代码。
- usbtick.c:USB 库中的基本时钟单元控制函数与接口函数源代码。
- usblibpriv.h:USB 库中不同模块之间共享变量与定义的头文件,主要是 USB 主机与设备的中断处理函数、tick 中断函数中使用此头文件。

图 4-11 为 device 文件夹,在 USBlib/device 路径下包含以下头文件及源代码:

名称	大小	类型	修改日期
usbdaudio.c	51 KB	C 文件	2012-2-19 18:58
usbdaudio.h	12 KB	H 文件	2012-2-19 18:58
usbdbulk.c	49 KB	C 文件	2012-2-19 18:58
usbdbulk.h	10 KB	H 文件	2012-2-19 18:58
usbdcdc.c	99 KB	C 文件	2012-2-19 18:58
usbdcdc.h	13 KB	H 文件	2012-2-19 18:58
usbdcdesc.c	25 KB	C 文件	2012-2-19 18:58
usbdcomp.c	38 KB	C 文件	2012-2-19 18:58
usbdcomp.h	10 KB	H 文件	2012-2-19 18:58
usbdconfig.c	25 KB	C 文件	2012-2-19 18:58
usbdenum.c	95 KB	C 文件	2012-2-19 18:58
usbdevice.h	7 KB	H 文件	2012-2-19 18:58
usbdevicepriv.h	3 KB	H 文件	2012-2-19 18:58
usbdhandler.c	3 KB	C 文件	2012-2-19 18:58
usbdhid.c	86 KB	C 文件	2012-2-19 18:58
usbdhid.h	41 KB	H 文件	2012-2-19 18:58
usbdhidkeyb.c	38 KB	C 文件	2012-2-19 18:58
usbdhidkeyb.h	15 KB	H 文件	2012-2-19 18:58
usbdhidmouse.c	28 KB	C 文件	2012-2-19 18:58
usbdhidmouse.h	12 KB	H 文件	2012-2-19 18:58
usbdmsc.c	65 KB	C 文件	2012-2-19 18:58
usbdmsc.h	15 KB	H 文件	2012-2-19 18:58

图 4-11 device 文件夹

- usbdevice.h:包含 USB 设备的数据类型与相关协议。
- usbdbulk.h:包含 Bulk 设备的 API 接口函数声明。
- usbdcdc.h:包含 CDC 设备的 API 接口函数声明。
- usbdhid.h:包含 HID 设备的 API 接口函数声明。
- usbdhidkeyb.h:包含键盘 API 接口函数声明。
- usbdhidmouse.h:包含鼠标 API 接口函数声明。
- usbdenum.c:枚举源代码。
- usbdhandler.c:中断处理函数源代码。

➤ usbdconfig.c：配置功能函数源代码。
➤ usbdcdesc.c：解析描述符的源代码。
➤ usbdbulk.c：bulk 相关函数的源代码。
➤ usbdhid.c：人机接口的源代码。
➤ usbdhidkeyb.c：键盘设备类。
➤ usbdhidmouse.c：鼠标设备类。

图 4-12 为 host 文件夹，在 USBlib / host 路径下包含以下头文件及源代码：
➤ usbhost.h：USB 主机模式下的所有宏定义与数据类型。
➤ usbhostenum.c：USB 主机枚举解析的源代码。
➤ usbhscsi.c：USB SCSI 接口，是大容量数据存储设备的驱动。
➤ usbhhid.c：hid 人机接口驱动。
➤ usbhhid.h：hid 人机接口驱动的头文件。
➤ usbhhidkeyboard.c：键盘设备访问源代码。
➤ usbhhidkeyboard.h：键盘设备头文件。
➤ usbhhidmouse.c：鼠标设备访问源代码。
➤ usbhhidmouse.h：鼠标设备头文件。

名称	大小	类型	修改日期
usbhaudio.c	5...	C 文件	2012-2-19 18:58
usbhaudio.h	7 KB	H 文件	2012-2-19 18:58
usbhhid.c	2...	C 文件	2012-2-19 18:58
usbhhid.h	6 KB	H 文件	2012-2-19 18:58
usbhhidkeyboard.c	2...	C 文件	2012-2-19 18:58
usbhhidkeyboard.h	3 KB	H 文件	2012-2-19 18:58
usbhhidmouse.c	1...	C 文件	2012-2-19 18:58
usbhhidmouse.h	3 KB	H 文件	2012-2-19 18:58
usbhmsc.c	2...	C 文件	2012-2-19 18:58
usbhmsc.h	4 KB	H 文件	2012-2-19 18:58
usbhost.h	1...	H 文件	2012-2-19 18:58
usbhostenum.c	1...	C 文件	2012-2-19 18:58
usbhscsi.c	2...	C 文件	2012-2-19 18:58
usbhscsi.h	5 KB	H 文件	2012-2-19 18:58

图 4-12 host 文件夹

➤ usbhmsc.c：大容量数据存储设备的驱动。
➤ usbhmsc.h：大容量数据存储设备的驱动函数头文件。

在 USB 系统设计时，开发人员不必详细了解所有函数具体的处理过程，只需要了解各个函数的功能与调用关系，就能完成各种 USB 设备与主机的开发。在系统设计中会用到的各个函数，将会在后面章节讲解。

4.3 使用底层驱动开发

使用最底层 USB 驱动开发，要求开发人员对 USB 协议及相关事务彻底了解，开发 USB 产品有一定难度，但是这种开发模式占用内存少，运行效率高，但也容易出现 bug。

例如，使用底层 USB 驱动开发，开发一个音频设备。

(1) 初始化 USB 处理器，包括内核电压、CPU 主频、USB 外设资源等。

```
SysCtlLDOSet(SYSCTL_LDO_2_75V);
// 主频 50 MHz
SysCtlClockSet(SYSCTL_XTAL_8MHZ | SYSCTL_SYSDIV_4 | SYSCTL_USE_PLL | SYSCTL_OSC_
                MAIN );
//打开 USB 外设
SysCtlPeripheralEnable(SYSCTL_PERIPH_USB0);
//打开 USB 主时钟
SysCtlUSBPLLEnable();
//清除中断标志，并重新打开中断
USBIntStatusControl(USB0_BASE);
USBIntStatusEndpoint(USB0_BASE);
USBIntEnableControl(USB0_BASE, USB_INTCTRL_RESET |
                    USB_INTCTRL_DISCONNECT |
                    USB_INTCTRL_RESUME |
                    USB_INTCTRL_SUSPEND |
                    USB_INTCTRL_SOF);
USBIntEnableEndpoint(USB0_BASE, USB_INTEP_ALL);
//断开 -> 连接
USBDevDisconnect(USB0_BASE);
SysCtlDelay(SysCtlClockGet() / 30);
USBDevConnect(USB0_BASE);
//使能总中断
IntEnable(INT_USB0);
//音频设备会传输大量数据，最好打开 DMA
uDMAChannelControlSet(psDevice->psPrivateData->ucOUTDMA,
                    (UDMA_SIZE_32 | UDMA_SRC_INC_NONE|
                    UDMA_DST_INC_32 | UDMA_ARB_16));
USBEndpointDMAChannel(USB0_BASE, psDevice->psPrivateData->ucOUTEndpoint,
                    psDevice->psPrivateData->ucOUTDMA);
```

(2) USB 音频设备描述符。

//语言描述符

```c
const unsigned char g_pLangDescriptor[] =
{
    4,
    USB_DTYPE_STRING,
    USBShort(USB_LANG_EN_US)
};
//制造商 字符串 描述符
const unsigned char g_pManufacturerString[] =
{
    (17 + 1) * 2,
    USB_DTYPE_STRING,
    'T', 0, 'e', 0, 'x', 0, 'a', 0, 's', 0, ' ', 0, 'I', 0, 'n', 0, 's', 0,
    't', 0, 'r', 0, 'u', 0, 'm', 0, 'e', 0, 'n', 0, 't', 0, 's', 0
};
//产品 字符串 描述符
const unsigned char g_pProductString[] =
{
    (13 + 1) * 2,
    USB_DTYPE_STRING,
    'A', 0, 'u', 0, 'd', 0, 'i', 0, 'o', 0, ' ', 0, 'E', 0, 'x', 0, 'a', 0,
    'm', 0, 'p', 0, 'l', 0, 'e', 0
};
//产品 序列号 描述符
const unsigned char g_pSerialNumberString[] =
{
    (8 + 1) * 2,
    USB_DTYPE_STRING,
    '1', 0, '2', 0, '3', 0, '4', 0, '5', 0, '6', 0, '7', 0, '8', 0
};
//设备接口字符串描述符
const unsigned char g_pInterfaceString[] =
{
    (15 + 1) * 2,
    USB_DTYPE_STRING,
    'A', 0, 'u', 0, 'd', 0, 'i', 0, 'o', 0, ' ', 0, 'I', 0, 'n', 0,
    't', 0, 'e', 0, 'r', 0, 'f', 0, 'a', 0, 'c', 0, 'e', 0
};
//设备配置字符串描述符
const unsigned char g_pConfigString[] =
{
    (20 + 1) * 2,
    USB_DTYPE_STRING,
```

```c
    'A',0,'u',0,'d',0,'i',0,'o',0,'',0,'',0,'C',0,
    'o',0,'n',0,'f',0,'i',0,'g',0,'u',0,'r',0,'a',0,
    't',0,'i',0,'o',0,'n',0
};
//字符串描述符集合
const unsigned char * const g_pStringDescriptors[] =
{
    g_pLangDescriptor,
    g_pManufacturerString,
    g_pProductString,
    g_pSerialNumberString,
    g_pInterfaceString,
    g_pConfigString
};
//音频设备描述符
static unsigned char g_pAudioDeviceDescriptor[] =
{
    18,                          // Size of this structure
    USB_DTYPE_DEVICE,            // Type of this structure
    USBShort(0x110),             // USB version 1.1 (if we say 2.0, hosts assume
                                 // high-speed - see USB 2.0 spec 9.2.6.6)
    USB_CLASS_AUDIO,             // USB Device Class (spec 5.1.1)
    USB_SUBCLASS_UNDEFINED,      // USB Device Sub-class (spec 5.1.1)
    USB_PROTOCOL_UNDEFINED,      // USB Device protocol (spec 5.1.1)
    64,                          // Maximum packet size for default pipe.
    USBShort(0x1111),            // Vendor ID (filled in during USBDAudioInit)
    USBShort(0xffee),            // Product ID (filled in during USBDAudioInit)
    USBShort(0x100),             // Device Version BCD
    1,                           // Manufacturer string identifier
    2,                           // Product string identifier
    3,                           // Product serial number
    1                            // Number of configurations
};
//音频配置描述符
static unsigned char g_pAudioDescriptor[] =
{
    9,                           // Size of the configuration descriptor
    USB_DTYPE_CONFIGURATION,     // Type of this descriptor
    USBShort(32),                // The total size of this full structure
    2,                           // The number of interfaces in this
                                 // configuration
    1,                           // The unique value for this configuration
```

```c
    0,                              // The string identifier that describes this
                                    // configuration
    USB_CONF_ATTR_BUS_PWR,          // Bus Powered, Self Powered, remote wake up
    250,                            // The maximum power in 2 mA increments
};
//音频接口描述符
unsigned char g_pIADAudioDescriptor[] =
{
    8,                              // Size of the interface descriptor
    USB_DTYPE_INTERFACE_ASC,        // Interface Association Type
    0x0,                            // Default starting interface is 0
    0x2,                            // Number of interfaces in this association
    USB_CLASS_AUDIO,                // The device class for this association
    USB_SUBCLASS_UNDEFINED,         // The device subclass for this association
    USB_PROTOCOL_UNDEFINED,         // The protocol for this association
    0                               // The string index for this association
};
//音频控制接口描述符
const unsigned char g_pAudioControlInterface[] =
{
    9,                              // Size of the interface descriptor
    USB_DTYPE_INTERFACE,            // Type of this descriptor
    AUDIO_INTERFACE_CONTROL,        // The index for this interface
    0,                              // The alternate setting for this interface
    0,                              // The number of endpoints used by this
                                    // interface
    USB_CLASS_AUDIO,                // The interface class
    USB_ASC_AUDIO_CONTROL,          // The interface sub-class
    0,                              // The interface protocol for the sub-class
                                    // specified above
    0,                              // The string index for this interface

                                    // Audio Header Descriptor
    9,                              // The size of this descriptor
    USB_DTYPE_CS_INTERFACE,         // Interface descriptor is class specific
    USB_ACDSTYPE_HEADER,            // Descriptor sub-type is HEADER
    USBShort(0x0100),               // Audio Device Class Specification Release
                                    // Number in Binary-Coded Decimal
                                    // Total number of bytes in
                                    // g_pAudioControlInterface
    USBShort((9 + 9 + 12 + 13 + 9)),
```

```
    1,                              // Number of streaming interfaces
    1,                              // Index of the first and only streaming
                                    // interface

                                    // Audio Input Terminal Descriptor
    12,                             // The size of this descriptor
    USB_DTYPE_CS_INTERFACE,         // Interface descriptor is class specific
    USB_ACDSTYPE_IN_TERMINAL,       // Descriptor sub-type is INPUT_TERMINAL
    AUDIO_IN_TERMINAL_ID,           // Terminal ID for this interface
                                    // USB streaming interface
    USBShort(USB_TTYPE_STREAMING),
    0,                              // ID of the Output Terminal to which this
                                    // Input Terminal is associated
    2,                              // Number of logical output channels in the
                                    // Terminal 扣 output audio channel cluster
    USBShort((USB_CHANNEL_L |       // Describes the spatial location of the
              USB_CHANNEL_R)),      // logical channels
    0,                              // Channel Name string index
    0,                              // Terminal Name string index

    // Audio Feature Unit Descriptor
    13,                             // The size of this descriptor
    USB_DTYPE_CS_INTERFACE,         // Interface descriptor is class specific
    USB_ACDSTYPE_FEATURE_UNIT,      // Descriptor sub-type is FEATURE_UNIT
    AUDIO_CONTROL_ID,               // Unit ID for this interface
    AUDIO_IN_TERMINAL_ID,           // ID of the Unit or Terminal to which this
                                    // Feature Unit is connected
    2,                              // Size in bytes of an element of the
                                    // bmaControls() array that follows
                                    // Master Mute control
    USBShort(USB_ACONTROL_MUTE),
                                    // Left channel volume control
    USBShort(USB_ACONTROL_VOLUME),
                                    // Right channel volume control
    USBShort(USB_ACONTROL_VOLUME),
    0,                              // Feature unit string index

    // Audio Output Terminal Descriptor
    9,                              // The size of this descriptor
    USB_DTYPE_CS_INTERFACE,         // Interface descriptor is class specific
    USB_ACDSTYPE_OUT_TERMINAL,      // Descriptor sub-type is INPUT_TERMINAL
    AUDIO_OUT_TERMINAL_ID,          // Terminal ID for this interface
```

```c
                                // Output type is a generic speaker
    USBShort(USB_ATTYPE_SPEAKER),
    AUDIO_IN_TERMINAL_ID,       // ID of the input terminal to which this
                                // output terminal is connected
    AUDIO_CONTROL_ID,           // ID of the feature unit that this output
                                // terminal is connected to
    0,                          // Output terminal string index
};
//音频流接口描述符
const unsigned char g_pAudioStreamInterface[] =
{
    9,                          // Size of the interface descriptor
    USB_DTYPE_INTERFACE,        // Type of this descriptor
    AUDIO_INTERFACE_OUTPUT,     // The index for this interface
    0,                          // The alternate setting for this interface
    0,                          // The number of endpoints used by this
                                // interface
    USB_CLASS_AUDIO,            // The interface class
    USB_ASC_AUDIO_STREAMING,    // The interface sub-class
    0,                          // Unused must be 0
    0,                          // The string index for this interface

    // Vendor-specific Interface Descriptor
    9,                          // Size of the interface descriptor
    USB_DTYPE_INTERFACE,        // Type of this descriptor
    1,                          // The index for this interface
    1,                          // The alternate setting for this interface
    1,                          // The number of endpoints used by this
                                // interface
    USB_CLASS_AUDIO,            // The interface class
    USB_ASC_AUDIO_STREAMING,    // The interface sub-class
    0,                          // Unused must be 0
    0,                          // The string index for this interface

    // Class specific Audio Streaming Interface descriptor
    7,                          // Size of the interface descriptor
    USB_DTYPE_CS_INTERFACE,     // Interface descriptor is class specific
    USB_ASDSTYPE_GENERAL,       // General information
    AUDIO_IN_TERMINAL_ID,       // ID of the terminal to which this streaming
                                // interface is connected
    1,                          // One frame delay
    USBShort(USB_ADF_PCM),      //
```

```c
// Format type Audio Streaming descriptor
11,                                  // Size of the interface descriptor
USB_DTYPE_CS_INTERFACE,              // Interface descriptor is class specific
USB_ASDSTYPE_FORMAT_TYPE,            // Audio Streaming format type
USB_AF_TYPE_TYPE_I,                  // Type I audio format type
2,                                   // Two audio channels
2,                                   // Two bytes per audio sub-frame
16,                                  // 16 bits per sample
1,                                   // One sample rate provided
USB3Byte(48000),                     // Only 48 000 sample rate supported

// Endpoint Descriptor
9,                                   // The size of the endpoint descriptor
USB_DTYPE_ENDPOINT,                  // Descriptor type is an endpoint
                                     // OUT endpoint with address
                                     // ISOC_OUT_ENDPOINT
USB_EP_DESC_OUT | USB_EP_TO_INDEX(ISOC_OUT_ENDPOINT),
USB_EP_ATTR_ISOC |                   // Endpoint is an adaptive isochronous data
USB_EP_ATTR_ISOC_ADAPT |             // endpoint
USB_EP_ATTR_USAGE_DATA,
USBShort(ISOC_OUT_EP_MAX_SIZE),      // The maximum packet size
1,                                   // The polling interval for this endpoint
0,                                   // Refresh is unused
0,                                   // Synch endpoint address

// Audio Streaming Isochronous Audio Data Endpoint Descriptor
7,                                   // The size of the descriptor
USB_ACSDT_ENDPOINT,                  // Audio Class Specific Endpoint Descriptor
USB_ASDSTYPE_GENERAL,                // This is a general descriptor
USB_EP_ATTR_ACG_SAMPLING,            // Sampling frequency is supported
USB_EP_LOCKDELAY_UNDEF,              // Undefined lock delay units
USBShort(0),                         // No lock delay
};
```

(3) USB 音频设备枚举。

```c
//枚举用到的函数
static void USBDGetStatus(tUSBRequest * pUSBRequest);
static void USBDClearFeature(tUSBRequest * pUSBRequest);
static void USBDSetFeature(tUSBRequest * pUSBRequest);
static void USBDSetAddress(tUSBRequest * pUSBRequest);
static void USBDGetDescriptor(tUSBRequest * pUSBRequest);
```

```c
static void USBDSetDescriptor(tUSBRequest * pUSBRequest);
static void USBDGetConfiguration(tUSBRequest * pUSBRequest);
static void USBDSetConfiguration(tUSBRequest * pUSBRequest);
static void USBDGetInterface(tUSBRequest * pUSBRequest);
static void USBDSetInterface(tUSBRequest * pUSBRequest);
static void USBDEP0StateTx(void);
static long USBDStringIndexFromRequest(unsigned short usLang,
                                       unsigned short usIndex);
//该结构能够完整表述 USB 设备枚举过程
typedef struct
{
    //当前 USB 设备地址,也可以通过 DEV_ADDR_PENDING 最高位改变
    volatile unsigned long ulDevAddress;
    //保存设备当前生效的配置
    unsigned long ulConfiguration;
    //当前设备的接口设置
    unsigned char ucAltSetting;
    //ApEPoData 数据指针指向端点 0 正发收或者发送的数据组
    unsigned char * pEP0Data;
    //指示端点 0 待接收或者发送数据的剩下数据字节数
    volatile unsigned long ulEP0DataRemain;
    //端点 0 待接收或者发送数据的字节总数
    unsigned long ulOUTDataSize;
    //当前设备状态
    unsigned char ucStatus;
    //在处理过程中是否使用 wakeup 信号
    tBoolean bRemoteWakeup;
    //bRemoteWakeup 信号计数
    unsigned char ucRemoteWakeupCount;
}
tDeviceState;

//定义端点输出/输入

#define HALT_EP_IN                   0
#define HALT_EP_OUT                  1
//端点 0 的状态,在枚举过程中使用
typedef enum
{
    //等待主机请求
    USB_STATE_IDLE,
```

第 4 章 USB 库介绍

```c
        //通过 IN 端口 0 给主机发送数据块
        USB_STATE_TX,
        //通过 OUT 端口 0 从主机接收数据块
        USB_STATE_RX,
        //端点 0 发送/接收完成,等待主机应答
        USB_STATE_STATUS,
        //端点 0STALL,等待主机响应 STALL
        USB_STATE_STALL
}
tEP0State;
//端点 0 最大传输包大小
#define EP0_MAX_PACKET_SIZE     64
//用于指示设备地址改变
#define DEV_ADDR_PENDING        0x80000000
//总线复位后,默认的配置编号
#define DEFAULT_CONFIG_ID       1
//REMOTE_WAKEUP 的信号毫秒数,在协议中定义为 1~15 ms
#define REMOTE_WAKEUP_PULSE_MS 10
//REMOTE_WAKEUP 保持 20 ms
#define REMOTE_WAKEUP_READY_MS 20
//端点 0 的读数据缓存
static unsigned char g_pucDataBufferIn[EP0_MAX_PACKET_SIZE];
//定义当前设备状态信息实例
static volatile tDeviceState g_sUSBDeviceState;
//定义当前端点 0 的状态
static volatile tEP0State g_eUSBDEP0State = USB_STATE_IDLE;
//请求函数表
static const tStdRequest g_psUSBDStdRequests[] =
{
    USBDGetStatus,
    USBDClearFeature,
    0,
    USBDSetFeature,
    0,
    USBDSetAddress,
    USBDGetDescriptor,
    USBDSetDescriptor,
    USBDGetConfiguration,
    USBDSetConfiguration,
    USBDGetInterface,
    USBDSetInterface,
};
```

第4章 USB库介绍

```c
//在读取usb中断时合并使用
#define USB_INT_RX_SHIFT            8
#define USB_INT_STATUS_SHIFT        24
#define USB_RX_EPSTATUS_SHIFT       16
//端点控制状态寄存器转换
#define EP_OFFSET(Endpoint)         (Endpoint - 0x10)


//从端点0的FIFO中获取数据
long USBEndpoint0DataGet(unsigned char * pucData, unsigned long * pulSize)
{
    unsigned long ulByteCount;
    //判断端点0的数据是否接收完成
    if((HWREGH(USB0_BASE + USB_O_CSRL0) & USB_CSRL0_RXRDY) == 0)
    {
        * pulSize = 0;
        return(-1);
    }
    //USB_O_COUNT0指示端点0收到的数据量
    ulByteCount = HWREGH(USB0_BASE + USB_O_COUNT0 + USB_EP_0);
    //确定读回的数据量
    ulByteCount = (ulByteCount < * pulSize) ? ulByteCount : * pulSize;
    * pulSize = ulByteCount;
    //从FIFO中读取数据
    for(; ulByteCount > 0; ulByteCount--)
    {
        * pucData++ = HWREGB(USB0_BASE + USB_O_FIFO0 + (USB_EP_0 >> 2));
    }
    return(0);
}
//端点0应答
void USBDevEndpoint0DataAck(tBoolean bIsLastPacket)
{
    HWREGB(USB0_BASE + USB_O_CSRL0) =
        USB_CSRL0_RXRDYC | (bIsLastPacket ? USB_CSRL0_DATAEND : 0);
}
// 向端点0中放入数据
long USBEndpoint0DataPut(unsigned char * pucData, unsigned long ulSize)
{
    if(HWREGB(USB0_BASE + USB_O_CSRL0 + USB_EP_0) & USB_CSRL0_TXRDY)
    {
        return(-1);
    }
```

第 4 章 USB 库介绍

```c
    for(; ulSize > 0; ulSize--)
    {
        HWREGB(USB0_BASE + USB_O_FIFO0 + (USB_EP_0 >> 2)) = *pucData++;
    }
    return(0);
}
//向端点 0 中写入数据
long USBEndpoint0DataSend(unsigned long ulTransType)
{
    //判断是否已经有数据准备好
    if(HWREGB(USB0_BASE + USB_O_CSRL0 + USB_EP_0) & USB_CSRL0_TXRDY)
    {
        return(-1);
    }
    HWREGB(USB0_BASE + USB_O_CSRL0 + USB_EP_0) = ulTransType & 0xff;
    return(0);
}
//端点 0 从主机上获取数据
void USBRequestDataEP0(unsigned char *pucData, unsigned long ulSize)
{
    g_eUSBDEP0State = USB_STATE_RX;
    g_sUSBDeviceState.pEP0Data = pucData;
    g_sUSBDeviceState.ulOUTDataSize = ulSize;
    g_sUSBDeviceState.ulEP0DataRemain = ulSize;
}
//端点 0 请求发送数据
void USBSendDataEP0(unsigned char *pucData, unsigned long ulSize)
{
    g_sUSBDeviceState.pEP0Data = pucData;
    g_sUSBDeviceState.ulEP0DataRemain = ulSize;
    g_sUSBDeviceState.ulOUTDataSize = ulSize;
    USBDEP0StateTx();
}
//端点 0 处于停止状态
void USBStallEP0(void)
{
    HWREGB(USB0_BASE + USB_O_CSRL0) |= (USB_CSRL0_STALL | USB_CSRL0_RXRDYC);
    g_eUSBDEP0State = USB_STATE_STALL;
}
//从端点 0 中获取一个请求
static void USBDReadAndDispatchRequest(void)
{
```

```c
    unsigned long ulSize;
    tUSBRequest * pRequest;
    pRequest = (tUSBRequest *)g_pucDataBufferIn;
    ulSize = EP0_MAX_PACKET_SIZE;
    USBEndpoint0DataGet(g_pucDataBufferIn, &ulSize);
    if(! ulSize)
    {
        return;
    }
    //判断是否是标准请求
    if((pRequest->bmRequestType & USB_RTYPE_TYPE_M) != USB_RTYPE_STANDARD)
    {
        UARTprintf("非标准请求..............\r\n");
    }
    else
    {
        //调用标准请求处理函数中
        if((pRequest->bRequest <
            (sizeof(g_psUSBDStdRequests) / sizeof(tStdRequest))) &&
            (g_psUSBDStdRequests[pRequest->bRequest] != 0))
        {
            g_psUSBDStdRequests[pRequest->bRequest](pRequest);
        }
        else
        {
            USBStallEP0();
        }
    }
}
//枚举过程
// USB_STATE_IDLE - * - - > USB_STATE_TX - * - > USB_STATE_STATUS - * - >
// USB_STATE_IDLE
//                  |                  |                         |
//                  |-- > USB_STATE_RX -                         |
//                  |                                            |
//                  |-- > USB_STATE_STALL ---------->---|
//
// ---------------------------------------------------------------
// | Current State      | State 0          | State 1             |
// | ------------------ | ---------------- | ------------------- |
// | USB_STATE_IDLE     | USB_STATE_TX/RX  | USB_STATE_STALL     |
// | USB_STATE_TX       | USB_STATE_STATUS |                     |
```

```
//  | USB_STATE_RX              | USB_STATE_STATUS  |
//  | USB_STATE_STATUS          | USB_STATE_IDLE    |
//  | USB_STATE_STALL           | USB_STATE_IDLE    |
//  --------------------------------------------------
void USBDeviceEnumHandler(void)
{
    unsigned long ulEPStatus;
    //获取中断状态
    ulEPStatus = HWREGH(USB0_BASE + EP_OFFSET(USB_EP_0) + USB_O_TXCSRL1);
    ulEPStatus |= ((HWREGH(USB0_BASE + EP_OFFSET(USB_EP_0) + USB_O_RXCSRL1)) <<
            USB_RX_EPSTATUS_SHIFT);

    //端点 0 的状态
    switch(g_eUSBDEP0State)
    {
        case USB_STATE_STATUS:
        {
            UARTprintf("USB_STATE_STATUS...............\r\n");
            g_eUSBDEP0State = USB_STATE_IDLE;
            //判断地址改变
            if(g_sUSBDeviceState.ulDevAddress & DEV_ADDR_PENDING)
            {
                //设置地址
                g_sUSBDeviceState.ulDevAddress &= ~DEV_ADDR_PENDING;
                HWREGB(USB0_BASE + USB_O_FADDR) =
                    (unsigned char)g_sUSBDeviceState.ulDevAddress;
            }
            //端点 0 接收包准备好
            if(ulEPStatus & USB_DEV_EP0_OUT_PKTRDY)
            {
                USBDReadAndDispatchRequest();
            }
            break;
        }
        //等待从主机接收数据
        case USB_STATE_IDLE:
        {
            if(ulEPStatus & USB_DEV_EP0_OUT_PKTRDY)
            {
                USBDReadAndDispatchRequest();
            }
            break;
```

```c
        }
        //数据处理好,准备发送
        case USB_STATE_TX:
        {
            USBDEP0StateTx();
            break;
        }
        //接收数据
        case USB_STATE_RX:
        {
            unsigned long ulDataSize;
            if(g_sUSBDeviceState.ulEP0DataRemain > EP0_MAX_PACKET_SIZE)
            {
                ulDataSize = EP0_MAX_PACKET_SIZE;
            }
            else
            {
                ulDataSize = g_sUSBDeviceState.ulEP0DataRemain;
            }
            USBEndpoint0DataGet(g_sUSBDeviceState.pEP0Data, &ulDataSize);
            if(g_sUSBDeviceState.ulEP0DataRemain < EP0_MAX_PACKET_SIZE)
            {
                USBDevEndpoint0DataAck(true);
                g_eUSBDEP0State =   USB_STATE_IDLE;
                if(g_sUSBDeviceState.ulOUTDataSize != 0)
                {
                }
            }
            else
            {
                USBDevEndpoint0DataAck(false);
            }
            g_sUSBDeviceState.pEP0Data += ulDataSize;
            g_sUSBDeviceState.ulEP0DataRemain -= ulDataSize;
            break;
        }
        //停止状态
        case USB_STATE_STALL:
        {
            if(ulEPStatus & USB_DEV_EP0_SENT_STALL)
            {
                HWREGB(USB0_BASE + USB_O_CSRL0) &= ~(USB_DEV_EP0_SENT_STALL);
```

```
                g_eUSBDEP0State = USB_STATE_IDLE;
            }
            break;
        }

        default:
        {
            break;
        }
    }
}
```

设备枚举过程中 USBDSetAddress 和 USBDGetDescriptor 很重要，下面只列出这两个函数的具体内容。

```
// SET_ADDRESS 标准请求
static void USBDSetAddress(tUSBRequest * pUSBRequest)
{
    USBDevEndpoint0DataAck(true);
    g_sUSBDeviceState.ulDevAddress = pUSBRequest->wValue | DEV_ADDR_PENDING;
    g_eUSBDEP0State = USB_STATE_STATUS;
    //HandleSetAddress();
}
//GET_DESCRIPTOR 标准请求
static void USBDGetDescriptor(tUSBRequest * pUSBRequest)
{
    USBDevEndpoint0DataAck(false);
    switch(pUSBRequest->wValue >> 8)
    {
        case USB_DTYPE_DEVICE:
        {
            g_sUSBDeviceState.pEP0Data =
                (unsigned char *)g_pDFUDeviceDescriptor;
            g_sUSBDeviceState.ulEP0DataRemain = g_pDFUDeviceDescriptor[0];
            break;
        }
        case USB_DTYPE_CONFIGURATION:
        {
            unsigned char ucIndex;
            ucIndex = (unsigned char)(pUSBRequest->wValue & 0xFF);
            if(ucIndex != 0)
            {
                USBStallEP0();
```

```c
            g_sUSBDeviceState.pEP0Data = 0;
            g_sUSBDeviceState.ulEP0DataRemain = 0;
        }
        else
        {
            g_sUSBDeviceState.pEP0Data =
                (unsigned char *)g_pDFUConfigDescriptor;
            g_sUSBDeviceState.ulEP0DataRemain =
                *(unsigned short *)&(g_pDFUConfigDescriptor[2]);
        }
        break;
    }
    case USB_DTYPE_STRING:
    {
        long lIndex;
        lIndex = USBDStringIndexFromRequest(pUSBRequest->wIndex,
                                            pUSBRequest->wValue & 0xFF);
        if(lIndex == -1)
        {
            USBStallEP0();
            break;
        }
        g_sUSBDeviceState.pEP0Data =
            (unsigned char *)g_pStringDescriptors[lIndex];
        g_sUSBDeviceState.ulEP0DataRemain = g_pStringDescriptors[lIndex][0];
        break;
    }
    case 0x22:
    {
        //USBDevEndpoint0DataAck(false);
        g_sUSBDeviceState.pEP0Data = (unsigned char *)ReportDescriptor;
        g_sUSBDeviceState.ulEP0DataRemain = sizeof(&ReportDescriptor[0]);
        //USBDEP0StateTx();
    }
    default:
    {
        USBStallEP0();
        break;
    }
}
if(g_sUSBDeviceState.pEP0Data)
{
```

第 4 章　USB 库介绍

```
        if(g_sUSBDeviceState.ulEP0DataRemain > pUSBRequest->wLength)
        {
            g_sUSBDeviceState.ulEP0DataRemain = pUSBRequest->wLength;
        }
        USBDEP0StateTx();
    }
}
```

（4）USB 音频数据处理与控制。此过程包括数据处理，音量控制，静音控制等，控制过程较为复杂，在此不再一一讲解，可以参考相关 USB 音频设备书籍。在第 6 章有讲解使用其他方法进行 USB 音频设备开发的内容。

4.4　使用 USB 库开发

使用 USB 库函数进行 USB 设备设计，开发人员可以不深入研究 USB 协议，包括枚举过程、中断处理、数据处理等，直接使用库函数提供的 API 接口函数就可以完成开发工作。使用 USB 库函数方便、快捷、缩短开发周期、不易出现 bug，但占用存储空间、内存较大，由于 Stellaris USB 处理器的存储空间达 128 KB，远远超过程序需要的存储空间，所以使用 USB 库函数开发是一个比较好的软件设计方案。

例如：使用库函数开发一个简单的 USB 鼠标，程序如下所示。

（1）完成字符串描述符。

```
//语言描述符
const unsigned char g_pLangDescriptor[] =
{
    4,
    USB_DTYPE_STRING,
    USBShort(USB_LANG_EN_US)
};
//制造商 字符串 描述符
const unsigned char g_pManufacturerString[] =
{
    (17 + 1) * 2,
    USB_DTYPE_STRING,
    'T', 0, 'e', 0, 'x', 0, 'a', 0, 's', 0, ' ', 0, 'I', 0, 'n', 0, 's', 0,
    't', 0, 'r', 0, 'u', 0, 'm', 0, 'e', 0, 'n', 0, 't', 0, 's', 0,
};
//产品 字符串 描述符
const unsigned char g_pProductString[] =
{
```

```c
    (13 + 1) * 2,
    USB_DTYPE_STRING,
    'M', 0, 'o', 0, 'u', 0, 's', 0, 'e', 0, ' ', 0, 'E', 0, 'x', 0, 'a', 0,
    'm', 0, 'p', 0, 'l', 0, 'e', 0
};
//产品 序列号 描述符
const unsigned char g_pSerialNumberString[] =
{
    (8 + 1) * 2,
    USB_DTYPE_STRING,
    '1', 0, '2', 0, '3', 0, '4', 0, '5', 0, '6', 0, '7', 0, '8', 0
};
//设备接口字符串描述
const unsigned char g_pHIDInterfaceString[] =
{
    (19 + 1) * 2,
    USB_DTYPE_STRING,
    'H', 0, 'I', 0, 'D', 0, ' ', 0, 'M', 0, 'o', 0, 'u', 0, 's', 0,
    'e', 0, ' ', 0, 'I', 0, 'n', 0, 't', 0, 'e', 0, 'r', 0, 'f', 0,
    'a', 0, 'c', 0, 'e', 0
};
// 配置字符串描述
const unsigned char g_pConfigString[] =
{
    (23 + 1) * 2,
    USB_DTYPE_STRING,
    'H', 0, 'I', 0, 'D', 0, ' ', 0, 'M', 0, 'o', 0, 'u', 0, 's', 0,
    'e', 0, ' ', 0, 'C', 0, 'o', 0, 'n', 0, 'f', 0, 'i', 0, 'g', 0,
    'u', 0, 'r', 0, 'a', 0, 't', 0, 'i', 0, 'o', 0, 'n', 0
};
//字符串描述符集合
const unsigned char * const g_pStringDescriptors[] =
{
    g_pLangDescriptor,
    g_pManufacturerString,
    g_pProductString,
    g_pSerialNumberString,
    g_pHIDInterfaceString,
    g_pConfigString
};
#define NUM_STRING_DESCRIPTORS (sizeof(g_pStringDescriptors) /        \
                                sizeof(unsigned char *))
```

(2) 了解鼠标设备 tUSBDHIDMouseDevice 在库函数中的定义,并完成鼠标设备实例。

```
//定义 USB 鼠标实例
tHIDMouseInstance g_sMouseInstance;
//定义 USB 鼠标相关信息
const tUSBDHIDMouseDevice g_sMouseDevice =
{
    USB_VID_STELLARIS,
    USB_PID_MOUSE,
    500,
    USB_CONF_ATTR_SELF_PWR,
    MouseHandler,
    (void *)&g_sMouseDevice,
    g_pStringDescriptors,
    NUM_STRING_DESCRIPTORS,
    &g_sMouseInstance
};
```

(3) 初始化 USB 鼠标设备,并进行数据处理。

```
//初始化 USB 鼠标设备,只需用这个函数就可完成配置,包括枚举配置
USBDHIDMouseInit(0, (tUSBDHIDMouseDevice *)&g_sMouseDevice);
//数据处理,改变鼠标位置和按键状态
USBDHIDMouseStateChange((void *)&g_sMouseDevice,
                        (char)lDeltaX, (char)lDeltaY,
                        ucButtons);
```

从使用 USB 库函数开发 USB 鼠标设备过程中可以看出,使用 USB 库函数开发非常简单、方便、快捷,不用考虑底层的驱动、类协议。在 HID 类中,报告符本身就很复杂,但是使用 USB 库函数开发完全屏蔽报告符配置过程。从第 5 章将会开始逐步介绍使用 USB 库函数开发 USB 设备与主机。

4.5 小 结

本章介绍了 USB 库的基本内容,使用 USB 库进行编程相对简单,容易理解,同时也列举了几种 USB 编程模式,开发人员可以根据自己的情况,具有选择性的编程。从第 5 章开始,将会逐一讲解几种常用的 USB 设备和主机开发。

第 5 章

HID 设备

前面 4 章对 USB 开发做了初步介绍,从本章起会逐一讲解 USB 几种常用设备的开发、USB 主机开发、USB OTG 开发过程与系统详细编程。本章主要介绍 HID 人机接口类的 USB 系统开发,包含 USB 键盘与 USB 鼠标开发。

5.1 HID 介绍

为简化 USB 设备的开发过程,USB IF 将 USB 设备分成了若干设备类。所有设备类都必须支持标准 USB 描述符和标准 USB 设备请求。如果有必要,设备类还可以自行定义其专用的描述符和设备请求,分别被称为设备类定义描述符和设备类定义请求。另外,一个完整的设备类还将指明其接口和端点的使用方法,如接口所包含端点的个数、端点的最大数据包长度等。

HID 设备类就是设备类的一类,HID 是 Human Interface Device 的缩写,人机交互设备,例如键盘、鼠标与游戏杆等。不过 HID 设备并不一定要有人机接口,只要符合 HID 类别规范的设备都是 HID 设备。

HID 设备既可以是低速设备也可以是全速设备,其典型的数据传输类型为中断 IN 传输,即它适用于主机接收 USB 设备发来的小量到中等量的数据。HID 具有以下功能特点:适用于传输少量或中量的数据;传输的数据具有突发性;传输的最大速率有限制;无固定的传输率。

HID 设备类除支持标准 USB 描述符外(设备描述符、配置描述符、接口描述符、端点描述符和字符串描述符),还自行定义了 3 种类描述符,分别为 HID 描述符(主要用于识别 HID 设备所包含的其他类描述符)、报告描述符(提供 HID 设备和主机间交换数据的格式)和物理描述符。一个 HID 设备只能支持一个 HID 描述符;可以支持一个或多个报告描述符;物理描述符是可选的,大多数 HID 设备不需要使用它。

5.2 HID 类描述符

除了标准 USB 描述符外,HID 设备自行定义了 3 种描述符:HID 描述符、报告描述符、物理描述符,其中物理描述符不常用,不在此讲解。3 种描述符分别定义为:

第 5 章 HID 设备

```
#define USB_HID_DTYPE_HID        0x21
#define USB_HID_DTYPE_REPORT     0x22
#define USB_HID_DTYPE_PHYSICAL   0x23
```

HID 设备只有一个 HID 描述符，HID 设备描述符主要描述 HID 规范的版本号、HID 通信所使用的额外描述符、报告描述符的长度等。表 5-1 为 HID 描述符表。

表 5-1 HID 描述符

偏移量	域	大小	值	描述
0	bLength	1	数字	字节数
1	bDescriptorType	1	常量	配置描述符类型
2	bcdHID	2	数字	版本号（BCD 码）
4	bCountryCode	1	数字	国家语言代码
5	bNumDescriptors	1	数字	描述符个数
6	bType	1	索引	下一个描述符类型
7	wLength	2	位图	下一个描述符长度

C 语言 HID 描述符结构体如下所示：

```
typedef struct
{
    //HID 描述符长度
    unsigned char bLength;
    // USB_HID_DTYPE_HID (0x21)
    unsigned char bDescriptorType;
    //HID 协议版本号
    unsigned short bcdHID;
    //国家语言
    unsigned char bCountryCode;
    //描述符个数，至少为 1
    unsigned char bNumDescriptors;
    //下一个描述符类型
    unsigned char bType;
    //下一个描述符长度
    unsigned short wLength;
}
tHIDDescriptor;
```

例如：定义一个实际的 HID 设备描述符，程序如下所示。

```
static const tHIDDescriptor g_sKeybHIDDescriptor =
```

```
{
    9,                              // bLength
    USB_HID_DTYPE_HID,              // bDescriptorType
    0x111,                          // bcdHID (version 1.11 compliant)
    USB_HID_COUNTRY_US,             // bCountryCode (not localized)
    1,                              // bNumDescriptors
    USB_HID_DTYPE_REPORT,           // Report descriptor
    sizeof(Report)                  // Size of report descriptor
};
```

国家语言定义如下：

```
#define USB_HID_COUNTRY_NONE                0x00
#define USB_HID_COUNTRY_ARABIC              0x01
#define USB_HID_COUNTRY_BELGIAN             0x02
#define USB_HID_COUNTRY_CANADA_BI           0x03
#define USB_HID_COUNTRY_CANADA_FR           0x04
#define USB_HID_COUNTRY_CZECH_REPUBLIC      0x05
#define USB_HID_COUNTRY_DANISH              0x06
#define USB_HID_COUNTRY_FINNISH             0x07
#define USB_HID_COUNTRY_FRENCH              0x08
#define USB_HID_COUNTRY_GERMAN              0x09
#define USB_HID_COUNTRY_GREEK               0x0A
#define USB_HID_COUNTRY_HEBREW              0x0B
#define USB_HID_COUNTRY_HUNGARY             0x0C
#define USB_HID_COUNTRY_INTERNATIONAL_ISO   0x0D
#define USB_HID_COUNTRY_ITALIAN             0x0E
#define USB_HID_COUNTRY_JAPAN_KATAKANA      0x0F
#define USB_HID_COUNTRY_KOREAN              0x10
#define USB_HID_COUNTRY_LATIN_AMERICAN      0x11
#define USB_HID_COUNTRY_NETHERLANDS         0x12
#define USB_HID_COUNTRY_NORWEGIAN           0x13
#define USB_HID_COUNTRY_PERSIAN             0x14
#define USB_HID_COUNTRY_POLAND              0x15
#define USB_HID_COUNTRY_PORTUGUESE          0x16
#define USB_HID_COUNTRY_RUSSIA              0x17
#define USB_HID_COUNTRY_SLOVAKIA            0x18
#define USB_HID_COUNTRY_SPANISH             0x19
#define USB_HID_COUNTRY_SWEDISH             0x1A
#define USB_HID_COUNTRY_SWISS_FRENCH        0x1B
#define USB_HID_COUNTRY_SWISS_GERMAN        0x1C
#define USB_HID_COUNTRY_SWITZERLAND         0x1D
#define USB_HID_COUNTRY_TAIWAN              0x1E
```

第 5 章 HID 设备

```
#define USB_HID_COUNTRY_TURKISH_Q        0x1F
#define USB_HID_COUNTRY_UK               0x20
#define USB_HID_COUNTRY_US               0x21
#define USB_HID_COUNTRY_YUGOSLAVIA       0x22
#define USB_HID_COUNTRY_TURKISH_F        0x23
```

中国使用 HID 设备较多的是美国语言 USB_HID_COUNTRY_US。

USB HID 设备通过报告来传送数据，报告有输入报告和输出报告。输入报告(input)是 USB 设备发送给主机的，例如 USB 鼠标将鼠标移动和鼠标单击等信息返回给计算机，键盘将按键数据返回给计算机等；输出报告(output)是主机发送给 USB 设备的，例如键盘上的数字键盘锁定灯和大写字母锁定灯等。报告是一个数据包，里面包含的是所要传送的数据。输入报告是通过中断输入端点输入的，而输出报告有点区别，当没有中断输出端点时，可以通过控制输出端点 0 发送，当有中断输出端点时，通过中断输出端点发出。

报告描述符是描述一个报告以及报告里面的数据的作用的。通过它，USB HOST 可以分析出报告里面的数据所表示的意思。它通过控制输入端点 0 返回，主机使用获取报告描述符命令来获取报告描述符，注意这个请求是发送到接口的，而不是到设备。一个报告描述符可以描述多个报告，不同的报告通过报告 ID 来识别，报告 ID 在报告的最前面，即第一个字节。当报告描述符中没有规定报告 ID 时，报告中就没有 ID 字段，报告的一开始就是数据。更详细的说明请参看 USB HID 协议。

下面以一个实例介绍鼠标报告描述符：

```
static const unsigned char g_pucMouseReportDescriptor[] =
{
    UsagePage(USB_HID_GENERIC_DESKTOP),    //通用桌面
    Usage(USB_HID_MOUSE),                  //HID 鼠标
    Collection(USB_HID_APPLICATION),       //应用集合,以 EndCollection 结束
        Usage(USB_HID_POINTER),            //指针设备
        Collection(USB_HID_PHYSICAL),      //集合
            UsagePage(USB_HID_BUTTONS),    //按键
            UsageMinimum(1),               //最小值
            UsageMaximum(3),               //最大值
            LogicalMinimum(0),             //逻辑最小值
            LogicalMaximum(1),             //逻辑最大值
            ReportSize(1),                 //Report 大小为 1bit
            ReportCount(3),                //Report 3 个位
            Input(USB_HID_INPUT_DATA | USB_HID_INPUT_VARIABLE |
                  USB_HID_INPUT_ABS),      //发送给主机的报告格式
            //剩余 5 位填满
            ReportSize(5),
```

```
                ReportCount(1),
                Input(USB_HID_INPUT_CONSTANT | USB_HID_INPUT_ARRAY |
                    USB_HID_INPUT_ABS),
                UsagePage(USB_HID_GENERIC_DESKTOP),    //通用桌面
                Usage(USB_HID_X),
                Usage(USB_HID_Y),
                LogicalMinimum(-127),
                LogicalMaximum(127),
                ReportSize(8),
                ReportCount(2),
                Input(USB_HID_INPUT_DATA | USB_HID_INPUT_VARIABLE |
                    USB_HID_INPUT_RELATIVE),
                ReportSize(8),
                ReportCount(MOUSE_REPORT_SIZE - 3),
                Input(USB_HID_INPUT_CONSTANT | USB_HID_INPUT_ARRAY |
                    USB_HID_INPUT_ABS),
            EndCollection,
        EndCollection,
};
```

UsagePage 常用参数如下所示。

```
#define USB_HID_GENERIC_DESKTOP     0x01
#define USB_HID_BUTTONS             0x09
#define USB_HID_USAGE_POINTER       0x0109
#define USB_HID_USAGE_BUTTONS       0x0509
#define USB_HID_USAGE_LEDS          0x0508
#define USB_HID_USAGE_KEYCODES      0x0507
#define USB_HID_X                   0x30
#define USB_HID_Y                   0x31
```

Usage 常用参数如下所示。

```
#define USB_HID_X                   0x30
#define USB_HID_Y                   0x31
#define USB_HID_POINTER             0x01
#define USB_HID_MOUSE               0x02
#define USB_HID_KEYBOARD            0x06
```

Collection 常用参数如下所示。

```
#define USB_HID_APPLICATION         0x00
#define USB_HID_PHYSICAL            0x01
```

第 5 章　HID 设备

Input 常用参数如下所示。

```
#define USB_HID_INPUT_DATA          0x0000
#define USB_HID_INPUT_CONSTANT      0x0001
#define USB_HID_INPUT_ARRAY         0x0000
#define USB_HID_INPUT_VARIABLE      0x0002
#define USB_HID_INPUT_ABS           0x0000
#define USB_HID_INPUT_RELATIVE      0x0004
#define USB_HID_INPUT_NOWRAP        0x0000
#define USB_HID_INPUT_WRAP          0x0008
#define USB_HID_INPUT_LINEAR        0x0000
#define USB_HID_INPUT_NONLINEAR     0x0010
#define USB_HID_INPUT_PREFER        0x0000
#define USB_HID_INPUT_NONPREFER     0x0020
#define USB_HID_INPUT_NONULL        0x0000
#define USB_HID_INPUT_NULL          0x0040
#define USB_HID_INPUT_BITF          0x0100
#define USB_HID_INPUT_BYTES         0x0000
```

Output 常用参数如下所示。

```
#define USB_HID_OUTPUT_DATA         0x0000
#define USB_HID_OUTPUT_CONSTANT     0x0001
#define USB_HID_OUTPUT_ARRAY        0x0000
#define USB_HID_OUTPUT_VARIABLE     0x0002
#define USB_HID_OUTPUT_ABS          0x0000
#define USB_HID_OUTPUT_RELATIVE     0x0004
#define USB_HID_OUTPUT_NOWRAP       0x0000
#define USB_HID_OUTPUT_WRAP         0x0008
#define USB_HID_OUTPUT_LINEAR       0x0000
#define USB_HID_OUTPUT_NONLINEAR    x0010
#define USB_HID_OUTPUT_PREFER       0x0000
#define USB_HID_OUTPUT_NONPREFER    0x0020
#define USB_HID_OUTPUT_NONULL       0x0000
#define USB_HID_OUTPUT_NULL         0x0040
#define USB_HID_OUTPUT_BITF         0x0100
#define USB_HID_OUTPUT_BYTES        0x0000
```

报告描述符使用复杂，可从 Http://www.usb.org 下载 HID Descriptor tool（如图 5-1 所示）来生成。HID Descriptor tool 是由 USB 官方编写的专用报告符生成工具。有关报告符具体内容请参阅 Http://www.usb.org 网上的 HID Usage Tables 文档。使用 USB 库函数开发 USB HID 设备可以不用考虑报告符，其在库中已经定义了。

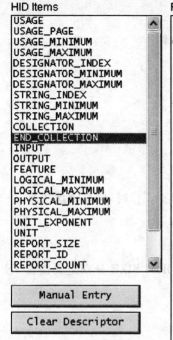

图 5-1　HID Descriptor tool

HID Descriptor tool 界面简洁、操作方便，集结了所有 HID 设备报告定义，如键盘、鼠标、操作杆、手写设备等，对于报告符不熟悉的开发者相当方便。

5.3　USB 键盘

在 USB 库中已经定义好 USB 键盘的数据类型、API 函数，开发 USB 键盘非常快捷方便。相关定义和数据类型放在 usbdhidkeyb.h 中。

5.3.1　数据类型

usbdhidkeyb.h 中已经定义好 USB 键盘使用的所有数据类型和函数，下面介绍 USB 键盘使用的数据类型。

```
#define KEYB_MAX_CHARS_PER_REPORT     6
```

KEYB_MAX_CHARS_PER_REPORT 定义 USB 键盘一次发送 6 个字节数据给主机，如果每次发送的数据多于定义的 6 个，将会通过 USBDHIDKeyboardKeyStateChange() 函数将返回发送数据太多的错误：KEYB_ERR_TOO_MANY_KEYS。KEYB_MAX_CHARS_PER_REPORT 这个值的定义由键盘报告符决定。

```c
typedef enum
{
    //状态还没有定义
    HID_KEYBOARD_STATE_UNCONFIGURED,
    //空闲状态,没有按键被按下或者没有等待数据
    HID_KEYBOARD_STATE_IDLE,
    //等待主机发送数据
    HID_KEYBOARD_STATE_WAIT_DATA,
    //等待数据发送
    HID_KEYBOARD_STATE_SEND
}
tKeyboardState;
```

tKeyboardState 用于定义键盘状态,用于在 USB 键盘工作时,保存键盘状态。

```c
#define KEYB_IN_REPORT_SIZE 8
#define KEYB_OUT_REPORT_SIZE 1
```

KEYB_IN_REPORT_SIZE 和 KEYB_OUT_REPORT_SIZE 分别定义键盘 IN 报告符和 OUT 报告符长度。

```c
typedef struct
{
    // 指示当前 USB 设备是否配置成功
    unsigned char ucUSBConfigured;
    // USB 键盘使用的子协议:USB_HID_PROTOCOL_BOOT 或者 USB_HID_PROTOCOL_REPORT
    // ucProtocol 值会自动传给该 USB 设备的设备描述符和端点描述符
    unsigned char ucProtocol;
    // 键盘 LED 灯当前状态
    volatile unsigned char ucLEDStates;
    // 记录有几个键被按下
    unsigned char ucKeyCount;
    // 中断 IN 端点状态
    volatile tKeyboardState eKeyboardState;
    // 指示当前是否有键状态改变
    volatile tBoolean bChangeMade;
    // 用于接收 OUT 报告
    unsigned char pucDataBuffer[KEYB_OUT_REPORT_SIZE];
    // 用于收送 IN 报告,保存最新一次报告
    unsigned char pucReport[KEYB_IN_REPORT_SIZE];
    // 按下键的 usage 代码
    unsigned char pucKeysPressed[KEYB_MAX_CHARS_PER_REPORT];
    // 为 IN 报告定义时间
    tHIDReportIdle sReportIdle;
```

```
    // HID 设备实例,保存键盘 HID 信息
    tHIDInstance sHIDInstance;
    // HID 设备驱动
    tUSBDHIDDevice sHIDDevice;
}
tHIDKeyboardInstance;
```

tHIDKeyboardInstance 表示键盘实例。在 tHIDInstance 和 tUSBDHIDDevice 的基础上,用于保存全部 USB 键盘的配置信息,包括描述符、callback 函数、按键事件等。

如图 5-2 所示,为 tHIDKeyboardInstance 的合作图,可以看出 tHIDKeyboardInstance 用于存放键盘实例,包括其 VID、PID、描述符等,通过调用其他函数,完成数据通信和数据传递。

在图 5-2 中,tUSBDHIDDevice 结构体定义在 usbdhid.h 中,描述源码如下:

```
typedef struct
{
    //VID
    unsigned short   usVID
    //PID
    unsigned short   usPID
    //该装置的最大功率消耗,以 mA 表示
    unsigned short   usMaxPowermA
    unsigned char    ucPwrAttributes
    //子类的接口
    unsigned char    ucSubclass
    //协议
    unsigned char    ucProtocol
    //该设备支持输入的报告数量
    unsigned char    ucNumInputReports
    tHIDReportIdle * psReportIdle
    tUSBCallback     pfnRxCallback
    //提供指针,回调 pfnRxCallback
    void  *   pvRxCBData
    //函数指针,指向 pfnTxCallback 回调函数
    tUSBCallback     pfnTxCallback
    //数据指针,为 pfnTxCallback 回调函数提供入口参数
    void  *   pvTxCBData
    tBoolean  bUseOutEndpoint
    //HID 的描述
    const tHIDDescriptor * psHIDDescriptor
    const unsigned char * const    ppClassDescriptors
```

第5章 HID 设备

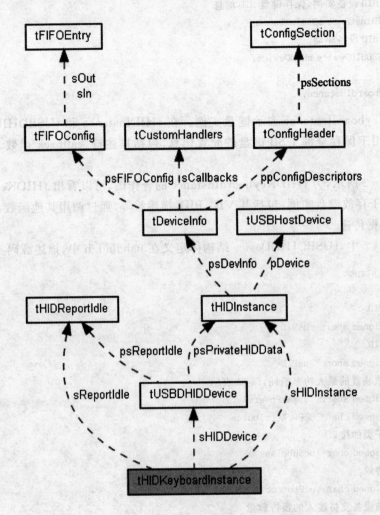

图 5-2 tHIDKeyboardInstance 合作图

```
    const unsigned char * const *   ppStringDescriptors
    //该描述符提供字符串描述符长度。即(1 + (5 * (num languages)))。
    unsigned long   ulNumStringDescriptors
    tHIDInstance *   psPrivateHIDData
}
tUSBDHIDDevice;
```

　　tUSBDHIDDevice 为 HID 设备类的结构体定义,HID 设备开发都会把更高层参数传递给该结构体,使用本结构体进行数据传输、控制,完成 HID 设备功能,如图 5-3 所示,为 tUSBDHIDDevice 的调用图。

第 5 章　HID 设备

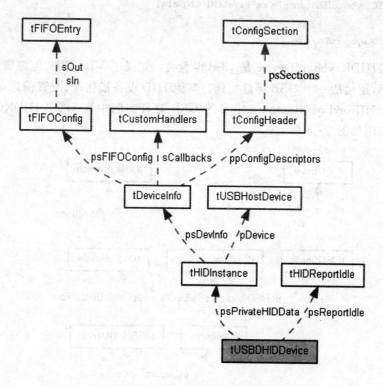

图 5-3　tUSBDHIDDevice 调用图

```
typedef struct
{
    //VID
    unsigned short usVID;
    //PID
    unsigned short usPID;
    //设备最大耗电量
    unsigned short usMaxPowermA;
    //电源属性
    unsigned char ucPwrAttributes;
    //函数指针,处理返回事务
    tUSBCallback pfnCallback;
    //Callback 第一个入口参数
    void * pvCBData;
    //指向字符串描述符集合
    const unsigned char * const * ppStringDescriptors;
    //字符串描述符个数为(1 + (5 * (num languages)))
    unsigned long ulNumStringDescriptors;
    //键盘实例,保存 USB 键盘的相关信息
```

```
    tHIDKeyboardInstance * psPrivateHIDKbdData;
}
tUSBDHIDKeyboardDevice;
```

tUSBDHIDKeyboardDevice 是 USB 键盘类。定义了 VID、PID、电源属性、字符串描述符等,还包括一个 USB 键盘实例。其他 HID 设备描述符、配置信息通过 API 函数输入 tHIDKeyboardInstance 定义的 USB 键盘实例中。tUSBDHIDKeyboardDevice 调用图如图 5-4 所示。

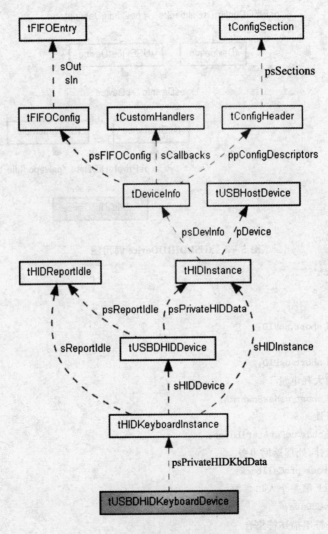

图 5-4 tUSBDHIDKeyboardDevice 调用图

tUSBDHIDKeyboardDevice 定义键盘相关的描述符,在 usbdhidkeyb.c 中通过 tHIDKeyboardInstance 把更详细的 hidkeyboard 参数传递给 USBHIDDevice,从而

完成键盘参数的初始化。如果要修改报告符,就必须进入 usbdhidkeyb.c 中,修改 g_pucKeybReportDescriptor 中的参数。

5.3.2 API 函数

在 USB 键盘 API 库中定义了 7 个函数,完成 USB 键盘初始化、配置及数据处理。下面为 usbdhidkeyb.h 中定义的 API 函数:

```
void * USBDHIDKeyboardInit(unsigned long ulIndex, const tUSBDHIDKeyboardDevice * psDevice);
void * USBDHIDKeyboardCompositeInit(unsigned long ulIndex,
                        const tUSBDHIDKeyboardDevice * psDevice);
unsigned long USBDHIDKeyboardKeyStateChange(void * pvInstance, unsigned char ucModifiers, unsigned char ucUsageCode, tBoolean bPressed);
void USBDHIDKeyboardTerm(void * pvInstance);
void * USBDHIDKeyboardSetCBData(void * pvInstance, void * pvCBData);
void USBDHIDKeyboardPowerStatusSet(void * pvInstance, unsigned char ucPower);
tBoolean USBDHIDKeyboardRemoteWakeupRequest(void * pvInstance);
```

(1) void * USBDHIDKeyboardInit(unsigned long ulIndex,
                          const tUSBDHIDKeyboardDevice * psDevice);

作用:初始化键盘硬件、协议,把其他配置参数填入 psDevice 的键盘实例中。
参数:ulIndex,USB 模块代码,固定值:USB_BASE0。psDevice,USB 键盘类。
返回:指向配置后的 tUSBDHIDKeyboardDevice。

(2) void * USBDHIDKeyboardCompositeInit(unsigned long ulIndex,
                          const tUSBDHIDKeyboardDevice * psDevice);

作用:初始化键盘协议,本函数在 USBDHIDKeyboardInit 中已经调用。
参数:ulIndex,USB 模块代码,固定值:USB_BASE0。psDevice,USB 键盘类。
返回:指向配置后的 tUSBDHIDKeyboardDevice。

(3) unsigned long USBDHIDKeyboardKeyStateChange(void * pvInstance,
                          unsigned char ucModifiers,
                          unsigned char ucUsageCode, tBoolean bPressed);

作用:键盘状态改变,并发送报告给主机。
参数:pvInstance,指向 tUSBDHIDKeyboardDevice,本函数将修改其按键状态等。ucModifiers,功能按键代码。ucUsageCode,普通按键代码。bPressed,是否加入到报告中并发送给主机。这个函数从当前按下的按键列表增加和移除一个按键代码,并产生一个调度来发送一个按键改变的状态到 USB 主机。如果按键超过最大的个数,则报告给主机时会有溢出错误代码,即 HID_KEYB_USAGE_ROLLOVER,调用者将会返回 KEYB_ERR_TOO_MANY_KEYS。

返回:程序错误代码。

(4) void USBDHIDKeyboardTerm(void * pvInstance);

作用:结束 USB 键盘。

参数:pvInstance,指向 tUSBDHIDKeyboardDevice。

返回:无。

(5) void * USBDHIDKeyboardSetCBData(void * pvInstance, void * pvCBData);

作用:修改 tUSBDHIDKeyboardDevice 中的 pvCBData 指针。

参数:pvInstance,指向 tUSBDHIDKeyboardDevice。pvCBData,数据指针,用于替换 tUSBDHIDKeyboardDevice 中的 pvCBData 指针。

返回:以前 tUSBDHIDKeyboardDevice 的 pvCBData 的指针。

(6) void USBDHIDKeyboardPowerStatusSet(void * pvInstance, unsigned char ucPower);

作用:设置键盘电源模式。

参数:pvInstance,指向 tUSBDHIDKeyboardDevice。ucPower,电源工作模式,USB_STATUS_SELF_PWR 或者 USB_STATUS_BUS_PWR。

返回:无。

(7) tBoolean USBDHIDKeyboardRemoteWakeupRequest(void * pvInstance);

作用:唤醒请求。

参数:pvInstance,指向 tUSBDHIDKeyboardDevice。

返回:是否成功唤醒。

这些 API 中使用最多是 USBDHIDKeyboardInit 和 USBDHIDKeyboardPowerStatusSet 两个函数,在首次使用 USB 键盘时,要初始化 USB 设备,使用 USBDHIDKeyboardInit 完成 USB 键盘初始化、打开 USB 中断、枚举设备、描述符补全等;USBDHIDKeyboardPowerStatusSet 设置按键状态,并通过报告发送给主机,这是键盘与主机进行数据通信最主要的接口函数,使用频率最高。

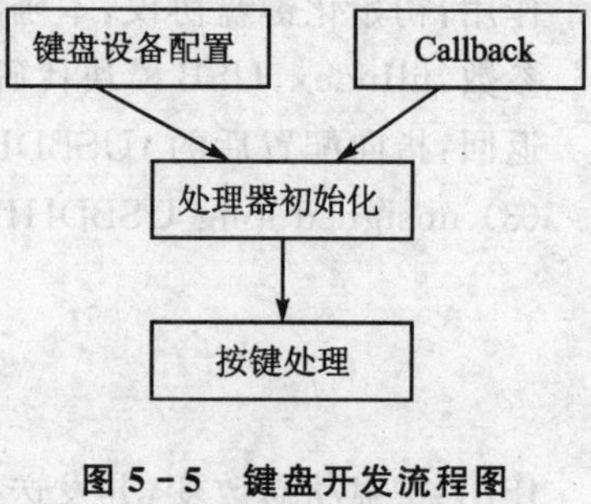

5.3.3 USB 键盘开发

USB 键盘开发只需要 4 步就能完成。如图 5-5 所示,为键盘开发流程图。键盘设备配置(主要是字符串描述符)、Callback 函数编写、USB 处理器初始化、按键处理。

图 5-5 键盘开发流程图

(1) 键盘设备配置(主要是字符串描述符),按字符串描述符标准完成串描述符配置,进而完成键盘设备配置。

```c
#include "inc/hw_ints.h"
#include "inc/hw_memmap.h"
#include "inc/hw_gpio.h"
#include "inc/hw_types.h"
#include "driverlib/debug.h"
#include "driverlib/gpio.h"
#include "driverlib/interrupt.h"
#include "driverlib/sysctl.h"
#include "driverlib/systick.h"
#include "driverlib/usb.h"
#include "usblib/usblib.h"
#include "usblib/usbhid.h"
#include "usblib/usb-ids.h"
#include "usblib/device/usbdevice.h"
#include "usblib/device/usbdhid.h"
#include "usblib/device/usbdhidkeyb.h"
//声明函数原型
unsigned long KeyboardHandler(void * pvCBData,
                              unsigned long ulEvent,
                              unsigned long ulMsgData,
                              void * pvMsgData);
//****************************************************************
// 语言描述符
//****************************************************************
const unsigned char g_pLangDescriptor[] =
{
    4,
    USB_DTYPE_STRING,
    USBShort(USB_LANG_EN_US)
};
//****************************************************************
// 制造商 字符串 描述符
//****************************************************************
const unsigned char g_pManufacturerString[] =
{
    (17 + 1) * 2,
    USB_DTYPE_STRING,
    'T', 0, 'e', 0, 'x', 0, 'a', 0, 's', 0, ' ', 0, 'I', 0, 'n', 0, 's', 0,
    't', 0, 'r', 0, 'u', 0, 'm', 0, 'e', 0, 'n', 0, 't', 0, 's', 0,
};
```

```c
// ****************************************************************
// 产品 字符串 描述符
// ****************************************************************
const unsigned char g_pProductString[] =
{
    (16 + 1) * 2,
    USB_DTYPE_STRING,
    'K', 0, 'e', 0, 'y', 0, 'b', 0, 'o', 0, 'a', 0, 'r', 0, 'd', 0, ' ', 0,
    'E', 0, 'x', 0, 'a', 0, 'm', 0, 'p', 0, 'l', 0, 'e', 0
};
// ****************************************************************
// 产品 序列号 描述符
// ****************************************************************
const unsigned char g_pSerialNumberString[] =
{
    (8 + 1) * 2,
    USB_DTYPE_STRING,
    '1', 0, '2', 0, '3', 0, '4', 0, '5', 0, '6', 0, '7', 0, '8', 0
};
// ****************************************************************
// 设备接口字符串描述符
// ****************************************************************
const unsigned char g_pHIDInterfaceString[] =
{
    (22 + 1) * 2,
    USB_DTYPE_STRING,
    'H', 0, 'I', 0, 'D', 0, ' ', 0, 'K', 0, 'e', 0, 'y', 0, 'b', 0,
    'o', 0, 'a', 0, 'r', 0, 'd', 0, ' ', 0, 'I', 0, 'n', 0, 't', 0,
    'e', 0, 'r', 0, 'f', 0, 'a', 0, 'c', 0, 'e', 0
};
// ****************************************************************
// 设备配置字符串描述符
// ****************************************************************
const unsigned char g_pConfigString[] =
{
    (26 + 1) * 2,
    USB_DTYPE_STRING,
    'H', 0, 'I', 0, 'D', 0, ' ', 0, 'K', 0, 'e', 0, 'y', 0, 'b', 0,
    'o', 0, 'a', 0, 'r', 0, 'd', 0, ' ', 0, 'C', 0, 'o', 0, 'n', 0,
    'f', 0, 'i', 0, 'g', 0, 'u', 0, 'r', 0, 'a', 0, 't', 0, 'i', 0,
    'o', 0, 'n', 0
};
```

```c
// ***************************************************************
// 字符串描述符集合,一定要按这个顺序排列,因为在描述符中已经定义好描述索引了
// ***************************************************************
const unsigned char * const g_pStringDescriptors[] =
{
    g_pLangDescriptor,
    g_pManufacturerString,
    g_pProductString,
    g_pSerialNumberString,
    g_pHIDInterfaceString,
    g_pConfigString
};
#define NUM_STRING_DESCRIPTORS (sizeof(g_pStringDescriptors) /     \
                               sizeof(unsigned char *))
// ***************************************************************
//键盘实例,键盘配置并为键盘设备信息提供空间
// ***************************************************************
tHIDKeyboardInstance g_KeyboardInstance;
// ***************************************************************
//键盘设备配置
// ***************************************************************
const tUSBDHIDKeyboardDevice g_sKeyboardDevice =
{
    USB_VID_STELLARIS,                    //自行定义 VIP
    USB_PID_KEYBOARD,                     //自行定义 PID
    500,
    USB_CONF_ATTR_SELF_PWR | USB_CONF_ATTR_RWAKE,
    KeyboardHandler,
    (void *)&g_sKeyboardDevice,
    g_pStringDescriptors,
    NUM_STRING_DESCRIPTORS,
    &g_KeyboardInstance
};
```

(2) 完成 Callback 函数。Callback 函数用于处理按键事务。可能是主机发出,也可能是状态信息。USB 键盘设备中包含了以下事务:USB_EVENT_CONNECTED、USB_EVENT_DISCONNECTED、USBD_HID_KEYB_EVENT_SET_LEDS、USB_EVENT_SUSPEND、USB_EVENT_RESUME、USB_EVENT_TX_COMPLETE。如表 5 - 2 所列,为 USB 键盘事务。

第 5 章 HID 设备

表 5-2 USB 键盘事务

名　称	说　明
USB_EVENT_CONNECTED	USB 设备已经连接到主机
USB_EVENT_DISCONNECTED	USB 设备已经与主机断开
USBD_HID_KEYB_EVENT_SET_LEDS	USB 键盘状态灯设置,通过查询功能按键状态来设置状态灯
USB_EVENT_SUSPEND	挂起
USB_EVENT_RESUME	唤醒
USB_EVENT_TX_COMPLETE	发送完成

根据以上事务编写 Callback 函数如下：

```
unsigned long KeyboardHandler(void * pvCBData, unsigned long ulEvent,
                unsigned long ulMsgData, void * pvMsgData)
{
    switch (ulEvent)
    {
        case USB_EVENT_CONNECTED:
        {
            //连接成功时,点亮 LED1
            GPIOPinWrite(GPIO_PORTF_BASE,0x10,0x10);
            break;
        }
        case USB_EVENT_DISCONNECTED:
        {
            //断开连接时,LED1 灭
            GPIOPinWrite(GPIO_PORTF_BASE,0x10,0x00);
            break;
        }
        case USB_EVENT_TX_COMPLETE:
        {
            //发送完成时,LED2 亮
            GPIOPinWrite(GPIO_PORTF_BASE,0x20,0x20);
            break;
        }
        case USB_EVENT_SUSPEND:
        {
            //USB 处于挂起模式,LED2 灭
            GPIOPinWrite(GPIO_PORTF_BASE,0x20,0x0);
            break;
        }
```

```
case USB_EVENT_RESUME:
{
    break;
}
case USBD_HID_KEYB_EVENT_SET_LEDS:
{
    //HID_KEYB_CAPS_LOCK 灯
    GPIOPinWrite(GPIO_PORTF_BASE,0x80,((char)ulMsgData &
                HID_KEYB_CAPS_LOCK)? 0 : 0x80);
    break;
}
default:
{
    break;
}
}
return (0);
}
```

（3）系统初始化，配置内核电压、系统主频、使能端口、配置按键端口、LED 控制等，本例中使用 4 个按键模拟普通键盘，使用 4 个 LED 进行指示。如图 5-6 所示，为 LED 与按键电路图，当 PF4～PF7 输入高电平，可以驱动 LED 亮；处理器内部配置 PF0～PF3 上拉，当 PF0～PF3 有按键被按下时，对应引脚会检测到低电平。

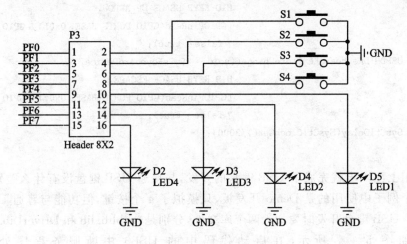

图 5-6 LED 与按键电路

系统初始化程序如下：

```
//设置系统内核电压与主频
SysCtlLDOSet(SYSCTL_LDO_2_75V);
```

第 5 章 HID 设备

```
SysCtlClockSet(SYSCTL_XTAL_8MHZ | SYSCTL_SYSDIV_4 | SYSCTL_USE_PLL    | SYSCTL_OSC
                _MAIN );         //使能端口
SysCtlPeripheralEnable(SYSCTL_PERIPH_GPIOF);
GPIOPinTypeGPIOOutput(GPIO_PORTF_BASE,0xf0);
GPIOPinTypeGPIOInput(GPIO_PORTF_BASE,0x0f);
HWREG(GPIO_PORTF_BASE + GPIO_O_PUR) |= 0x0f;
  //初始化键盘设备
USBDHIDKeyboardInit(0, &g_sKeyboardDevice);
```

(4) 按键处理。主要使用 USBDHIDKeyboardPowerStatusSet 设置按键状态，并通过报告发送给主机。

```
while(1)
{
    USBDHIDKeyboardKeyStateChange((void * )&g_sKeyboardDevice, HID_KEYB_CAPS_LOCK,
                        HID_KEYB_USAGE_A,
                        (GPIOPinRead(GPIO_PORTF_BASE, 0x0f) & GPIO_PIN_0)
                        ? false : true);
    USBDHIDKeyboardKeyStateChange((void * )&g_sKeyboardDevice, 0,
                        HID_KEYB_USAGE_DOWN_ARROW,
                        (GPIOPinRead(GPIO_PORTF_BASE, 0x0f) & GPIO_PIN_1)
                        ? false : true);
    USBDHIDKeyboardKeyStateChange((void * )&g_sKeyboardDevice, 0,
                        HID_KEYB_USAGE_UP_ARROW,
                        (GPIOPinRead(GPIO_PORTF_BASE, 0x0f) & GPIO_PIN_2)
                        ? false : true);
    USBDHIDKeyboardKeyStateChange((void * )&g_sKeyboardDevice, 0,
                        HID_KEYB_USAGE_ESCAPE,
                        (GPIOPinRead(GPIO_PORTF_BASE, 0x0f) & GPIO_PIN_3)
                        ? false : true);
    SysCtlDelay(SysCtlClockGet()/3000);
}
```

使用上面 4 步就完成了 USB 键盘的开发，与普通 USB 键盘没有什么差别。由于在这个例子中使用的是 Demo 开发板，只模拟了 4 个按键，但功能与普通 USB 键盘一样。USB 键盘开发时要加入两个库文件，分别是：usblib.lib 和 DriverLib.lib 库文件。如图 5-7 所示，在启动代码中的 USB0 中断服务程序处加入 USB0DeviceIntHandler 中断服务函数，此函数已经包含在 USB 库中，不必由开发人员自己编写。

USB 键盘源码如下：

```
#include "inc/hw_ints.h"
```

```
DCD     IntDefaultHandler           ; GPIO Port F
DCD     IntDefaultHandler           ; GPIO Port G
DCD     IntDefaultHandler           ; GPIO Port H
DCD     IntDefaultHandler           ; UART2 Rx and Tx
DCD     IntDefaultHandler           ; SSI1 Rx and Tx
DCD     IntDefaultHandler           ; Timer 3 subtimer A
DCD     IntDefaultHandler           ; Timer 3 subtimer B
DCD     IntDefaultHandler           ; I2C1 Master and Slave
DCD     IntDefaultHandler           ; Quadrature Encoder 1
DCD     IntDefaultHandler           ; CAN0
DCD     IntDefaultHandler           ; CAN1
DCD     IntDefaultHandler           ; CAN2
DCD     IntDefaultHandler           ; Ethernet
DCD     IntDefaultHandler           ; Hibernate
DCD     USB0DeviceIntHandler        ; USB0
DCD     IntDefaultHandler           ; PWM Generator 3
DCD     IntDefaultHandler           ; uDMA Software Transfer
DCD     IntDefaultHandler           ; uDMA Error
DCD     IntDefaultHandler           ; ADC1 Sequence 0
DCD     IntDefaultHandler           ; ADC1 Sequence 1
DCD     IntDefaultHandler           ; ADC1 Sequence 2
DCD     IntDefaultHandler           ; ADC1 Sequence 3
DCD     IntDefaultHandler           ; I2S0
DCD     IntDefaultHandler           ; External Bus Interface 0
DCD     IntDefaultHandler           ; GPIO Port J
```

图 5-7 USB0 中断服务程序

```c
#include "inc/hw_memmap.h"
#include "inc/hw_gpio.h"
#include "inc/hw_types.h"
#include "driverlib/debug.h"
#include "driverlib/gpio.h"
#include "driverlib/interrupt.h"
#include "driverlib/sysctl.h"
#include "driverlib/systick.h"
#include "driverlib/usb.h"
#include "usblib/usblib.h"
#include "usblib/usbhid.h"
#include "usblib/usb-ids.h"
#include "usblib/device/usbdevice.h"
#include "usblib/device/usbdhid.h"
#include "usblib/device/usbdhidkeyb.h"
//声明函数原型
unsigned long KeyboardHandler(void * pvCBData,
                unsigned long ulEvent,
```

第 5 章　HID 设备

```c
                            unsigned long ulMsgData,
                            void * pvMsgData);
//*************************************************************
// 语言描述符
//*************************************************************
const unsigned char g_pLangDescriptor[] =
{
    4,
    USB_DTYPE_STRING,
    USBShort(USB_LANG_EN_US)
};
//*************************************************************
// 制造商 字符串 描述符
//*************************************************************
const unsigned char g_pManufacturerString[] =
{
    (17 + 1) * 2,
    USB_DTYPE_STRING,
    'T', 0, 'e', 0, 'x', 0, 'a', 0, 's', 0, ' ', 0, 'I', 0, 'n', 0, 's', 0,
    't', 0, 'r', 0, 'u', 0, 'm', 0, 'e', 0, 'n', 0, 't', 0, 's', 0,
};
//*************************************************************
//产品 字符串 描述符
//*************************************************************
const unsigned char g_pProductString[] =
{
    (16 + 1) * 2,
    USB_DTYPE_STRING,
    'K', 0, 'e', 0, 'y', 0, 'b', 0, 'o', 0, 'a', 0, 'r', 0, 'd', 0, ' ', 0,
    'E', 0, 'x', 0, 'a', 0, 'm', 0, 'p', 0, 'l', 0, 'e', 0
};
//*************************************************************
// 产品 序列号 描述符
//*************************************************************
const unsigned char g_pSerialNumberString[] =
{
    (8 + 1) * 2,
    USB_DTYPE_STRING,
    '1', 0, '2', 0, '3', 0, '4', 0, '5', 0, '6', 0, '7', 0, '8', 0
};
//*************************************************************
// 设备接口字符串描述符
```

```c
//****************************************************************
const unsigned char g_pHIDInterfaceString[] =
{
    (22 + 1) * 2,
    USB_DTYPE_STRING,
    'H', 0, 'I', 0, 'D', 0, ' ', 0, 'K', 0, 'e', 0, 'y', 0, 'b', 0,
    'o', 0, 'a', 0, 'r', 0, 'd', 0, ' ', 0, 'I', 0, 'n', 0, 't', 0,
    'e', 0, 'r', 0, 'f', 0, 'a', 0, 'c', 0, 'e', 0
};
//****************************************************************
// 设备配置字符串描述符
//****************************************************************
const unsigned char g_pConfigString[] =
{
    (26 + 1) * 2,
    USB_DTYPE_STRING,
    'H', 0, 'I', 0, 'D', 0, ' ', 0, 'K', 0, 'e', 0, 'y', 0, 'b', 0,
    'o', 0, 'a', 0, 'r', 0, 'd', 0, ' ', 0, 'C', 0, 'o', 0, 'n', 0,
    'f', 0, 'i', 0, 'g', 0, 'u', 0, 'r', 0, 'a', 0, 't', 0, 'i', 0,
    'o', 0, 'n', 0
};
//****************************************************************
// 字符串描述符集合
//****************************************************************
const unsigned char * const g_pStringDescriptors[] =
{
    g_pLangDescriptor,
    g_pManufacturerString,
    g_pProductString,
    g_pSerialNumberString,
    g_pHIDInterfaceString,
    g_pConfigString
};

#define NUM_STRING_DESCRIPTORS (sizeof(g_pStringDescriptors) / \
                                sizeof(unsigned char *))
//****************************************************************
//键盘实例,键盘配置并为键盘设备信息提供空间
//****************************************************************
tHIDKeyboardInstance g_KeyboardInstance;
//****************************************************************
//键盘设备配置
```

```c
//*****************************************************************
const tUSBDHIDKeyboardDevice g_sKeyboardDevice =
{
    USB_VID_STELLARIS,
    USB_PID_KEYBOARD,
    500,
    USB_CONF_ATTR_SELF_PWR | USB_CONF_ATTR_RWAKE,
    KeyboardHandler,
    (void *)&g_sKeyboardDevice,
    g_pStringDescriptors,
    NUM_STRING_DESCRIPTORS,
    &g_KeyboardInstance
};
//*****************************************************************
//键盘 Callback 函数
//*****************************************************************
unsigned long KeyboardHandler(void * pvCBData, unsigned long ulEvent,
                              unsigned long ulMsgData, void * pvMsgData)
{
    switch (ulEvent)
    {
        case USB_EVENT_CONNECTED:
        {
            //连接成功时,点亮 LED1
            GPIOPinWrite(GPIO_PORTF_BASE,0x10,0x10);
            break;
        }
        case USB_EVENT_DISCONNECTED:
        {
            //断开连接时,LED1 灭
            GPIOPinWrite(GPIO_PORTF_BASE,0x10,0x00);
            break;
        }
        case USB_EVENT_TX_COMPLETE:
        {
            //发送完成时,LED2 亮
            GPIOPinWrite(GPIO_PORTF_BASE,0x20,0x20);
            break;
        }
        case USB_EVENT_SUSPEND:
        {
            //USB 处于挂起模式时,LED2 灭
```

```c
            GPIOPinWrite(GPIO_PORTF_BASE,0x20,0x0);
            break;
        }
        case USB_EVENT_RESUME:
        {
            break;
        }
        case USBD_HID_KEYB_EVENT_SET_LEDS:
        {
            //HID_KEYB_CAPS_LOCK 灯
            GPIOPinWrite(GPIO_PORTF_BASE,0x80,((char)ulMsgData & HID_KEYB_CAPS_
                    LOCK)? 0 : 0x80);
            break;
        }
        default:
        {
            break;
        }
    }
    return (0);
}
//****************************************************************
//主函数
//****************************************************************
int main(void)
{
    //系统初始化
    SysCtlLDOSet(SYSCTL_LDO_2_75V);
    SysCtlClockSet(SYSCTL_XTAL_8MHZ | SYSCTL_SYSDIV_4 | SYSCTL_USE_PLL   | SYSCTL_OSC
            _MAIN );
    SysCtlPeripheralEnable(SYSCTL_PERIPH_GPIOF);
    GPIOPinTypeGPIOOutput(GPIO_PORTF_BASE,0xf0);
    GPIOPinTypeGPIOInput(GPIO_PORTF_BASE,0x0f);
    HWREG(GPIO_PORTF_BASE + GPIO_O_PUR) |= 0x0f;
    USBDHIDKeyboardInit(0, &g_sKeyboardDevice);
    while(1)
    {
        USBDHIDKeyboardKeyStateChange((void *)&g_sKeyboardDevice, HID_KEYB_CAPS_LOCK,
                        HID_KEYB_USAGE_A,
                        (GPIOPinRead(GPIO_PORTF_BASE, 0x0f) & GPIO_PIN_0)
                        ? false : true);
```

```
                USBDHIDKeyboardKeyStateChange((void *)&g_sKeyboardDevice, 0,
                                HID_KEYB_USAGE_DOWN_ARROW,
                                (GPIOPinRead(GPIO_PORTF_BASE, 0x0f) & GPIO_PIN_1)
                                ? false : true);
                USBDHIDKeyboardKeyStateChange((void *)&g_sKeyboardDevice, 0,
                                HID_KEYB_USAGE_UP_ARROW,
                                (GPIOPinRead(GPIO_PORTF_BASE, 0x0f) & GPIO_PIN_2)
                                ? false : true);
                USBDHIDKeyboardKeyStateChange((void *)&g_sKeyboardDevice, 0,
                                HID_KEYB_USAGE_ESCAPE,
                                (GPIOPinRead(GPIO_PORTF_BASE, 0x0f) & GPIO_PIN_3)
                                ? false : true);
                SysCtlDelay(SysCtlClockGet()/3000);
        }
    }
```

5.4 USB 鼠标

在 USB 库中已经定义好了 USB 鼠标的数据类型和 API 函数，使开发 USB 鼠标非常方便快捷。相关定义和数据类型放在 usbdhidmouse.h 中。

5.4.1 数据类型

usbdhidmouse.h 中已经定义好 USB 鼠标使用的所有数据类型和函数，下面介绍 USB 鼠标使用的数据类型。

```
#define MOUSE_REPORT_SIZE           3
```

MOUSE_REPORT_SIZE 定义鼠标 IN 报告的大小，可用于判断 IN 报告是否正确，如果超出 3 个字节会返回错误代码。

```
typedef enum
{
    //状态还没有定义
    HID_MOUSE_STATE_UNCONFIGURED,
    //空闲状态,没有按键按下或者没有等待数据
    HID_MOUSE_STATE_IDLE,
    //等待主机发送数据
    HID_MOUSE_STATE_WAIT_DATA,
    //等待数据发送
    HID_MOUSE_STATE_SEND
}
```

```
tMouseState;
```

tMouseState，定义 USB 鼠标状态，USB 鼠标正常工作时，保存鼠标工作状态。

```
typedef struct
{
    // 指示当前 USB 设备是否配置成功
    unsigned char ucUSBConfigured;
    // USB 键盘使用的子协议：USB_HID_PROTOCOL_BOOT 或者 USB_HID_PROTOCOL_REPORT
    // 将会传给设备描述符和端点描述符
    unsigned char ucProtocol;
    // 用于收送 IN 报告，保存最新一次报告
    unsigned char pucReport[MOUSE_REPORT_SIZE];
    // 中断 IN 端点状态
    volatile tMouseState eMouseState;
    // 为 IN 报告定义时间
    tHIDReportIdle sReportIdle;
    // HID 设备实例，保存键盘 HID 信息
    tHIDInstance sHIDInstance;
    // HID 设备驱动
    tUSBDHIDDevice sHIDDevice;
}
tHIDMouseInstance;
```

tHIDMouseInstance，USB 鼠标实例。在 tHIDInstance 和 tUSBDHIDDevice 的基础上，用于保存全部 USB 鼠标的配置信息，包括描述符、Callback 函数和按键事件等。

```
typedef struct
{
    //VID
    unsigned short usVID;
    //PID
    unsigned short usPID;
    //设备最大耗电量
    unsigned short usMaxPowermA;
    //电源属性
    unsigned char ucPwrAttributes;
    //函数指针，处理返回事务
    tUSBCallback pfnCallback;
    //Callback 第一个入口参数
    void * pvCBData;
    //指向字符串描述符集合
```

第5章 HID设备

```
    const unsigned char * const * ppStringDescriptors;
    //字符串描述符个数为(1 + (5 * (num languages)))
    unsigned long ulNumStringDescriptors;
    //鼠标实例,保存USB鼠标的相关信息
    tHIDMouseInstance * psPrivateHIDMouseData;
}
tUSBDHIDMouseDevice;
```

tUSBDHIDMouseDevice,USB鼠标类。定义了VID、PID、电源属性以及字符串描述符等,还包括一个USB鼠标实例。其他HID设备描述符、配置信息通过API函数输入tHIDMouseInstance定义的USB鼠标实例中。tUSBDHIDMouseDevice调用图如图5-8所示。

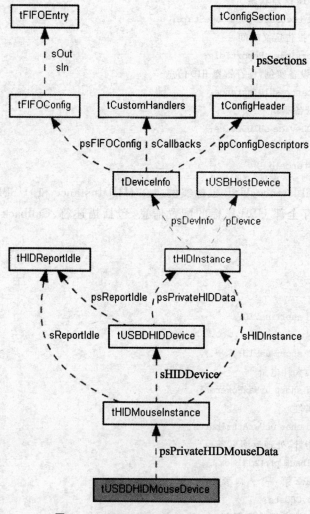

图5-8 tUSBDHIDMouseDevice调用图

由图 5-8 和图 5-4 对比可以看出，USB 键盘与鼠标同属于 HID 设备，都是通过 tUSBHIDDevice 传递参数给底层驱动，当然也可以直接使用 tUSBHIDDevice 定义键盘、鼠标类，完成鼠标、键盘的设计。在 USB 库中，已经通过 HID 设备把键盘与鼠标独立出来，做为 HID 设备的两个分类设备。

```
#define MOUSE_ERR_TX_ERROR      2
```

MOUSE_ERR_TX_ERROR 表示 USB 鼠标 API 函数 USBDHIDMouseStateChange 返回的错误代码。

5.4.2 API 函数

在 USB 鼠标 API 库中定义了 7 个函数，用于完成 USB 键盘初始化、配置及数据处理。下面为 usbdhidkeyb.h 中定义的 API 函数：

```
void * USBDHIDMouseInit(unsigned long ulIndex,
                       const tUSBDHIDMouseDevice * psDevice);
void * USBDHIDMouseCompositeInit(unsigned long ulIndex,
                       const tUSBDHIDMouseDevice * psDevice);
unsigned long USBDHIDMouseStateChange(void * pvInstance, char cDeltaX,
                       char cDeltaY,
                       unsigned char ucButtons);
void USBDHIDMouseTerm(void * pvInstance);
void * USBDHIDMouseSetCBData(void * pvInstance, void * pvCBData);
void USBDHIDMousePowerStatusSet(void * pvInstance,
                       unsigned char ucPower);
tBoolean USBDHIDMouseRemoteWakeupRequest(void * pvInstance);
```

(1) void * USBDHIDMouseInit(unsigned long ulIndex,
                       const tUSBDHIDMouseDevice * psDevice);
作用：初始化鼠标硬件、协议，把其他配置参数填入 psDevice 的鼠标实例中。
参数：ulIndex，USB 模块代码，固定值：USB_BASE0。psDevice，USB 鼠标类。
返回：指向配置后的 tUSBDHIDMouseDevice。

(2) void * USBDHIDMouseCompositeInit(unsigned long ulIndex,
                       const tUSBDHIDMouseDevice * psDevice);
作用：初始化鼠标协议，此函数在 USBDHIDMouseInit 中被调用一次。
参数：ulIndex，USB 模块代码，固定值：USB_BASE0。psDevice，USB 鼠标类。
返回：指向配置后的 tUSBDHIDMouseDevice。

(3) unsigned long USBDHIDMouseStateChange(void * pvInstance, char cDeltaX,
                       char cDeltaY,
                       unsigned char ucButtons);

作用:鼠标状态改变,并发送报告给主机。

参数:pvInstance,指向 tUSBDHIDMouseDevice,本函数将修改其 X、Y 及按键状态等。cDeltaX,X 值。cDeltaY,Y 值。ucButtons,鼠标按键。

返回:程序错误代码。

(4) void USBDHIDMouseTerm(void * pvInstance);

作用:结束 USB 鼠标。

参数:pvInstance,指向 tUSBDHIDMouseDevice。

返回:无。

(5) void * USBDHIDMouseSetCBData(void * pvInstance, void * pvCBData);

作用:修改 tUSBDHIDMouseDevice 中的 pvCBData 指针。

参数:pvInstance,指向 tUSBDHIDMouseDevice。pvCBData,数据指针,用于替换 tUSBDHIDMouseDevice 中的 pvCBData 指针。

返回:以前 tUSBDHIDMouseDevice 的 pvCBData 的指针。

(6) void USBDHIDMousePowerStatusSet(void * pvInstance,
                    unsigned char ucPower);

作用:设置鼠标电源模式。

参数:pvInstance,指向 tUSBDHIDMouseDevice。ucPower,电源工作模式,USB_STATUS_SELF_PWR 或者 USB_STATUS_BUS_PWR。

返回:无。

(7) tBoolean USBDHIDMouseRemoteWakeupRequest(void * pvInstance);

作用:唤醒请求。

参数:pvInstance,指向 tUSBDHIDMouseDevice。

返回:是否成功唤醒。

这些 API 中使用最多的是 USBDHIDMouseInit 和 USBDHIDMouseStateChange 两个函数,在首次使用 USB 鼠标时,要初始化 USB 设备,使用 USBDHIDMouseInit 完成 USB 鼠标初始化、打开 USB 中断、枚举设备、描述符补全等;USBDHIDMouseStateChange 设置鼠标 X、Y、按键状态,并通过 IN 报告发送给主机,这是 USB 鼠标与主机进行数据通信最主要的接口函数,使用频率最高。

5.4.3 USB 鼠标开发

USB 鼠标开发只需要 4 步就能完成。如图 5-9 所示,为鼠标开发流程图。鼠标设备配置(主要是字符串描述符)、Callback 函数编写、USB 处理器初始化、X、Y、按键处理。

(1) 鼠标设备配置(主要是字符串描述符),按字符串描述符标准完成串描述符配置,进而完

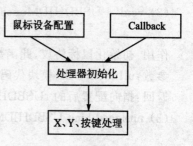

图 5-9 鼠标开发流程图

成鼠标设备配置。

```c
#include "inc/hw_types.h"
#include "driverlib/usb.h"
#include "usblib/usblib.h"
#include "usblib/usbhid.h"
#include "usblib/usb-ids.h"
#include "usblib/device/usbdevice.h"
#include "usblib/device/usbdhid.h"
#include "usblib/device/usbdhidmouse.h"
#include "usb_mouse_structs.h"
//声明函数原型
unsigned long MouseHandler(void * pvCBData,
                           unsigned long ulEvent,
                           unsigned long ulMsgData,
                           void * pvMsgData);
//***************************************************************
// 语言描述符
//***************************************************************
const unsigned char g_pLangDescriptor[] =
{
    4,
    USB_DTYPE_STRING,
    USBShort(USB_LANG_EN_US)
};
//***************************************************************
// 制造商 字符串 描述符
//***************************************************************
const unsigned char g_pManufacturerString[] =
{
    (17 + 1) * 2,
    USB_DTYPE_STRING,
    'T', 0, 'e', 0, 'x', 0, 'a', 0, 's', 0, '', 0, 'I', 0, 'n', 0, 's', 0,
    't', 0, 'r', 0, 'u', 0, 'm', 0, 'e', 0, 'n', 0, 't', 0, 's', 0,
};
//***************************************************************
//产品 字符串 描述符
//***************************************************************
const unsigned char g_pProductString[] =
{
    (13 + 1) * 2,
    USB_DTYPE_STRING,
```

```c
    'M', 0, 'o', 0, 'u', 0, 's', 0, 'e', 0, ' ', 0, 'E', 0, 'x', 0, 'a', 0,
    'm', 0, 'p', 0, 'l', 0, 'e', 0
};
//***************************************************************
//产品 序列号 描述符
//***************************************************************
const unsigned char g_pSerialNumberString[] =
{
    (8 + 1) * 2,
    USB_DTYPE_STRING,
    '1', 0, '2', 0, '3', 0, '4', 0, '5', 0, '6', 0, '7', 0, '8', 0
};
//***************************************************************
// 设备接口字符串描述符
//***************************************************************
const unsigned char g_pHIDInterfaceString[] =
{
    (19 + 1) * 2,
    USB_DTYPE_STRING,
    'H', 0, 'I', 0, 'D', 0, ' ', 0, 'M', 0, 'o', 0, 'u', 0, 's', 0,
    'e', 0, ' ', 0, 'I', 0, 'n', 0, 't', 0, 'e', 0, 'r', 0, 'f', 0,
    'a', 0, 'c', 0, 'e', 0
};
//***************************************************************
//设备配置字符串描述符
//***************************************************************
const unsigned char g_pConfigString[] =
{
    (23 + 1) * 2,
    USB_DTYPE_STRING,
    'H', 0, 'I', 0, 'D', 0, ' ', 0, 'M', 0, 'o', 0, 'u', 0, 's', 0,
    'e', 0, ' ', 0, 'C', 0, 'o', 0, 'n', 0, 'f', 0, 'i', 0, 'g', 0,
    'u', 0, 'r', 0, 'a', 0, 't', 0, 'i', 0, 'o', 0, 'n', 0
};
//***************************************************************
// 字符串描述符集合
//***************************************************************
const unsigned char * const g_pStringDescriptors[] =
{
    g_pLangDescriptor,
    g_pManufacturerString,
    g_pProductString,
```

```
    g_pSerialNumberString,
    g_pHIDInterfaceString,
    g_pConfigString
};

#define NUM_STRING_DESCRIPTORS (sizeof(g_pStringDescriptors) /     \
                                sizeof(unsigned char *))
//*************************************************************
//鼠标实例,鼠标配置并为鼠标设备信息提供空间
//*************************************************************
tHIDMouseInstance g_sMouseInstance;
//*************************************************************
//鼠标设备配置
//*************************************************************
const tUSBDHIDMouseDevice g_sMouseDevice =
{
    USB_VID_STELLARIS,              //开发者自己定义
    USB_PID_MOUSE,                  //开发者自己定义
    500,
    USB_CONF_ATTR_SELF_PWR,
    MouseHandler,
    (void *)&g_sMouseDevice,
    g_pStringDescriptors,
    NUM_STRING_DESCRIPTORS,
    &g_sMouseInstance
};
```

(2) 完成 Callback 函数。Callback 函数用于处理 X、Y 和按键事务。可能是主机发出,也可能是状态信息。USB 鼠标设备中包含以下事务:USB_EVENT_CONNECTED、USB_EVENT_DISCONNECTED、USB_EVENT_SUSPEND、USB_EVENT_RESUME、USB_EVENT_TX_COMPLETE。如表 5-3 所列,为 USB 鼠标事务表。

表 5-3 USB 鼠标事务

名 称	说 明
USB_EVENT_CONNECTED	USB 设备已经连接到主机
USB_EVENT_DISCONNECTED	USB 设备已经与主机断开
USB_EVENT_SUSPEND	挂起
USB_EVENT_RESUME	唤醒
USB_EVENT_TX_COMPLETE	发送完成

根据以上事务编写 Callback 函数：

```c
unsigned long MouseHandler(void * pvCBData, unsigned long ulEvent,
                unsigned long ulMsgData, void * pvMsgData)
{
    switch (ulEvent)
    {
        case USB_EVENT_CONNECTED:
        {
            //连接成功时,点亮 LED1
            GPIOPinWrite(GPIO_PORTF_BASE,0x10,0x10);
            break;
        }
        case USB_EVENT_DISCONNECTED:
        {
            //断开连接时,LED1 灭
            GPIOPinWrite(GPIO_PORTF_BASE,0x10,0x00);
            break;
        }
        case USB_EVENT_TX_COMPLETE:
        {
            //发送完成时,LED2 亮
            GPIOPinWrite(GPIO_PORTF_BASE,0x20,0x20);
            break;
        }
        case USB_EVENT_SUSPEND:
        {
            //USB 处于挂起模式时,LED2 灭
            GPIOPinWrite(GPIO_PORTF_BASE,0x20,0x0);
            break;
        }
        case USB_EVENT_RESUME:
        {
            break;
        }
        default:
        {
            break;
        }
    }
    return (0);
}
```

(3) 系统初始化,配置内核电压、系统主频、使能端口、配置按键端口以及 LED 控制等,本例中使用 4 个按键控制鼠标移动,使用 4 个 LED 进行指示动作。LED 与键盘电路如图 5-6 所示。

系统初始化如下所示。

```
//设置系统内核电压与主频
SysCtlLDOSet(SYSCTL_LDO_2_75V);
SysCtlClockSet(SYSCTL_XTAL_8MHZ | SYSCTL_SYSDIV_4 | SYSCTL_USE_PLL | SYSCTL_OSC_
            MAIN );
//使能端口
SysCtlPeripheralEnable(SYSCTL_PERIPH_GPIOF);
GPIOPinTypeGPIOOutput(GPIO_PORTF_BASE,0xf0);
GPIOPinTypeGPIOInput(GPIO_PORTF_BASE,0x0f);
HWREG(GPIO_PORTF_BASE + GPIO_O_PUR) |= 0x0f;
//初始化鼠标设备
USBDHIDMouseInit (0, &g_sMouseDevice);
```

(4) X、Y 及按键处理。主要使用 USBDHIDMouseStateChange 设置 X、Y 及按键状态,并通过报告发送给主机。

```
while(1)
{
    ulTemp = (~GPIOPinRead(GPIO_PORTF_BASE, 0x0f)) & 0x0f;
    switch(ulTemp)
    {
        case 0x01:
                x = x + 1;
                key = 0;
                break;
        case 0x02:
                x = x - 1;
                key = 0;
                break;
        case 0x04:
                y = y + 1;
                key = 0;
                break;
        case 0x08:
                y = y - 1;
                key = 0;
                break;
        case 0x03:
                key = 1;
```

```
                    break;
          case 0x0c:
                    key = 2;
                    break;
          case 0x09:
                    key = 4;
                    break;
          default:
                    key = 0;
                    break;
    }
    if(ulTemp)
          USBDHIDMouseStateChange((void *)&g_sMouseDevice,x,y,key);
    SysCtlDelay(SysCtlClockGet()/30);
}
```

使用上面4步就能完成USB鼠标开发,与普通USB鼠标一样操作。由于在这个例子中使用的是Demo开发板,只能用4个按键,模拟鼠标移动。USB鼠标开发时也要加入两个库文件,分别是usblib.lib和DriverLib.lib,在启动代码中加入USB0DeviceIntHandler中断服务函数。如图5-10所示,为发现新硬件图。此时计算机正在枚举USB鼠标设备。

图5-10 发现新硬件

如图5-11所示,为枚举成功图。可以看出刚才设计USB鼠标枚举成功,在"设备管理器"中可以看到"USB人体学输入设备",而且可以在"鼠标和其他指针设备"中找到HID-compliant mouse。其中有一个HID-compliant mouse就是刚才开发的USB鼠标,查看"属性"可以看到如图5-12所示的内容,其中VID和PID都与USB鼠标之前配置的VID和PID相同。

USB鼠标源码如下所示。

```
#include "inc/hw_ints.h"
#include "inc/hw_memmap.h"
#include "inc/hw_gpio.h"
#include "inc/hw_types.h"
```

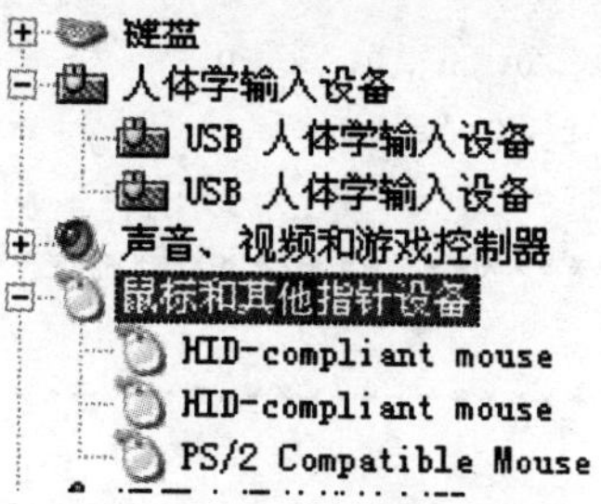

第 5 章　HID 设备

图 5-11　枚举成功

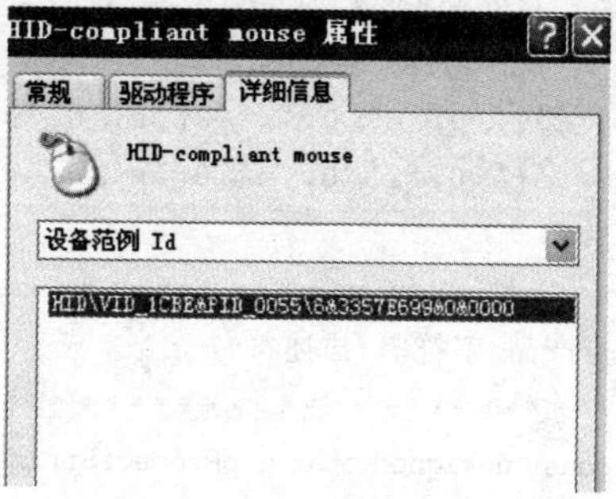

图 5-12　PID 和 VID

```c
#include "driverlib/debug.h"
#include "driverlib/gpio.h"
#include "inc/hw_types.h"
#include "driverlib/usb.h"
#include "inc/hw_sysctl.h"
#include "driverlib/sysctl.h"
#include "usblib/usblib.h"
#include "usblib/usbhid.h"
#include "usblib/usb-ids.h"
#include "usblib/device/usbdevice.h"
#include "usblib/device/usbdhid.h"
#include "usblib/device/usbdhidmouse.h"
//声明函数原型
unsigned long MouseHandler(void * pvCBData,
                           unsigned long ulEvent,
                           unsigned long ulMsgData,
                           void * pvMsgData);
//****************************************************************
// 语言描述符
//****************************************************************
const unsigned char g_pLangDescriptor[] =
{
    4,
    USB_DTYPE_STRING,
    USBShort(USB_LANG_EN_US)
};
//****************************************************************
//制造商 字符串 描述符
//****************************************************************
const unsigned char g_pManufacturerString[] =
{
```

```c
    (17 + 1) * 2,
    USB_DTYPE_STRING,
    'T', 0, 'e', 0, 'x', 0, 'a', 0, 's', 0, ' ', 0, 'I', 0, 'n', 0, 's', 0,
    't', 0, 'r', 0, 'u', 0, 'm', 0, 'e', 0, 'n', 0, 't', 0, 's', 0,
};
//****************************************************************
//产品 字符串 描述符
//****************************************************************
const unsigned char g_pProductString[] =
{
    (13 + 1) * 2,
    USB_DTYPE_STRING,
    'M', 0, 'o', 0, 'u', 0, 's', 0, 'e', 0, ' ', 0, 'E', 0, 'x', 0, 'a', 0,
    'm', 0, 'p', 0, 'l', 0, 'e', 0
};
//****************************************************************
//产品 序列号 描述符
//****************************************************************
const unsigned char g_pSerialNumberString[] =
{
    (8 + 1) * 2,
    USB_DTYPE_STRING,
    '1', 0, '2', 0, '3', 0, '4', 0, '5', 0, '6', 0, '7', 0, '8', 0
};
//****************************************************************
//设备接口字符串描述符
//****************************************************************
const unsigned char g_pHIDInterfaceString[] =
{
    (19 + 1) * 2,
    USB_DTYPE_STRING,
    'H', 0, 'I', 0, 'D', 0, ' ', 0, 'M', 0, 'o', 0, 'u', 0, 's', 0,
    'e', 0, ' ', 0, 'I', 0, 'n', 0, 't', 0, 'e', 0, 'r', 0, 'f', 0,
    'a', 0, 'c', 0, 'e', 0
};
//****************************************************************
//设备配置字符串描述符
//****************************************************************
const unsigned char g_pConfigString[] =
{
    (23 + 1) * 2,
    USB_DTYPE_STRING,
```

```c
    'H',0,'I',0,'D',0,' ',0,'M',0,'o',0,'u',0,'s',0,
    'e',0,' ',0,'C',0,'o',0,'n',0,'f',0,'i',0,'g',0,
    'u',0,'r',0,'a',0,'t',0,'i',0,'o',0,'n',0
};
//*************************************************************
//字符串描述符集合
//*************************************************************
const unsigned char * const g_pStringDescriptors[] =
{
    g_pLangDescriptor,
    g_pManufacturerString,
    g_pProductString,
    g_pSerialNumberString,
    g_pHIDInterfaceString,
    g_pConfigString
};
#define NUM_STRING_DESCRIPTORS (sizeof(g_pStringDescriptors) /  \
                                sizeof(unsigned char *))
//*************************************************************
//键盘实例,键盘配置并为键盘设备信息提供空间
//*************************************************************
tHIDMouseInstance g_sMouseInstance;
//*************************************************************
//键盘设备配置
//*************************************************************
const tUSBDHIDMouseDevice g_sMouseDevice =
{
    USB_VID_STELLARIS,
    USB_PID_MOUSE,
    500,
    USB_CONF_ATTR_SELF_PWR,
    MouseHandler,
    (void *)&g_sMouseDevice,
    g_pStringDescriptors,
    NUM_STRING_DESCRIPTORS,
    &g_sMouseInstance
};
//*************************************************************
//键盘 Callback 回调
//*************************************************************
unsigned long MouseHandler(void * pvCBData, unsigned long ulEvent,
                           unsigned long ulMsgData, void * pvMsgData)
```

```
    {
        switch (ulEvent)
        {
            case USB_EVENT_CONNECTED:
            {
                //连接成功时,点亮 LED1
                GPIOPinWrite(GPIO_PORTF_BASE,0x10,0x10);
                break;
            }
            case USB_EVENT_DISCONNECTED:
            {
                //断开连接时,LED1 灭
                GPIOPinWrite(GPIO_PORTF_BASE,0x10,0x00);
                break;
            }
            case USB_EVENT_TX_COMPLETE:
            {
                //发送完成时,LED2 亮
                GPIOPinWrite(GPIO_PORTF_BASE,0x20,0x20);
                break;
            }
            case USB_EVENT_SUSPEND:
            {
                //USB 处于挂起模式时,LED2 灭
                GPIOPinWrite(GPIO_PORTF_BASE,0x20,0x0);
                break;
            }
            case USB_EVENT_RESUME:
            {
                break;
            }
            default:
            {
                break;
            }
        }
        return (0);
    }
//************************************************************
//主函数
//************************************************************
int main(void)
```

```
{
    //系统初始化
    unsigned long x = 0, y = 0, key = 0, ulTemp = 0;
    SysCtlLDOSet(SYSCTL_LDO_2_75V);
    SysCtlClockSet(SYSCTL_XTAL_8MHZ | SYSCTL_SYSDIV_4 | SYSCTL_USE_PLL  | SYSCTL_OSC
                _MAIN );
    SysCtlPeripheralEnable(SYSCTL_PERIPH_GPIOF);
    GPIOPinTypeGPIOOutput(GPIO_PORTF_BASE,0xf0);
    GPIOPinTypeGPIOInput(GPIO_PORTF_BASE,0x0f);
    HWREG(GPIO_PORTF_BASE + GPIO_O_PUR) |= 0x0f;
    USBDHIDMouseInit(0, &g_sMouseDevice);
    while(1)
    {
        ulTemp = (~GPIOPinRead(GPIO_PORTF_BASE, 0x0f)) & 0x0f;
        switch(ulTemp)
        {
            case 0x01:
                x = x + 1;
                key = 0;
                break;
            case 0x02:
                x = x - 1;
                key = 0;
                break;
            case 0x04:
                y = y + 1;
                key = 0;
                break;
            case 0x08:
                y = y - 1;
                key = 0;
                break;
            case 0x03:
                key = 1;
                break;
            case 0x0c:
                key = 2;
                break;
            case 0x09:
                key = 4;
                break;
            default:
```

第 5 章　HID 设备

```
            key = 0;
            break;
    }
    if(ulTemp)
            USBDHIDMouseStateChange((void *)&g_sMouseDevice,x,y,key);
    SysCtlDelay(SysCtlClockGet()/30);
    }
}
```

5.5 小　结

经过本章介绍，读者对 HID 设备有了初步了解，如果想了解更深层次的 HID 设备类，可以参考官方数据手册。本章主要介绍了 HID 设备类的结构与编程、USB 库函数编程、USB 键盘开发与 USB 鼠标开发。当然这些代码都只是简单地实现了功能，真正的产品还需要在此基础之上进一步完善与优化。

第 6 章

Audio 设备

本章主要介绍 USB Audio 设备开发，便携式多媒体设备在当前生活中非常流行，是专门针对 USB 音频设备定义的一种专用类别，采用 USB 传输模式中的 Isochronous transfers 模式。USB 接口可以有很多种方式实现数字音频数据传输，不同的开发者可以根据自己的喜好和需求，定义任意的控制方式、传输模式、音频格式等参数，USB 接口拥有远远高于音频需求的带宽，可以传输极高品质（高采样率、高编码率、多声道）的音频数据。

6.1 Audio 设备介绍

USB 协议制定时，为了方便不同设备的开发商基于 USB 进行设计，定义了不同的设备类来支持不同类型的设备。虽然在 USB 标准中定义了 USB_DEVICE_CLASS_AUDIO——AUDIO 设备。但是很少有此类设备问世。目前称为 USB 音箱的设备，大都使用 USB_DEVICE_CLASS_POWER，仅仅将 USB 接口作为电源使用，完全基于 USB 协议的 USB_DEVICE_CLASS_AUDIO 设备，采用一根 USB 连接线，在 USB 音频设备中每一个端点都可用于音频信号传输，音频信号传输不仅包括音频流数据传输，还包括声音控制信号传输。

AUDIO 设备是专门针对 USB 音频设备定义的一种专用类别，不仅定义了音频输入、输出端点的标准，还提供了音量控制、混音器配置、左右声道平衡，甚至包括对支持杜比音效解码设备的支持，功能相当强大。不同的开发者可以根据不同的需求对主机枚举自己的设备结构，主机则根据枚举的不同设备结构提供相应的服务。

AUDIO 设备采用 USB 传输模式中的 Isochronous transfers 模式，Isochronous transfers 传输模式是专门针对流媒体特点的传输方法。它依照设备在链接初始化时枚举的参数，保证提供稳定的带宽给采用该模式的设备或端点。由于实时性的要求较严格，它不提供相应的接收/应答和握手协议。这很好地适应了音频数据流量稳定、对差错相对不敏感的特点。

6.2 Audio 描述符

Audio 设备定义了3种接口描述符子类,子协议恒为 0x00。3种接口描述符子类分别定义为:

```
// Audio Interface Subclass Codes
#define USB_ASC_AUDIO_CONTROL    0x01
#define USB_ASC_AUDIO_STREAMING  0x02
#define USB_ASC_MIDI_STREAMING   0x03
```

USB_ASC_AUDIO_CONTROL,音频设备控制类;USB_ASC_AUDIO_STREAMING,音频流类;USB_ASC_MIDI_STREAMING,MIDI 流类。以上3种用于描述接口描述符的子类,用于确定被描述接口的功能。Audio 设备没有定义子协议,所以为定值 0x00。

Audio 设备接口描述符中要包含 Class-Specific AC Interface Header Descriptor,用于定义接口的其他功能端口。Class-Specific AC Interface Header Descriptor(AC 接口头)描述符如表6-1所列。

表6-1 AC 接口头描述符

偏移量	域	大小	值	描述
0	bLength	1	数字	字节数
1	bDescriptorType	1	常量	配置描述符类型 USB_DTYPE_CS_INTERFACE(36)
2	bDescriptorSubtype	1	数字	AC 接口头:USB_DSUBTYPE_HEADER
3	bcdADC	2	数字	Audio 设备发行号,BCD 码
5	wTotalLength	2	数字	接口描述符总长度,包括标准接口描述符
6	bInCollection	1	索引	AudioStreaming、MIDIStreaming 使用代码
7	baInterfaceNr	2	位图	使用 AudioStreaming 或者 MIDIStreaming 的接口号

C 语言 AC 接口头描述符结构体为:

```
typedef struct
{
    //本描述符长度
    unsigned char bLength;
    //描述类型 USB_DTYPE_CS_INTERFACE (36)
    unsigned char bDescriptorSubtype;
```

```
    //发行号(BCD 码)
    unsigned short bcdADC;
    //接口描述符总长度,包括标准接口描述符
    unsigned short wTotalLength;
    //AudioStreaming 和 MIDIStreaming 使用代码
    unsigned char bInCollection;
    //使用 AudioStreaming 或者 MIDIStreaming 的接口号
    unsigned char baInterfaceNr;
}
tACHeader;
```

例如:定义一个实际的 AC 接口头描述符,程序代码如下。

```
const unsigned char g_pAudioInterfaceHeader[] =
{
    9,                              // The size of this descriptor
    USB_DTYPE_CS_INTERFACE,         // Interface descriptor is class specific
    USB_ACDSTYPE_HEADER,            // Descriptor sub-type is HEADER
    USBShort(0x0100),               // Audio Device Class Specification Release
                                    // Number in Binary-Coded Decimal
                                    // Total number of bytes in
                                    // g_pAudioControlInterface
    USBShort((9 + 9 + 12 + 13 + 9)),
    1,                              // Number of streaming interfaces
    1,                              // Index of the first and only streaming interface
}
```

Audio 设备的输入端口描述符(Input Terminal Descriptor),用于对输入端定义。输入端口描述符如表 6-2 所列。

表 6-2 输入端口描述符

偏移量	域	大小	值	描述
0	bLength	1	数字	字节数
1	bDescriptorType	1	常量	配置描述符类型 USB_DTYPE_CS_INTERFACE(36)
2	bDescriptorSubtype	1	常量	USB_ACDSTYPE_IN_TERMINAL
3	bTerminalID	1	常量	本端口 ID 号
4	wTerminalType	2	常量	端口类型
6	bAssocTerminal	1	常量	对应输出端口 ID
7	bNrChannels	2	位图	通道数量

第 6 章 Audio 设备

续表 6-2

偏移量	域	大小	值	描述
8	wChannelConfig	2	数字	通道配置
10	iChannelNames	1	常量	通道名字
11	iTerminal	1	数字	端口字符串描述符索引

C 语言输入端口描述符结构体为：

```
typedef struct
{
    unsigned char bLength;
    unsigned char bDescriptorType;
    unsigned char bDescriptorSubtype;
    unsigned char bTerminalID;
    unsigned short wTerminalType;
    unsigned char bAssocTerminal;
    unsigned char bNrChannels;
    unsigned short wChannelConfig;
    unsigned char iChannelNames;
    unsigned char iTerminal;
}
tACInputTerminal;
```

Audio 设备的输出端口描述符（Output Terminal Descriptor），对于输出端定义。输出端口描述符如表 6-3 所列。

表 6-3 输出端口描述符

偏移量	域	大小	值	描述
0	bLength	1	数字	字节数
1	bDescriptorType	1	常量	配置描述符类型 USB_DTYPE_CS_INTERFACE(36)
2	bDescriptorSubtype	1	常量	USB_ACDSTYPE_OUT_TERMINAL
3	bTerminalID	1	常量	本端口 ID 号
4	wTerminalType	2	常量	端口类型
6	bAssocTerminal	1	常量	对应输入端口 ID
7	bSourceID	1	常量	连接端口的 ID 号
8	iTerminal	1	数字	端口字符串描述符索引

C 语言输出端口描述符结构体为:

```c
typedef struct
{
    unsigned char bLength;
    unsigned char bDescriptorType;
    unsigned char bDescriptorSubtype;
    unsigned char bTerminalID;
    unsigned short wTerminalType;
    unsigned char bAssocTerminal;
    unsigned char bSourceID;
    unsigned char iTerminal;
}
tACOutputTerminal;
```

其他与 Audio 设备相关的描述符,请读者参阅 USB_Audio_Class 手册。下面列出一个 Audio 设备的接口描述符:

```c
const unsigned char g_pAudioControlInterface[] =
{
    //标准接口描述符
    9,                              // Size of the interface descriptor
    USB_DTYPE_INTERFACE,            // Type of this descriptor
    AUDIO_INTERFACE_CONTROL,        // 接口编号,从 0 开发编。AUDIO_INTERFACE_CONTROL = 0
    0,                              // The alternate setting for this interface
    0,                              // The number of endpoints used by this interface
    USB_CLASS_AUDIO,                // AUDIO 设备
    USB_ASC_AUDIO_CONTROL,          // 子类,USB_ASC_AUDIO_CONTROL 用于 Audio 控制
    0,                              // 无协议,定值 0
    0,                              // The string index for this interface
    // Audio 接口头描述符
    9,                              // The size of this descriptor
    USB_DTYPE_CS_INTERFACE,         // 描述符类型
    USB_ACDSTYPE_HEADER,            // 子类型,本描述为描述头
    USBShort(0x0100),               // Audio Device Class Specification Release
                                    // Number in Binary-Coded Decimal
                                    // Total number of bytes in g_pAudioControlInterface
    USBShort((9 + 9 + 12 + 13 + 9)),
    1,                              // Number of streaming interfaces
    1,                              // Index of the first and only streaming interface
    // Audio 设备输入端口描述符
    12,                             // The size of this descriptor
    USB_DTYPE_CS_INTERFACE,         // Interface descriptor is class specific
```

```
    USB_ACDSTYPE_IN_TERMINAL,       // 本描述符为输入端口
    AUDIO_IN_TERMINAL_ID,           // Terminal ID for this interface
                                    // USB streaming interface
    USBShort(USB_TTYPE_STREAMING),
    0,                              // ID of the Output Terminal to which this
                                    // Input Terminal is associated
    2,                              // Number of logical output channels in the
                                    // Terminal output audio channel cluster
    USBShort((USB_CHANNEL_L |       // Describes the spatial location of the
            USB_CHANNEL_R)),        // logical channels
    0,                              // Channel Name string index
    0,                              // Terminal Name string index

// Audio 设备特征单元描述符
    13,                             // The size of this descriptor
    USB_DTYPE_CS_INTERFACE,         // Interface descriptor is class specific
    USB_ACDSTYPE_FEATURE_UNIT,      // Descriptor sub-type is FEATURE_UNIT
    AUDIO_CONTROL_ID,               // Unit ID for this interface
    AUDIO_IN_TERMINAL_ID,           // ID of the Unit or Terminal to which this
                                    // Feature Unit is connected
    2,                              // Size in bytes of an element of the
                                    // bmaControls() array that follows
                                    // Master Mute control
    USBShort(USB_ACONTROL_MUTE),
                                    // Left channel volume control
    USBShort(USB_ACONTROL_VOLUME),
                                    // Right channel volume control
    USBShort(USB_ACONTROL_VOLUME),
    0,                              // Feature unit string index
//输出端口描述符
    9,                              // The size of this descriptor
    USB_DTYPE_CS_INTERFACE,         // Interface descriptor is class specific
    USB_ACDSTYPE_OUT_TERMINAL,      // 输出端口描述符
    AUDIO_OUT_TERMINAL_ID,          // Terminal ID for this interface
                                    // Output type is a generic speaker
    USBShort(USB_ATTYPE_SPEAKER),
    AUDIO_IN_TERMINAL_ID,           // ID of the input terminal to which this
                                    // output terminal is connected
    AUDIO_CONTROL_ID,               // ID of the feature unit that this output
                                    // terminal is connected to
    0,                              // Output terminal string index
```

```c
};
//音频流接口
const unsigned char g_pAudioStreamInterface[] =
{
    //标准接口描述符
    9,                              // Size of the interface descriptor
    USB_DTYPE_INTERFACE,            // Type of this descriptor
    AUDIO_INTERFACE_OUTPUT,         // The index for this interface
    0,                              // The alternate setting for this interface
    0,                              // The number of endpoints used by this interface
    USB_CLASS_AUDIO,                // The interface class
    USB_ASC_AUDIO_STREAMING,        // The interface sub-class
    0,                              // Unused must be 0
    0,                              // The string index for this interface
    // Vendor-specific Interface Descriptor
    9,                              // Size of the interface descriptor
    USB_DTYPE_INTERFACE,            // Type of this descriptor
    1,                              // The index for this interface
    1,                              // The alternate setting for this interface
    1,                              // The number of endpoints used by this interface
    USB_CLASS_AUDIO,                // The interface class
    USB_ASC_AUDIO_STREAMING,        // The interface sub-class
    0,                              // Unused must be 0
    0,                              // The string index for this interface
    // Class specific Audio Streaming Interface descriptor
    7,                              // Size of the interface descriptor
    USB_DTYPE_CS_INTERFACE,         // Interface descriptor is class specific
    USB_ASDSTYPE_GENERAL,           // General information
    AUDIO_IN_TERMINAL_ID,           // ID of the terminal to which this streaming
                                    // interface is connected
    1,                              // One frame delay
    USBShort(USB_ADF_PCM),
    // Format type Audio Streaming descriptor
    11,                             // Size of the interface descriptor
    USB_DTYPE_CS_INTERFACE,         // Interface descriptor is class specific
    USB_ASDSTYPE_FORMAT_TYPE,       // Audio Streaming format type
    USB_AF_TYPE_TYPE_I,             // Type I audio format type
    2,                              // Two audio channels
    2,                              // Two bytes per audio sub-frame
    16,                             // 16 bits per sample
    1,                              // One sample rate provided
    USB3Byte(48000),                // Only 48 000 sample rate supported
```

第 6 章 Audio 设备

```
//端点描述符
9,                                  // The size of the endpoint descriptor
USB_DTYPE_ENDPOINT,                 // Descriptor type is an endpoint
                                    // OUT endpoint with address ISOC_OUT_ENDPOINT
USB_EP_DESC_OUT | USB_EP_TO_INDEX(ISOC_OUT_ENDPOINT),
USB_EP_ATTR_ISOC |                  // Endpoint is an adaptive isochronous data
USB_EP_ATTR_ISOC_ADAPT |            // endpoint
USB_EP_ATTR_USAGE_DATA,
USBShort(ISOC_OUT_EP_MAX_SIZE),     // The maximum packet size
1,                                  // The polling interval for this endpoint
0,                                  // Refresh is unused
0,                                  // Synch endpoint address
// Audio Streaming Isochronous Audio Data Endpoint Descriptor
7,                                  // The size of the descriptor
USB_ACSDT_ENDPOINT,                 // Audio Class Specific Endpoint Descriptor
USB_ASDSTYPE_GENERAL,               // This is a general descriptor
USB_EP_ATTR_ACG_SAMPLING,           // Sampling frequency is supported
USB_EP_LOCKDELAY_UNDEF,             // Undefined lock delay units
USBShort(0),                        // No lock delay
};
```

6.3 Audio 数据类型

usbdaudio.h 中已经定义好 Audio 设备类中使用的所有数据类型和函数，下面介绍 Audio 设备类使用的数据类型。

```
typedef struct
{
    unsigned long ulUSBBase;
    //设备信息指针
    tDeviceInfo * psDevInfo;
    //配置描述符
    tConfigDescriptor * psConfDescriptor;
    //最大音量值
    short sVolumeMax;
    //最小音量值
    short sVolumeMin;
    //音量控制阶梯值
    short sVolumeStep;
    struct
    {
```

```
        //Callback 入口参数
        void * pvData;
        // pvData 大小
        unsigned long ulSize;
        // 可用 pvData 大小
        unsigned long ulNumBytes;
        // Callback
        tUSBAudioBufferCallback pfnCallback;
    } sBuffer;
    //请求类型
    unsigned short usRequestType;
    //请求标志
    unsigned char ucRequest;
    //更新值
    unsigned short usUpdate;
    //当前音量设置
    unsigned short usVolume;
    //静音设置
    unsigned char ucMute;
    //采样率
    unsigned long ulSampleRate;
    // 使用输出端点
    unsigned char ucOUTEndpoint;
    // 输出端点 DMA 通道
    unsigned char ucOUTDMA;
    //控制接口
    unsigned char ucInterfaceControl;
    //Audio 接口
    unsigned char ucInterfaceAudio;
}
tAudioInstance;
```

tAudioInstance，Audio 设备类实例。用于保存全部 Audio 设备类的配置信息，包括描述符、Callback 函数及控制事件等。

```
#define USB_AUDIO_INSTANCE_SIZE sizeof(tAudioInstance);
#define COMPOSITE_DAUDIO_SIZE    (8 + 52 + 52)
```

USB_AUDIO_INSTANCE_SIZE，定义 Audio 设备类实例信息的大小。COMPOSITE_DAUDIO_SIZE 定义设备描述符与所有接口描述符总长度。

第 6 章　Audio 设备

```c
typedef struct
{
    //VID
    unsigned short usVID;
    //PID
    unsigned short usPID;
    //8 字节供应商字符串
    unsigned char pucVendor[8];
    //16 字节产品字符串
    unsigned char pucProduct[16];
    //4 字节版本号
    unsigned char pucVersion[4];
    //最大耗电量
    unsigned short usMaxPowermA;
    //电源属性:USB_CONF_ATTR_SELF_PWR、USB_CONF_ATTR_BUS_PWR、USB_CONF_ATTR_RWAKE
    unsigned char ucPwrAttributes;
    // Callback 函数
    tUSBCallback pfnCallback;
    //字符串描述符集合
    const unsigned char * const * ppStringDescriptors;
    //字符串描述符个数
    unsigned long ulNumStringDescriptors;
    //最大音量
    short sVolumeMax;
    //最小音量
    short sVolumeMin;
    //音量调节步进
    short sVolumeStep;
    //Audio 设备类实例
    tAudioInstance * psPrivateData;
}
tUSBDAudioDevice;
```

　　tUSBDAudioDevice，Audio 设备类。定义了 VID、PID、电源属性以及字符串描述符等，还包括一个 Audio 设备类实例。其他设备描述符、配置信息通过 API 函数输入 tAudioInstance 定义的 Audio 设备实例中。如图 6-1 所示，为 tUSBDAudioDevice 调用图。

　　如图 6-1 所示，tUSBDAudioDevice 通过 tAudioInstance 把参数传递给底层驱动函数，完成参数初始化。tUSBDAudioDevice 主要完成各个描述符与 API 函数接口，方便 USB 库调用。

第 6 章　Audio 设备

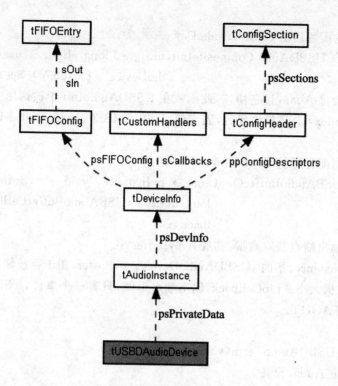

图 6-1　tUSBDAudioDevice 调用图

6.4　API 函数

在 Audio 设备类 API 库中定义了 4 个函数,完成 USB Audio 设备初始化、配置及数据处理。下面为 usbdaudio.h 中定义的 API 函数。

```
void * USBDAudioInit(unsigned long ulIndex,
                const tUSBDAudioDevice * psAudioDevice);
void * USBDAudioCompositeInit(unsigned long ulIndex,
                const tUSBDAudioDevice * psAudioDevice);
int USBAudioBufferOut(void * pvInstance, void * pvBuffer, unsigned long ulSize,
                tUSBAudioBufferCallback pfnCallback);
void USBDAudioTerm(void * pvInstance);
```

(1) void * USBDAudioInit(unsigned long ulIndex,
　　　　　　　　const tUSBDAudioDevice * psAudioDevice);

作用:初始化 Audio 设备硬件、协议,把其他配置参数填入 psAudioDevice 实例中。

参数:ulIndex,USB 模块代码,固定值:USB_BASE0。psAudioDevice,Audio 设

备类。

返回:指向配置后的 tUSBDAudioDevice。

(2) void * USBDAudioCompositeInit(unsigned long ulIndex,const tUSBDAudioDevice * psAudioDevice);

作用:初始化 Audio 设备协议,此函数在 USBDAudioInit 中被调用一次。

参数:ulIndex,USB 模块代码,固定值:USB_BASE0。psAudioDevice,Audio 设备类。

返回:指向配置后的 tUSBDAudioDevice。

(3) int USBAudioBufferOut(void * pvInstance, void * pvBuffer, unsigned long ulSize, tUSBAudioBufferCallback pfnCallback);

作用:从输出端点获取数据,并放入 pvBuffer 中。

参数:pvInstance,指向 tUSBDAudioDevice。pvBuffer,用于存放输出端点数据。ulSize,设置数据大小。pfnCallback,输出端点返回,只有一个事件,USBD_AUDIO_EVENT_DATAOUT。

返回:无。

(4) void USBDAudioTerm(void * pvInstance);

作用:结束 Audio 设备。

参数:pvInstance,指向 tUSBDAudioDevice。

返回:无。

在这些函数中 USBDAudioInit 和 USBAudioBufferOut 函数最重要并且使用最多,第一次使用 Audio 设备时 USBDAudioInit 函数用于初始化 Audio 设备的配置与控制。USBAudioBufferOut 函数从输出端点中获取数据,并放入数据缓存区内。

6.5 Audio 设备开发

Audio 设备开发只需要 5 步就能完成,如图 6-2 所示,为 Audio 设备开发流程图。Audio 设备配置(主要是字符串描述符)、Callback 函数编写、USB 处理器初始化、DMA 控制及数据处理。

(1) Audio 设备配置(主要是字符串描述符),按字符串描述符标准完成串描述符配置,进而完成 Audio 设备配置。

```
# include "inc/hw_ints.h"
# include "inc/hw_memmap.h"
# include "inc/hw_types.h"
```

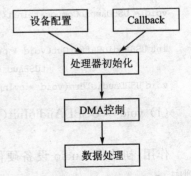

图 6-2　Audio 设备开发流程图

第 6 章 Audio 设备

```c
#include "inc/hw_sysctl.h"
#include "inc/hw_udma.h"
#include "inc/hw_gpio.h"
#include "driverlib/gpio.h"
#include "driverlib/interrupt.h"
#include "driverlib/sysctl.h"
#include "driverlib/udma.h"
#include "usblib/usblib.h"
#include "usblib/usb-ids.h"
#include "usblib/device/usbdevice.h"
#include "usblib/device/usbdaudio.h"
//根据具体 Audio 芯片修改
#define VOLUME_MAX          ((short)0x0C00)   // +12 db
#define VOLUME_MIN          ((short)0xDC80)   // -34.5 db
#define VOLUME_STEP         ((short)0x0180)   // 1.5 db
//Audio 设备
const tUSBDAudioDevice g_sAudioDevice;
//DMA 控制
tDMAControlTable sDMAControlTable[64] __attribute__ ((aligned(1024)));
//*************************************************************
// 缓存与标志
//*************************************************************
#define AUDIO_PACKET_SIZE        ((48000 * 4)/1000)
#define AUDIO_BUFFER_SIZE        (AUDIO_PACKET_SIZE * 20)
#define SBUFFER_FLAGS_PLAYING    0x00000001
#define SBUFFER_FLAGS_FILLING    0x00000002
struct
{
    //主要 buffer,USB audio class 和 sound driver 使用
    volatile unsigned char pucBuffer[AUDIO_BUFFER_SIZE];
    // play pointer
    volatile unsigned char * pucPlay;
    // USB fill pointer
    volatile unsigned char * pucFill;
    // 采样率调整
    volatile int iAdjust;
    // 播放状态
    volatile unsigned long ulFlags;
} g_sBuffer;
//*************************************************************
// 当前音量
//*************************************************************
```

第6章 Audio 设备

```c
short g_sVolume;
//**************************************************************
// 通过 USBDAudioInit() 函数,完善 Audio 设备配置信息
//**************************************************************
void * g_pvAudioDevice;
// 音量更新
#define FLAG_VOLUME_UPDATE          0x00000001
// 更新静音状态
#define FLAG_MUTE_UPDATE            0x00000002
// 静音状态
#define FLAG_MUTED                  0x00000004
// 连接成功
#define FLAG_CONNECTED              0x00000008
volatile unsigned long g_ulFlags;
extern unsigned long AudioMessageHandler(void * pvCBData, unsigned long ulEvent,
                              unsigned long ulMsgParam, void * pvMsgData);
//**************************************************************
// 设备语言描述符
//**************************************************************
const unsigned char g_pLangDescriptor[] =
{
    4,
    USB_DTYPE_STRING,
    USBShort(USB_LANG_EN_US)
};
//**************************************************************
// 制造商 字符串 描述符
//**************************************************************
const unsigned char g_pManufacturerString[] =
{
    (17 + 1) * 2,
    USB_DTYPE_STRING,
    'T', 0, 'e', 0, 'x', 0, 'a', 0, 's', 0, ' ', 0, 'I', 0, 'n', 0, 's', 0,
    't', 0, 'r', 0, 'u', 0, 'm', 0, 'e', 0, 'n', 0, 't', 0, 's', 0,
};
//**************************************************************
//产品 字符串 描述符
//**************************************************************
const unsigned char g_pProductString[] =
{
    (13 + 1) * 2,
    USB_DTYPE_STRING,
```

```c
    'A', 0, 'u', 0, 'd', 0, 'i', 0, 'o', 0, ' ', 0, 'E', 0, 'x', 0, 'a', 0,
    'm', 0, 'p', 0, 'l', 0, 'e', 0
};
//**************************************************************
//产品 序列号 描述符
//**************************************************************
const unsigned char g_pSerialNumberString[] =
{
    (8 + 1) * 2,
    USB_DTYPE_STRING,
    '1', 0, '2', 0, '3', 0, '4', 0, '5', 0, '6', 0, '7', 0, '8', 0
};
//**************************************************************
//设备接口字符串描述符
//**************************************************************
const unsigned char g_pInterfaceString[] =
{
    (15 + 1) * 2,
    USB_DTYPE_STRING,
    'A', 0, 'u', 0, 'd', 0, 'i', 0, 'o', 0, ' ', 0, 'I', 0, 'n', 0,
    't', 0, 'e', 0, 'r', 0, 'f', 0, 'a', 0, 'c', 0, 'e', 0
};
//**************************************************************
//设备配置字符串描述符
//**************************************************************
const unsigned char g_pConfigString[] =
{
    (20 + 1) * 2,
    USB_DTYPE_STRING,
    'A', 0, 'u', 0, 'd', 0, 'i', 0, 'o', 0, ' ', 0, ' ', 0, 'C', 0,
    'o', 0, 'n', 0, 'f', 0, 'i', 0, 'g', 0, 'u', 0, 'r', 0, 'a', 0,
    't', 0, 'i', 0, 'o', 0, 'n', 0
};
//**************************************************************
//字符串描述符集合
//**************************************************************
const unsigned char * const g_pStringDescriptors[] =
{
    g_pLangDescriptor,
    g_pManufacturerString,
    g_pProductString,
    g_pSerialNumberString,
```

```c
        g_pInterfaceString,
        g_pConfigString
};
#define NUM_STRING_DESCRIPTORS (sizeof(g_pStringDescriptors) /      \
                                sizeof(unsigned char *))
//***************************************************************
// 定义 Audio 设备实例
//***************************************************************
static tAudioInstance g_sAudioInstance;
//***************************************************************
// 定义 Audio 设备类
//***************************************************************
const tUSBDAudioDevice g_sAudioDevice =
{
    // VID
    USB_VID_STELLARIS,
    // PID
    USB_PID_AUDIO,
    //8 字节供应商字符串
    "TI        ",
    //16 字节产品字符串
    "Audio Device    ",
    //4 字节版本字符串
    "1.00",
    500,
    USB_CONF_ATTR_SELF_PWR,
    AudioMessageHandler,
    g_pStringDescriptors,
    NUM_STRING_DESCRIPTORS,
    VOLUME_MAX,
    VOLUME_MIN,
    VOLUME_STEP,
    &g_sAudioInstance
};
```

(2) 完成 Callback 函数。Callback 函数用于处理输出端点数据事务。主机发出的音频流数据，也可能是状态信息。Audio 设备中包含了以下事务：USBD_AUDIO_EVENT_IDLE、USBD_AUDIO_EVENT_ACTIVE、USBD_AUDIO_EVENT_MUTE、USBD_AUDIO_EVENT_VOLUME、USB_EVENT_DISCONNECTED、USB_EVENT_CONNECTED。如表 6-4 所列，为 Audio 事务，为 USB 音频设备的 Callback 回调函数提供事件标志。

表 6-4 Audio 事务

名 称	说 明
USB_EVENT_CONNECTED	USB 设备已经连接到主机
USB_EVENT_DISCONNECTED	USB 设备已经与主机断开
USBD_AUDIO_EVENT_VOLUME	更新音量
USBD_AUDIO_EVENT_MUTE	静音
USBD_AUDIO_EVENT_ACTIVE	Audio 处于活动状态
USBD_AUDIO_EVENT_IDLE	Audio 处于空闲状态

根据以上事务编写的 Callback 函数如下所示。

```
//*************************************************************
//USB Audio 设备类返回事件处理函数(Callback)
//*************************************************************
unsigned long AudioMessageHandler(void * pvCBData, unsigned long ulEvent,
                    unsigned long ulMsgParam, void * pvMsgData)
{
    switch(ulEvent)
    {
        //Audio 正处于空闲或者工作状态
        case USBD_AUDIO_EVENT_IDLE:
        case USBD_AUDIO_EVENT_ACTIVE:
        {
            GPIOPinWrite(GPIO_PORTF_BASE,0x10,0x10);
            g_ulFlags |= FLAG_CONNECTED;
            break;
        }
        // 静音控制
        case USBD_AUDIO_EVENT_MUTE:
        {
            // 检查是否静音
            if(ulMsgParam == 1)
            {
                //静音
                g_ulFlags |= FLAG_MUTE_UPDATE | FLAG_MUTED;
            }
            else
            {
                // 取消静音
```

第 6 章 Audio 设备

```c
            g_ulFlags &= ~(FLAG_MUTE_UPDATE | FLAG_MUTED);
            g_ulFlags |= FLAG_MUTE_UPDATE;
        }
        break;
    }
    //音量控制
    case USBD_AUDIO_EVENT_VOLUME:
    {
        g_ulFlags |= FLAG_VOLUME_UPDATE;
        //最大音量
        if(ulMsgParam == 0x8000)
        {
            //设置为最小
            g_sVolume = 0;
        }
        else
        {
            //声音控制器,设置音量
            g_sVolume = (short)ulMsgParam - (short)VOLUME_MIN;
        }
        break;
    }
    // 断开连接
    case USB_EVENT_DISCONNECTED:
    {
        GPIOPinWrite(GPIO_PORTF_BASE,0x10,0x00);
        GPIOPinWrite(GPIO_PORTF_BASE,0x80,0x00);
        g_ulFlags &= ~FLAG_CONNECTED;
        break;
    }

    //连接
    case USB_EVENT_CONNECTED:
    {
        GPIOPinWrite(GPIO_PORTF_BASE,0x80,0x80);
        g_ulFlags |= FLAG_CONNECTED;
        break;
    }
    default:
    {
        break;
    }
}
```

```
    return(0);
}
```

(3) 系统初始化,配置内核电压、系统主频、使能端口、配置按键端口、LED 控制等,本例中使用 4 个 LED 进行指示。如图 5-6 所示,为 LED 与键盘电路图。

系统初始化代码如下。

```
//设置内核电压、主频为 50 Mhz
SysCtlLDOSet(SYSCTL_LDO_2_75V);
SysCtlClockSet(SYSCTL_XTAL_8MHZ | SYSCTL_SYSDIV_4 | SYSCTL_USE_PLL | SYSCTL_OSC_
            MAIN);
SysCtlPeripheralEnable(SYSCTL_PERIPH_GPIOF);
GPIOPinTypeGPIOOutput(GPIO_PORTF_BASE,0xf0);
GPIOPinTypeGPIOInput(GPIO_PORTF_BASE,0x0f);
HWREG(GPIO_PORTF_BASE + GPIO_O_PUR) |= 0x0f;
// 全局状态标志
g_ulFlags = 0;
// 初始化 Audio 设备
g_pvAudioDevice = USBDAudioInit(0, (tUSBDAudioDevice *)&g_sAudioDevice);
```

(4) 使能、配置 DMA,Audio 设备要传输大量数据,所以 USB 库函数内部已经使用了 DMA,在使用前必须使能、配置 DMA。

```
//配置使能 DMA
SysCtlPeripheralEnable(SYSCTL_PERIPH_UDMA);
SysCtlDelay(10);
uDMAControlBaseSet(&sDMAControlTable[0]);
uDMAEnable();
```

(5) 数据处理。主要使用 USBAudioBufferOut 从输出端点中获取数据并处理,并且进行 Audio 设备控制。

```
while(1)
{
    //等待连接结束
    while((g_ulFlags & FLAG_CONNECTED) == 0)
    {
    }
    //初始化 Buffer
    g_sBuffer.pucFill = g_sBuffer.pucBuffer;
    g_sBuffer.pucPlay = g_sBuffer.pucBuffer;
    g_sBuffer.ulFlags = 0;
    //从 Audio 设备类中获取数据
    if(USBAudioBufferOut(g_pvAudioDevice,
```

```c
                        (unsigned char *)g_sBuffer.pucFill,
                        AUDIO_PACKET_SIZE, USBBufferCallback) == 0)
    {
        //标记数据放入 buffer 中
        g_sBuffer.ulFlags |= SBUFFER_FLAGS_FILLING;
    }
    //设备连接到主机
    while(g_ulFlags & FLAG_CONNECTED)
    {
        // 检查音量是否有改变
        if(g_ulFlags & FLAG_VOLUME_UPDATE)
        {
            // 清除更新音量标志
            g_ulFlags &= ~FLAG_VOLUME_UPDATE;
            // 修改音量,自行添加代码,在此以 LED 灯做指示
            //UpdateVolume();
            GPIOPinWrite(GPIO_PORTF_BASE,0x40,~GPIOPinRead(GPIO_PORTF_BASE,
                    0x40));
        }
        //是否静音
        if(g_ulFlags & FLAG_MUTE_UPDATE)
        {
            //修改静音状态,自行添加函数,在此以 LED 灯做指示
            //UpdateMute();
            if(g_ulFlags & FLAG_MUTED)
                GPIOPinWrite(GPIO_PORTF_BASE,0x20,0x20);
            else
                GPIOPinWrite(GPIO_PORTF_BASE,0x20,0x00);
            // 清除静音标志
            g_ulFlags &= ~FLAG_MUTE_UPDATE;
        }
    }
}
//*************************************************************
//USBAudioBufferOut 的 Callback 入口参数
//*************************************************************
void USBBufferCallback(void *pvBuffer, unsigned long ulParam, unsigned long ulEvent)
{
    //数据处理,自行加入代码
    // Your Codes ......
    //再一次获取数据
    USBAudioBufferOut(g_pvAudioDevice, (unsigned char *)g_sBuffer.pucFill,
```

第 6 章 Audio 设备

```
                     AUDIO_PACKET_SIZE, USBBufferCallback);
}
```

　　使用上面 5 步就完成了 Audio 设备开发。Audio 设备开发时要加入两个库函数，分别是 usblib.lib 和 DriverLib.lib，在启动代码中加入 USB0DeviceIntHandler 中断服务函数。以上 Audio 设备开发完成，在 Windows XP 下运行的效果如图 6-3 所示，Audio 正在枚举。可以看出 Windows XP 提示有 Audio 设备插入，并进行枚举。

　　在枚举过程中可以在计算机右下脚可以看到 Audio Example 字样，表示正在进行枚举。枚举成功后，在"设备管理器"的"声音、视频和游戏控制器"中看到 USB Audio Device 设备，如图 6-4 所示，Audio 设备枚举成功，现在 Audio 设备可以正式使用了。

图 6-3　Audio 正在枚举　　　　图 6-4　Audio 设备枚举成功

Audio 设备开发源码如下：

```
# include "inc/hw_ints.h"
# include "inc/hw_memmap.h"
# include "inc/hw_types.h"
# include "inc/hw_sysctl.h"
# include "inc/hw_udma.h"
# include "inc/hw_gpio.h"
# include "driverlib/gpio.h"
# include "driverlib/interrupt.h"
# include "driverlib/sysctl.h"
# include "driverlib/udma.h"
# include "usblib/usblib.h"
# include "usblib/usb-ids.h"
# include "usblib/device/usbdevice.h"
# include "usblib/device/usbdaudio.h"

//根据具体 Audio 芯片修改
# define VOLUME_MAX        ((short)0x0C00)   // + 12 db
```

```c
#define VOLUME_MIN           ((short)0xDC80)   // -34.5 db
#define VOLUME_STEP          ((short)0x0180)   // 1.5 db
//Audio 设备
const tUSBDAudioDevice g_sAudioDevice;
//DMA 控制
tDMAControlTable sDMAControlTable[64] __attribute__ ((aligned(1024)));
//****************************************************************
// 缓存与标志
//****************************************************************
#define AUDIO_PACKET_SIZE         ((48000 * 4)/1000)
#define AUDIO_BUFFER_SIZE         (AUDIO_PACKET_SIZE * 20)
#define SBUFFER_FLAGS_PLAYING     0x00000001
#define SBUFFER_FLAGS_FILLING     0x00000002
struct
{
    //主要 buffer,USB audio class 和 sound driver 使用
    volatile unsigned char pucBuffer[AUDIO_BUFFER_SIZE];
    // play pointer
    volatile unsigned char * pucPlay;
    // USB fill pointer
    volatile unsigned char * pucFill;
    // 采样率调整
    volatile int iAdjust;
    // 播放状态
    volatile unsigned long ulFlags;
} g_sBuffer;
//****************************************************************
// 当前音量
//****************************************************************
short g_sVolume;
//****************************************************************
// 通过 USBDAudioInit() 函数,完善 Audio 设备配置信息
//****************************************************************
void * g_pvAudioDevice;
// 音量更新
#define FLAG_VOLUME_UPDATE        0x00000001
// 更新静音状态
#define FLAG_MUTE_UPDATE          0x00000002
// 静音状态
#define FLAG_MUTED                0x00000004
// 连接成功
#define FLAG_CONNECTED            0x00000008
```

```c
volatile unsigned long g_ulFlags;
extern unsigned long
AudioMessageHandler(void * pvCBData, unsigned long ulEvent,
                    unsigned long ulMsgParam, void * pvMsgData);
//***************************************************************
// 设备语言描述符
//***************************************************************
const unsigned char g_pLangDescriptor[] =
{
    4,
    USB_DTYPE_STRING,
    USBShort(USB_LANG_EN_US)
};
//***************************************************************
// 制造商 字符串 描述符
//***************************************************************
const unsigned char g_pManufacturerString[] =
{
    (17 + 1) * 2,
    USB_DTYPE_STRING,
    'T', 0, 'e', 0, 'x', 0, 'a', 0, 's', 0, '', 0, 'I', 0, 'n', 0, 's', 0,
    't', 0, 'r', 0, 'u', 0, 'm', 0, 'e', 0, 'n', 0, 't', 0, 's', 0,
};
//***************************************************************
//产品 字符串 描述符
//***************************************************************
const unsigned char g_pProductString[] =
{
    (13 + 1) * 2,
    USB_DTYPE_STRING,
    'A', 0, 'u', 0, 'd', 0, 'i', 0, 'o', 0, '', 0, 'E', 0, 'x', 0, 'a', 0,
    'm', 0, 'p', 0, 'l', 0, 'e', 0
};
//***************************************************************
// 产品 序列号 描述符
//***************************************************************
const unsigned char g_pSerialNumberString[] =
{
    (8 + 1) * 2,
    USB_DTYPE_STRING,
    '1', 0, '2', 0, '3', 0, '4', 0, '5', 0, '6', 0, '7', 0, '8', 0
};
```

```c
//*****************************************************************
// 设备接口字符串描述符
//*****************************************************************
const unsigned char g_pInterfaceString[] =
{
    (15 + 1) * 2,
    USB_DTYPE_STRING,
    'A', 0, 'u', 0, 'd', 0, 'i', 0, 'o', 0, ' ', 0, 'I', 0, 'n', 0,
    't', 0, 'e', 0, 'r', 0, 'f', 0, 'a', 0, 'c', 0, 'e', 0
};
//*****************************************************************
// 设备配置字符串描述符
//*****************************************************************
const unsigned char g_pConfigString[] =
{
    (20 + 1) * 2,
    USB_DTYPE_STRING,
    'A', 0, 'u', 0, 'd', 0, 'i', 0, 'o', 0, ' ', 0, ' ', 0, 'C', 0,
    'o', 0, 'n', 0, 'f', 0, 'i', 0, 'g', 0, 'u', 0, 'r', 0, 'a', 0,
    't', 0, 'i', 0, 'o', 0, 'n', 0
};
//*****************************************************************
// 字符串描述符集合
//*****************************************************************
const unsigned char * const g_pStringDescriptors[] =
{
    g_pLangDescriptor,
    g_pManufacturerString,
    g_pProductString,
    g_pSerialNumberString,
    g_pInterfaceString,
    g_pConfigString
};
#define NUM_STRING_DESCRIPTORS (sizeof(g_pStringDescriptors) /          \
                                sizeof(unsigned char *))
//*****************************************************************
// 定义 Audio 设备实例
//*****************************************************************
static tAudioInstance g_sAudioInstance;
//*****************************************************************
// 定义 Audio 设备类
//*****************************************************************
```

```c
const tUSBDAudioDevice g_sAudioDevice =
{
    // VID
    USB_VID_STELLARIS,
    // PID
    USB_PID_AUDIO,
    // 8 字节供应商字符串
    "TI      ",
    //16 字节产品字符串
    "Audio Device    ",
    //4 字节版本字符串
    "1.00",
    500,
    USB_CONF_ATTR_SELF_PWR,
    AudioMessageHandler,
    g_pStringDescriptors,
    NUM_STRING_DESCRIPTORS,
    VOLUME_MAX,
    VOLUME_MIN,
    VOLUME_STEP,
    &g_sAudioInstance
};
//**************************************************************
//USB Audio 设备类返回事件处理函数(Callback)
//**************************************************************
unsigned long AudioMessageHandler(void * pvCBData, unsigned long ulEvent,
                                  unsigned long ulMsgParam, void * pvMsgData)
{
    switch(ulEvent)
    {
        //Audio 正处于空闲或者工作状态
        case USBD_AUDIO_EVENT_IDLE:
        case USBD_AUDIO_EVENT_ACTIVE:
        {
            GPIOPinWrite(GPIO_PORTF_BASE,0x10,0x10);
            g_ulFlags |= FLAG_CONNECTED;
            break;
        }
        //静音控制
        case USBD_AUDIO_EVENT_MUTE:
        {
            //检查是否静音
```

```c
            if(ulMsgParam == 1)
            {
                //静音
                g_ulFlags |= FLAG_MUTE_UPDATE | FLAG_MUTED;
            }
            else
            {
                // 取消静音
                g_ulFlags &= ~(FLAG_MUTE_UPDATE | FLAG_MUTED);
                g_ulFlags |= FLAG_MUTE_UPDATE;
            }
            break;
        }
        //音量控制
        case USBD_AUDIO_EVENT_VOLUME:
        {
            g_ulFlags |= FLAG_VOLUME_UPDATE;
            //最大音量
            if(ulMsgParam == 0x8000)
            {
                //设置为最小
                g_sVolume = 0;
            }
            else
            {
                //声音控制器,设置音量
                g_sVolume = (short)ulMsgParam - (short)VOLUME_MIN;
            }
            break;
        }
        // 断开连接
        case USB_EVENT_DISCONNECTED:
        {
                GPIOPinWrite(GPIO_PORTF_BASE,0x10,0x00);
                GPIOPinWrite(GPIO_PORTF_BASE,0x80,0x00);
                g_ulFlags &= ~FLAG_CONNECTED;
                break;
        }
        case USB_EVENT_CONNECTED:
        {
                GPIOPinWrite(GPIO_PORTF_BASE,0x80,0x80);
                g_ulFlags |= FLAG_CONNECTED;
```

```c
                break;
        }
            default:
            {
                break;
            }
    }
    return(0);
}
//*************************************************************
//USBAudioBufferOut 的 Callback 入口参数
//*************************************************************
void USBBufferCallback(void * pvBuffer, unsigned long ulParam, unsigned long ulEvent)
{
    //数据处理,自行加入代码
    // Your Codes ......
    //再一次获取数据
    USBAudioBufferOut(g_pvAudioDevice, (unsigned char *)g_sBuffer.pucFill,
                    AUDIO_PACKET_SIZE, USBBufferCallback);
}


//*************************************************************
// 应用主函数
//*************************************************************
int main(void)
{
    //设置内核电压、主频为 50 MHz
    SysCtlLDOSet(SYSCTL_LDO_2_75V);
    SysCtlClockSet(SYSCTL_XTAL_8MHZ | SYSCTL_SYSDIV_4 |
                    SYSCTL_USE_PLL | SYSCTL_OSC_MAIN );
    SysCtlPeripheralEnable(SYSCTL_PERIPH_GPIOF);
    GPIOPinTypeGPIOOutput(GPIO_PORTF_BASE,0xf0);
    GPIOPinTypeGPIOInput(GPIO_PORTF_BASE,0x0f);
    HWREG(GPIO_PORTF_BASE + GPIO_O_PUR) |= 0x0f;

    //配置使能 DMA
    SysCtlPeripheralEnable(SYSCTL_PERIPH_UDMA);
    SysCtlDelay(10);
    uDMAControlBaseSet(&sDMAControlTable[0]);
    uDMAEnable();

    //全局状态标志
```

```c
    g_ulFlags = 0;
    //初始化 Audio 设备
    g_pvAudioDevice = USBDAudioInit(0,(tUSBDAudioDevice *)&g_sAudioDevice);
    while(1)
    {
        //等待连接结束
        while((g_ulFlags & FLAG_CONNECTED) == 0)
        {
        }
        //初始化 Buffer
        g_sBuffer.pucFill = g_sBuffer.pucBuffer;
        g_sBuffer.pucPlay = g_sBuffer.pucBuffer;
        g_sBuffer.ulFlags = 0;
        //从 Audio 设备类中获取数据
        if(USBAudioBufferOut(g_pvAudioDevice,
                             (unsigned char *)g_sBuffer.pucFill,
                             AUDIO_PACKET_SIZE, USBBufferCallback) == 0)
        {
            //标记数据放入 buffer 中
            g_sBuffer.ulFlags |= SBUFFER_FLAGS_FILLING;
        }
        //设备连接到主机
        while(g_ulFlags & FLAG_CONNECTED)
        {
            // 检查音量是否有改变
            if(g_ulFlags & FLAG_VOLUME_UPDATE)
            {
                // 清除更新音量标志
                g_ulFlags &= ~FLAG_VOLUME_UPDATE;
                // 修改音量,自行添加代码,在此以 LED 灯做指示
                //UpdateVolume();
                GPIOPinWrite(GPIO_PORTF_BASE,0x40,~GPIOPinRead(GPIO_PORTF_BASE,
                             0x40));
            }
            //是否静音
            if(g_ulFlags & FLAG_MUTE_UPDATE)
            {
                //修改静音状态,自行添加函数,在此以 LED 灯做指示
                //UpdateMute();
                if(g_ulFlags & FLAG_MUTED)
                    GPIOPinWrite(GPIO_PORTF_BASE,0x20,0x20);
                else
```

```
            GPIOPinWrite(GPIO_PORTF_BASE,0x20,0x00);
            // 清除静音标志
            g_ulFlags &= ~FLAG_MUTE_UPDATE;
        }
    }
}
```

6.6 小　结

　　本章主要对 Audio 设备、描述符、数据类型、相关 API 函数、Audio 设备开发流程等进行了详细介绍，并通过一个 Audio 设备开发，让读者更深入地了解 Audio 设备的协议与开发。当然要开发功能更强大的 Audio 设备还需要在此基础上完善。

第 7 章

Bulk 设备

批量传输(Bulk)采用的是流状态传输,可以实现 PC 与 USB 设备的通信与控制。批量传输也是 USB 通信的基础,本章主要介绍 USB 使用 Bulk 与 PC 进行通信开发。

7.1 Bulk 设备介绍

USB 通道的数据传输格式有两种,而且这两种格式还是互斥的。有消息和流两种状态。对于流状态,不具有 USB 数据的格式,遵循的规则就是先进先出。对于消息通道,它的通信模式符合 USB 的数据格式,一般由 3 个阶段组成:分别是建立阶段,数据阶段和确认阶段。所有通信的开始都是由主机方面发起的。

USB 协议制定时,为了方便不同设备的开发商基于 USB 进行设计,定义了不同的设备类来支持不同类型的设备。Bulk 也是其中一种,Bulk 设备使用端点 Bulk 传输模式,可以快速传输大量数据。

批量传输(Bulk)采用的是流状态传输,批量传送的一个特点就是支持不确定时间内进行大量数据传输,能够保证数据一定可以传输,但是不能保证传输的带宽和传输的延迟。而且批量传输是一种单向的传输,要进行双向传输必须要使用两个通道。本章将介绍双向 Bulk 的批量传输。

7.2 Bulk 数据类型

usbdbulk.h 和 usblib.h 中已经定义好了 Bulk 设备类中使用的所有数据类型和函数,下面介绍 Bulk 设备类使用的数据类型。

```
typedef struct
{
    //定义本 Buffer 是用于发送还是接收,True 为接收,False 为发送
    tBoolean bTransmitBuffer;
    //Callback 函数,用于 Buffer 数据处理完成后
    tUSBCallback pfnCallback;
```

```c
    //Callback 第一个输入参数
    void * pvCBData;
    //数据传输或者接收时调用的函数,用于完成发送或者接收的函数
    tUSBPacketTransfer pfnTransfer;
    //数据传输或者接收时调用的函数
    //发送时,用于检查是否有足够的空间;接收时,用于检查可以接收的数量
    tUSBPacketAvailable pfnAvailable;
    //在设备模式下,设备类指针
    void * pvHandle;
    //用于存放发送或者接收的数据
    unsigned char * pcBuffer;
    //发送或者接收数据的大小
    unsigned long ulBufferSize;
    //RAM Buffer
    void * pvWorkspace;
}
tUSBBuffer;
```

tUSBBuffer,数据缓存控制结构体,定义在 usblib.h 中,用于在 Bulk 传输过程中,发送数据或者接收数据,是 Bulk 设备传输的主要载体,结构体内部包含数据发送、接收及处理函数等。

```c
typedef enum
{
    //Bulk 状态没定义
    BULK_STATE_UNCONFIGURED,
    //空闲状态
    BULK_STATE_IDLE,
    //等待数据发送或者结束
    BULK_STATE_WAIT_DATA,
    //等待数据处理
    BULK_STATE_WAIT_CLIENT
} tBulkState;
```

tBulkState,定义 Bulk 端点状态。定义在 usbdbulk.h 文件中。用于端点状态标记与控制,可以保证数据传输不相互冲突。

```c
typedef struct
{
    //USB 基地址
    unsigned long ulUSBBase;
    //设备信息
    tDeviceInfo * psDevInfo;
```

第 7 章 Bulk 设备

```c
    //配置信息
    tConfigDescriptor * psConfDescriptor;
    //Bulk 接收端点状态
    volatile tBulkState eBulkRxState;
    //Bulk 发送端点状态
    volatile tBulkState eBulkTxState;
    //标志位
    volatile unsigned short usDeferredOpFlags;
    //最后一次发送数据大小
    unsigned short usLastTxSize;
    //连接是否成功
    volatile tBoolean bConnected;
    //IN 端点号
    unsigned char ucINEndpoint;
    //OUT 端点号
    unsigned char ucOUTEndpoint;
    //接口号
    unsigned char ucInterface;
}
tBulkInstance;
```

tBulkInstance,Bulk 设备类实例。定义了 Bulk 设备类的 USB 基地址、设备信息、IN 端点以及 OUT 端点等信息。

```c
typedef struct
{
    //VID
    unsigned short usVID;
    //PID
    unsigned short usPID;
    //最大耗电量
    unsigned short usMaxPowermA;
    //电源属性
    unsigned char ucPwrAttributes;
    //接收回调函数,主要用于接收数据处理
    tUSBCallback pfnRxCallback;
    //接收回调函数的第一个参数
    void * pvRxCBData;
    //发送回调函数,主要用于发送数据处理
    tUSBCallback pfnTxCallback;
    //发送回调函数的第一个参数
    void * pvTxCBData;
    //字符串描述符集合
```

```
    const unsigned char * const * ppStringDescriptors;
    //字符串描述符个数
    unsigned long ulNumStringDescriptors;
    //Bulk 设备实例
    tBulkInstance * psPrivateBulkData;
}
tUSBDBulkDevice;
```

tUSBDBulkDevice，Bulk 设备类，定义了 VID、PID、电源属性、字符串描述符等，还包括了一个 Bulk 设备类实例。其他设备描述符、配置信息通过 API 函数输入 tBulkInstance 定义的 Bulk 设备实例中。图 7-1 为 tUSBDBulkDevice 调用图。

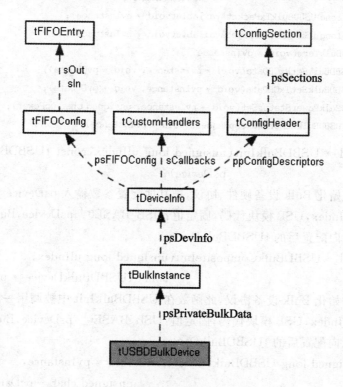

图 7-1 tUSBDBulkDevice 调用图

7.3 API 函数

在 Bulk 设备类 API 库中定义了 11 个函数，完成 USB Bulk 设备初始化、配置及数据处理。同时 Bulk 设备类还会用到 11 个 Buffer 操作函数，定义在 usblib.h 头文件中。下面为 usbdbulk.h 中定义的 API 函数。

```
    void * USBDBulkInit(unsigned long ulIndex,
```

第7章 Bulk 设备

```
                        const tUSBDBulkDevice * psDevice);
void * USBDBulkCompositeInit(unsigned long ulIndex,
                        const tUSBDBulkDevice * psDevice);
unsigned long USBDBulkPacketWrite(void * pvInstance,
                        unsigned char * pcData,
                        unsigned long ulLength,
                        tBoolean bLast);
unsigned long USBDBulkPacketRead(void * pvInstance,
                        unsigned char * pcData,
                        unsigned long ulLength,
                        tBoolean bLast);
unsigned long USBDBulkTxPacketAvailable(void * pvInstance);
unsigned long USBDBulkRxPacketAvailable(void * pvInstance);
void USBDBulkTerm(void * pvInstance);
void * USBDBulkSetRxCBData(void * pvInstance, void * pvCBData);
void * USBDBulkSetTxCBData(void * pvInstance, void * pvCBData);
void USBDBulkPowerStatusSet(void * pvInstance, unsigned char ucPower);
tBoolean USBDBulkRemoteWakeupRequest(void * pvInstance);
```

(1) void * USBDBulkInit(unsigned long ulIndex, const tUSBDBulkDevice * psDevice);

作用:初始化 Bulk 设备硬件、协议,把其他配置参数输入 psDevice 实例中。

参数:ulIndex,USB 模块代码,固定值:USB_BASE0。psDevice,Bulk 设备类。

返回:指向配置后的 tUSBDBulkDevice。

(2) void * USBDBulkCompositeInit(unsigned long ulIndex,
                        const tUSBDBulkDevice * psDevice);

作用:初始化 Bulk 设备协议,此函数在 USBDBulkInit 中被调用一次。

参数:ulIndex,USB 模块代码,固定值:USB_BASE0。psDevice,Bulk 设备类。

返回:指向配置后的 tUSBDBulkDevice。

(3) unsigned long USBDBulkPacketWrite(void * pvInstance,
                        unsigned char * pcData,
                        unsigned long ulLength,
                        tBoolean bLast);

作用:通过 Bulk 传输发送一个包数据,底层驱动,在 Buffer 中使用。

参数:pvInstance,tUSBDBulkDevice 设备指针。pcData,待写入的数据指针。ulLength,待写入数据的长度。bLast,是否传输结束包。

返回:成功发送长度,可能与 ulLength 长度不一样。

(4) unsigned long USBDBulkPacketRead(void * pvInstance,
                        unsigned char * pcData,

unsigned long ulLength,
tBoolean bLast);

作用：通过 Bulk 传输接收一个包数据，底层驱动，在 Buffer 中使用。

参数：pvInstance,tUSBDBulkDevice 设备指针。pcData,读出数据指针。ulLength,读出数据的长度。bLast,表明当前接收的数据是否是最后一个数据包。

返回：成功接收长度，可能与 ulLength 长度不一样。

(5) unsigned long USBDBulkTxPacketAvailable(void * pvInstance);

作用：获取可用发送数据的长度。

参数：pvInstance,tUSBDBulkDevice 设备指针。

返回：待发送数据的字节数，为数据包发送函数 USBDBulkPacketwrite 提供 ulLength 参数。

(6) unsigned long USBDBulkRxPacketAvailable(void * pvInstance);

作用：获取接收数据的长度。

参数：pvInstance,tUSBDBulkDevice 设备指针。

返回：可用接收数据个数，可读取的有效数据。

(7) void USBDBulkTerm(void * pvInstance);

作用：结束 Bulk 设备。

参数：pvInstance,指向 tUSBDBulkDevice。

返回：无。

(8) void * USBDBulkSetRxCBData(void * pvInstance, void * pvCBData);

作用：改变接收回调函数的第一个参数。

参数：pvInstance,指向 tUSBDBulkDevice。pvCBData,用于替换的参数。

返回：旧参数指针。

(9) void * USBDBulkSetTxCBData(void * pvInstance, void * pvCBData);

作用：改变发送回调函数的第一个参数。

参数：pvInstance,指向 tUSBDBulkDevice。pvCBData,用于替换的参数

返回：旧参数指针。

(10) void USBDBulkPowerStatusSet(void * pvInstance, unsigned char ucPower);

作用：修改电源属性、状态。

参数：pvInstance,指向 tUSBDBulkDevice。ucPower,电源属性。

返回：无。

(11) tBoolean USBDBulkRemoteWakeupRequest(void * pvInstance);

作用：唤醒请求。

参数：pvInstance,指向 tUSBDBulkDevice。

返回：无。

第 7 章 Bulk 设备

在这些函数中 USBDBulkInit 和 USBDBulkPacketWrite、USBDBulkPacketRead、USBDBulkTxPacketAvailable、USBDBulkRxPacketAvailable 函数最重要并且使用最多。第一次使用 Bulk 设备时，USBDBulkInit 用于初始化 Bulk 设备的配置与控制。USBDBulkPacketRead、USBDBulkPacketWrite、USBDBulkTxPacketAvailable、USBDBulkRxPacketAvailable 为 Bulk 传输数据的底层驱动函数用于驱动 Buffer。

usblib.h 中定义了 11 个 Buffer 操作函数，用于数据发送、接收、以及回调其他函数，下面介绍这 11 个 Buffer 操作函数。

```
const tUSBBuffer * USBBufferInit(const tUSBBuffer * psBuffer);
void USBBufferInfoGet(const tUSBBuffer * psBuffer,
                    tUSBRingBufObject * psRingBuf);
unsigned long USBBufferWrite(const tUSBBuffer * psBuffer,
                    const unsigned char * pucData,
                    unsigned long ulLength);
void USBBufferDataWritten(const tUSBBuffer * psBuffer,
                    unsigned long ulLength);
void USBBufferDataRemoved(const tUSBBuffer * psBuffer,
                    unsigned long ulLength);
void USBBufferFlush(const tUSBBuffer * psBuffer);
unsigned long USBBufferRead(const tUSBBuffer * psBuffer,
                    unsigned char * pucData,
                    unsigned long ulLength);
unsigned long USBBufferDataAvailable(const tUSBBuffer * psBuffer);
unsigned long USBBufferSpaceAvailable(const tUSBBuffer * psBuffer);
void * USBBufferCallbackDataSet(tUSBBuffer * psBuffer, void * pvCBData);
unsigned long USBBufferEventCallback(void * pvCBData,
                    unsigned long ulEvent,
                    unsigned long ulMsgValue,
                    void * pvMsgData);
```

（1）const tUSBBuffer * USBBufferInit(const tUSBBuffer * psBuffer);

作用：初始化 Buffer，把它加入到当前设备中。首次使用 Buffer 必须使用此函数。

参数：psBuffer，待初始化的 Buffer。

返回：指向配置后的 Buffer。

（2）void USBBufferInfoGet(const tUSBBuffer * psBuffer,
　　　　　　　　　　　tUSBRingBufObject * psRingBuf);

作用：直接写入数据到 US 缓冲区，而不是使用 USBBufferwrite()函数。

参数：psBuffer，操作的目标 Buffer。psRingBuf，声明一个 tUSBRingBufObject 变量。

返回:无。

(3) unsigned long USBBufferWrite(const tUSBBuffer * psBuffer,
　　　　　　　　　　　　　　 const unsigned char * pucData,
　　　　　　　　　　　　　　 unsigned long ulLength);

作用:写入一组数据。直接写入。
参数:psBuffer,目标 Buffer。pucData,待写入数据指针。ulLength,写入长度。
返回:写入的数据长度。

(4) void USBBufferDataWritten(const tUSBBuffer * psBuffer,
　　　　　　　　　　　　　　 unsigned long ulLength);

作用:写入一组数据。使用前要调用 USBBufferInfoGet。
参数:psBuffer,目标 Buffer。ulLength,写入长度。
返回:写入的数据长度。

(5) void USBBufferDataRemoved(const tUSBBuffer * psBuffer,
　　　　　　　　　　　　　　 unsigned long ulLength);

作用:从 Buffer 中移出数据。
参数:psBuffer,目标 Buffer。ulLength,移出数据个数。
返回:无。

(6) void USBBufferFlush(const tUSBBuffer * psBuffer);

作用:清除 Buffer 中的数据。
参数:psBuffer,目标 Buffer。
返回:无。

(7) unsigned long USBBufferRead(const tUSBBuffer * psBuffer,
　　　　　　　　　　　　　　 unsigned char * pucData,
　　　　　　　　　　　　　　 unsigned long ulLength);

作用:读取数据。
参数:psBuffer,目标 Buffer。pucData,数据存放指针。ulLength,读取个数。
返回:读取数据的个数。

(8) unsigned long USBBufferDataAvailable(const tUSBBuffer * psBuffer);

作用:可读取数据个数。
参数:psBuffer,目标 Buffer。
返回:可读取数据个数。

(9) unsigned long USBBufferSpaceAvailable(const tUSBBuffer * psBuffer);

作用:可用数据空间大小。
参数:psBuffer,目标 Buffer。
返回:数据空间大小。

(10) void * USBBufferCallbackDataSet(tUSBBuffer * psBuffer, void * pvCB-

Data);

作用：修改设备 Buffer。
参数：psBuffer,用于替换的新 Buffer 指针。
返回：旧 Buffer 指针。
(11) unsigned long USBBufferEventCallback(void * pvCBData,
　　　　　　　　　　　　　　　　　　　　unsigned long ulEvent,
　　　　　　　　　　　　　　　　　　　　unsigned long ulMsgValue,
　　　　　　　　　　　　　　　　　　　　void * pvMsgData);

作用：Buffer 事件调用函数。
参数：pvCBData,设备指针。ulEvent,Buffer 事务。ulMsgValue,数据长度。pvMsgData 数据指针。
返回：函数是否成功执行。

以上是 11 个 Buffer 处理函数，用于 Buffer 数据接收、发送及处理，在 Bulk 传输中大量使用。

7.4　Bulk 设备开发

Bulk 设备开发只需要 4 步就能完成。如图 7-2 所示，为 Bulk 开发流程图。Bulk 设备配置（主要是字符串描述符）、Callback 函数编写、USB 处理器初始化及数据处理。

(1) Bulk 设备配置（主要是字符串描述符），按字符串描述符标准完成串描述符配置，进而完成 Bulk 设备配置。

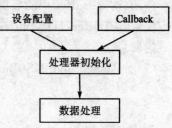

图 7-2　Bulk 开发流程图

```
# include "inc/hw_ints.h"
# include "inc/hw_memmap.h"
# include "inc/hw_types.h"
# include "inc/hw_sysctl.h"
# include "inc/hw_udma.h"
# include "inc/hw_gpio.h"
# include "driverlib/gpio.h"
# include "driverlib/interrupt.h"
# include "driverlib/sysctl.h"
# include "driverlib/usb.h"
# include "usblib/usblib.h"
# include "usblib/usb-ids.h"
# include "usblib/device/usbdevice.h"
# include "usblib/device/usbdbulk.h"
# include "uartstdio.h"
```

```c
#include "ustdlib.h"
//每次传输数据大小
#define BULK_BUFFER_SIZE 256
unsigned long RxHandler(void * pvCBData, unsigned long ulEvent,
                        unsigned long ulMsgValue, void * pvMsgData);
unsigned long TxHandler(void * pvlCBData, unsigned long ulEvent,
                        unsigned long ulMsgValue, void * pvMsgData);
unsigned long EchoNewDataToHost(tUSBDBulkDevice * psDevice, unsigned char * pcData,
                        unsigned long ulNumBytes);
#define COMMAND_PACKET_RECEIVED 0x00000001
#define COMMAND_STATUS_UPDATE   0x00000002
volatile unsigned long g_ulFlags = 0;
char * g_pcStatus;
static volatile tBoolean g_bUSBConfigured = false;
volatile unsigned long g_ulTxCount = 0;
volatile unsigned long g_ulRxCount = 0;
const tUSBBuffer g_sRxBuffer;
const tUSBBuffer g_sTxBuffer;
//*****************************************************************
// 设备语言描述符
//*****************************************************************
const unsigned char g_pLangDescriptor[] =
{
    4,
    USB_DTYPE_STRING,
    USBShort(USB_LANG_EN_US)
};
//*****************************************************************
// 制造商 字符串 描述符
//*****************************************************************
const unsigned char g_pManufacturerString[] =
{
    (17 + 1) * 2,
    USB_DTYPE_STRING,
    'T', 0, 'e', 0, 'x', 0, 'a', 0, 's', 0, ' ', 0, 'I', 0, 'n', 0, 's', 0,
    't', 0, 'r', 0, 'u', 0, 'm', 0, 'e', 0, 'n', 0, 't', 0, 's', 0,
};
//*****************************************************************
//产品 字符串 描述符
//*****************************************************************
const unsigned char g_pProductString[] =
{
```

```c
    (19 + 1) * 2,
    USB_DTYPE_STRING,
    'G', 0, 'e', 0, 'n', 0, 'e', 0, 'r', 0, 'i', 0, 'c', 0, ' ', 0, 'B', 0,
    'u', 0, 'l', 0, 'k', 0, ' ', 0, 'D', 0, 'e', 0, 'v', 0, 'i', 0, 'c', 0,
    'e', 0
};
//*****************************************************************
//产品 序列号 描述符
//*****************************************************************
const unsigned char g_pSerialNumberString[] =
{
    (8 + 1) * 2,
    USB_DTYPE_STRING,
    '1', 0, '2', 0, '3', 0, '4', 0, '5', 0, '6', 0, '7', 0, '8', 0
};
//*****************************************************************
//设备接口字符串描述符
//*****************************************************************
const unsigned char g_pDataInterfaceString[] =
{
    (19 + 1) * 2,
    USB_DTYPE_STRING,
    'B', 0, 'u', 0, 'l', 0, 'k', 0, ' ', 0, 'D', 0, 'a', 0, 't', 0,
    'a', 0, ' ', 0, 'I', 0, 'n', 0, 't', 0, 'e', 0, 'r', 0, 'f', 0,
    'a', 0, 'c', 0, 'e', 0
};
//*****************************************************************
//设备配置字符串描述符
//*****************************************************************
const unsigned char g_pConfigString[] =
{
    (23 + 1) * 2,
    USB_DTYPE_STRING,
    'B', 0, 'u', 0, 'l', 0, 'k', 0, ' ', 0, 'D', 0, 'a', 0, 't', 0,
    'a', 0, ' ', 0, 'C', 0, 'o', 0, 'n', 0, 'f', 0, 'i', 0, 'g', 0,
    'u', 0, 'r', 0, 'a', 0, 't', 0, 'i', 0, 'o', 0, 'n', 0
};
//*****************************************************************
// 字符串描述符集合
//*****************************************************************
const unsigned char * const g_pStringDescriptors[] =
{
```

```c
    g_pLangDescriptor,
    g_pManufacturerString,
    g_pProductString,
    g_pSerialNumberString,
    g_pDataInterfaceString,
    g_pConfigString
};
#define NUM_STRING_DESCRIPTORS (sizeof(g_pStringDescriptors) /      \
                               sizeof(unsigned char *))
//****************************************************************
// 定义 Bulk 设备实例
//****************************************************************
tBulkInstance g_sBulkInstance;
//****************************************************************
// 定义 Bulk 设备
//****************************************************************
const tUSBDBulkDevice g_sBulkDevice =
{
    0x1234,
    USB_PID_BULK,
    500,
    USB_CONF_ATTR_SELF_PWR,
    USBBufferEventCallback,
    (void *)&g_sRxBuffer,
    USBBufferEventCallback,
    (void *)&g_sTxBuffer,
    g_pStringDescriptors,
    NUM_STRING_DESCRIPTORS,
    &g_sBulkInstance
};
//****************************************************************
// 定义 Buffer
//****************************************************************
unsigned char g_pucUSBRxBuffer[BULK_BUFFER_SIZE];
unsigned char g_pucUSBTxBuffer[BULK_BUFFER_SIZE];
unsigned char g_pucTxBufferWorkspace[USB_BUFFER_WORKSPACE_SIZE];
unsigned char g_pucRxBufferWorkspace[USB_BUFFER_WORKSPACE_SIZE];
const tUSBBuffer g_sRxBuffer =
{
    false,                              // This is a receive buffer
    RxHandler,                          // pfnCallback
    (void *)&g_sBulkDevice,             // Callback data is our device pointer
```

第 7 章 Bulk 设备

```
    USBDBulkPacketRead,              // pfnTransfer
    USBDBulkRxPacketAvailable,       // pfnAvailable
    (void *)&g_sBulkDevice,          // pvHandle
    g_pucUSBRxBuffer,                // pcBuffer
    BULK_BUFFER_SIZE,                // ulBufferSize
    g_pucRxBufferWorkspace           // pvWorkspace
};
const tUSBBuffer g_sTxBuffer =
{
    true,                            // This is a transmit buffer
    TxHandler,                       // pfnCallback
    (void *)&g_sBulkDevice,          // Callback data is our device pointer
    USBDBulkPacketWrite,             // pfnTransfer
    USBDBulkTxPacketAvailable,       // pfnAvailable
    (void *)&g_sBulkDevice,          // pvHandle
    g_pucUSBTxBuffer,                // pcBuffer
    BULK_BUFFER_SIZE,                // ulBufferSize
    g_pucTxBufferWorkspace           // pvWorkspace
};
```

tUSBDBulkDevice g_sBulkDevice、tUSBBuffer g_sTxBuffer、tUSBBuffer g_sRxBuffer 是管理 Bulk 设备的主要结构体，它们 3 者的关系如图 7-3 所示。

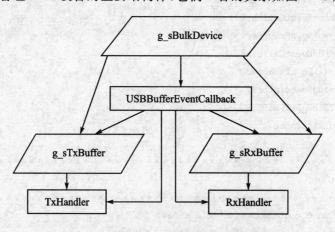

图 7-3 buffer 使用

tUSBDBulkDevice g_sBulkDevice 主要管理 Bulk 设备的 VID、PID、电源属性、字符串描述符、Bulk 设备实例，并将待发送数据和已经接收到的数据存放入 Buffer 中。其通过 USBBufferEventCallback 函数处理 Buffer 数据：数据发送、接收、控制事件，通过 g_sTxBuffer 中的 USBDBulkPacketWrite 和 USBDBulkTxPacketAvailable 实现底层数据发送，并通过 TxHandler 返回处理结果；通过 g_sRxBuffer 中的 USB-

DBulkPacketRead 和 USBDBulkRxPacketAvailable 实现底层数据接收，并通过 RxHandler 返回处理结果。在 g_sBulkDevice 层可以直接使用 Buffer 函数对 Buffer 层操作，并通过 TxHandler 和 RxHandler 返回处理结果。所有处理过程中的数据都保存在 Buffer 层的 g_pucUSBTxBuffer 或者 g_pucUSBRxBuffer 中，隶属于 g_sBulkDevice 的一部分。注意：USBDBulkPacketWrite、USBDBulkTxPacketAvailable、USBDBulkPacketRead、USBDBulkRxPacketAvailable 由 Bulk 设备类 API 定义，可以直接使用。USBBufferEventCallback 为 Buffer 层定义的标准 API，用于处理、调用 USBDBulkPacketWrite、USBDBulkTxPacketAvailable、USBDBulkPacketRead、USBDBulkRxPacketAvailable、TxHandler 和 RxHandler 完成 g_sBulkDevice 层发送的数据接收与发送命令。

（2）完成 Callback 函数。Callback 函数用于处理输出端点、输入端点数据事务。Bulk 设备接收回调函数包含以下事务：USB_EVENT_CONNECTED、USB_EVENT_DISCONNECTED、USB_EVENT_RX_AVAILABLE、USB_EVENT_SUSPEND、USB_EVENT_RESUME、USB_EVENT_ERROR。Bulk 设备发送回调函数包含了以下事务：USB_EVENT_TX_COMPLETE。如表 7-1 所列。

表 7-1 Bulk 事务

名 称	属 性	说 明
USB_EVENT_CONNECTED	接收	USB 设备已经连接到主机
USB_EVENT_DISCONNECTED	接收	USB 设备已经与主机断开
USB_EVENT_RX_AVAILABLE	接收	有接收数据
USB_EVENT_SUSPEND	接收	挂起
USB_EVENT_RESUME	接收	唤醒
USB_EVENT_ERROR	接收	错误
USB_EVENT_TX_COMPLETE	发送	发送完成

根据以上事务编写 Callback 函数如下所示。

```
//************************************************************
//USB Bulk 设备类返回事件处理函数(Callback)
//************************************************************
unsigned long   TxHandler(void * pvCBData, unsigned long ulEvent, unsigned long ulMs-
                gValue,void * pvMsgData)
{
    //发送完成事件
    if(ulEvent == USB_EVENT_TX_COMPLETE)
    {
        g_ulTxCount += ulMsgValue;
    }
```

```c
        return(0);
    }
    unsigned long RxHandler(void * pvCBData, unsigned long ulEvent,
                            unsigned long ulMsgValue, void * pvMsgData)
    {
        // 接收事件
        switch(ulEvent)
        {
            //连接成功
            case USB_EVENT_CONNECTED:
            {
                GPIOPinWrite(GPIO_PORTF_BASE,0x40,0x40);
                g_bUSBConfigured = true;
                g_pcStatus = "Host connected.";
                g_ulFlags |= COMMAND_STATUS_UPDATE;
                // Flush our buffers
                USBBufferFlush(&g_sTxBuffer);
                USBBufferFlush(&g_sRxBuffer);
                break;
            }

            // 断开连接
            case USB_EVENT_DISCONNECTED:
            {
                GPIOPinWrite(GPIO_PORTF_BASE,0x40,0x00);
                g_bUSBConfigured = false;
                g_pcStatus = "Host disconnected.";
                g_ulFlags |= COMMAND_STATUS_UPDATE;
                break;
            }

            // 数据缓冲区有可接收的数据,获取有效数据,并发送给计算机
            case USB_EVENT_RX_AVAILABLE:
            {
                tUSBDBulkDevice * psDevice;
                psDevice = (tUSBDBulkDevice *)pvCBData;
                // 把接收到的数据发送回去
                return(EchoNewDataToHost(psDevice, pvMsgData, ulMsgValue));
            }
            //挂起,唤醒
            case USB_EVENT_SUSPEND:
            case USB_EVENT_RESUME:break;
            default:break;
        }
```

```c
    return(0);
}
//****************************************************************
//EchoNewDataToHost 函数
//****************************************************************
unsigned long EchoNewDataToHost(tUSBDBulkDevice * psDevice, unsigned char * pcData,
                    unsigned long ulNumBytes)
{
    unsigned long ulLoop, ulSpace, ulCount;
    unsigned long ulReadIndex;
    unsigned long ulWriteIndex;
    tUSBRingBufObject sTxRing;
    // 获取 Buffer 信息
    USBBufferInfoGet(&g_sTxBuffer, &sTxRing);
    // 有多少可能的空间
    ulSpace = USBBufferSpaceAvailable(&g_sTxBuffer);
    // 改变数据
    ulLoop = (ulSpace < ulNumBytes) ? ulSpace : ulNumBytes;
    ulCount = ulLoop;
    // 更新接收到的数据个数
    g_ulRxCount += ulNumBytes;
    ulReadIndex = (unsigned long)(pcData - g_pucUSBRxBuffer);
    ulWriteIndex = sTxRing.ulWriteIndex;
    while(ulLoop)
    {
        //更新接收的数据
        if((g_pucUSBRxBuffer[ulReadIndex] >= 'a') &&
           (g_pucUSBRxBuffer[ulReadIndex] <= 'z'))
        {
            //转换
            g_pucUSBTxBuffer[ulWriteIndex] =
                (g_pucUSBRxBuffer[ulReadIndex] - 'a') + 'A';
        }
        else
        {
            //转换
            if((g_pucUSBRxBuffer[ulReadIndex] >= 'A') &&
               (g_pucUSBRxBuffer[ulReadIndex] <= 'Z'))
            {
                //转换
                g_pucUSBTxBuffer[ulWriteIndex] =
                    (g_pucUSBRxBuffer[ulReadIndex] - 'Z') + 'z';
```

```
            }
            else
            {
                //转换
                g_pucUSBTxBuffer[ulWriteIndex] = g_pucUSBRxBuffer[ulReadIndex];
            }
        }
        // 更新指针
        ulWriteIndex++;
        ulWriteIndex = (ulWriteIndex == BULK_BUFFER_SIZE) ? 0 : ulWriteIndex;
        ulReadIndex++;
        ulReadIndex = (ulReadIndex == BULK_BUFFER_SIZE) ? 0 : ulReadIndex;
        ulLoop--;
    }
    // 发送数据
    USBBufferDataWritten(&g_sTxBuffer, ulCount);
    return(ulCount);
}
```

(3) 系统初始化,配置内核电压、系统主频、使能端口、LED 控制等,本例中使用 4 个 LED 进行指示数据传输。在这个例子中,Bulk 传输接收的数据发送给主机。LED 与键盘电路图如图 5-6 所示。

系统初始化程序如下所示。

```
unsigned long ulTxCount = 0;
unsigned long ulRxCount = 0;
//char pcBuffer[16];
//设置内核电压、主频为 50 MHz
SysCtlLDOSet(SYSCTL_LDO_2_75V);
SysCtlClockSet(SYSCTL_XTAL_8MHZ | SYSCTL_SYSDIV_4 | SYSCTL_USE_PLL | SYSCTL_OSC
               _MAIN);
SysCtlPeripheralEnable(SYSCTL_PERIPH_GPIOF);
GPIOPinTypeGPIOOutput(GPIO_PORTF_BASE,0xf0);
GPIOPinTypeGPIOInput(GPIO_PORTF_BASE,0x0f);
HWREG(GPIO_PORTF_BASE + GPIO_O_PUR) |= 0x0f;

// 初始化发送与接收 Buffer
USBBufferInit((tUSBBuffer *)&g_sTxBuffer);
USBBufferInit((tUSBBuffer *)&g_sRxBuffer);
// 初始化 Bulk 设备
USBDBulkInit(0, (tUSBDBulkDevice *)&g_sBulkDevice);
```

(4) 数据处理。主要使用 11 个 Buffer 处理函数,用于 Buffer 数据的接收、发送及处理。

```c
while(1)
{
    //等待连接结束
    while((g_ulFlags & FLAG_CONNECTED) == 0)
    {
    }
    //初始化 Buffer
    g_sBuffer.pucFill = g_sBuffer.pucBuffer;
    g_sBuffer.pucPlay = g_sBuffer.pucBuffer;
    g_sBuffer.ulFlags = 0;
    //从 Bulk 设备类中获取数据
    if(USBBulkBufferOut(g_pvBulkDevice,
                        (unsigned char *)g_sBuffer.pucFill,
                        BULK_PACKET_SIZE, USBBufferCallback) == 0)
    {
        //标记数据放入 buffer 中
        g_sBuffer.ulFlags |= SBUFFER_FLAGS_FILLING;
    }
    //设备连接到主机
    while(g_ulFlags & FLAG_CONNECTED)
    {
        // 检查音量是否有改变
        if(g_ulFlags & FLAG_VOLUME_UPDATE)
        {
            // 清除更新音量标志
            g_ulFlags &= ~FLAG_VOLUME_UPDATE;
            // 修改音量,自行添加代码,在此以 LED 灯做指示
            //UpdateVolume();
            GPIOPinWrite(GPIO_PORTF_BASE,0x40,~GPIOPinRead(GPIO_PORTF_BASE,
                0x40));
        }
        //是否静音
        if(g_ulFlags & FLAG_MUTE_UPDATE)
        {
            //修改静音状态,自行添加函数.在此以 LED 灯做指示
            //UpdateMute();
              if(g_ulFlags & FLAG_MUTED)
                 GPIOPinWrite(GPIO_PORTF_BASE,0x20,0x20);
              else
                 GPIOPinWrite(GPIO_PORTF_BASE,0x20,0x00);
```

第7章 Bulk 设备

```
                // 清除静音标志
                g_ulFlags &= ~FLAG_MUTE_UPDATE;
            }
        }
}
//*****************************************************************
//USBBulkBufferOut 的 Callback 入口参数
//*****************************************************************
void USBBufferCallback(void * pvBuffer, unsigned long ulParam, unsigned long ulEvent)
{
    //数据处理,自行加入代码
    // Your Codes ......
    //再一次获取数据
    USBBulkBufferOut(g_pvBulkDevice, (unsigned char *)g_sBuffer.pucFill,
                     BULK_PACKET_SIZE, USBBufferCallback);
}
```

使用上面4步就可完成 Bulk 设备的开发。Bulk 设备开发时要加入两个库函数,分别是 usblib.lib 和 DriverLib.lib,在启动代码中加入 USB0DeviceIntHandler 中断服务函数。以上 Bulk 设备开发完成,在 Windows XP 下的运行效果如图7-4所示,Bulk 设备正在枚举,在计算机的右下脚可以看到 Generic Bulk Device 提示。

在枚举过程中,在计算机右下脚可以看到 Generic Bulk Device 字样,表示正在进行枚举,并手动安装驱动,如图7-5所示,驱动安装完成。枚举成功后,在"设备

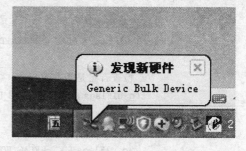

图7-4 Bulk 设备正在枚举

管理器"的 Stellaris Bulk Device 中看到 Generic Bulk Device 设备,如图7-6所示,可在"设备管理器"中找到 Bulk 设备,到此 Bulk 设备可以正式使用了,PC 与 USB 系统可以正常通信。

Bulk 设备必须配合上位机使用,上位机要发送的字符串通过 USB Bulk 设备传送到 USB 处理器中,处理器将数据转换后发送给主机(PC)。如图7-7所示,为上位机通信图,从图中可以看出,主机发送"Hello paulhyde!"到 USB 设备中,该字符串经设备转换后,上传至 PC 上,并在当前屏幕中显示转换结果为"hELLO PAULHYDE!"。

上位机源码如下:

```
#include <windows.h>
#include <strsafe.h>
```

第7章 Bulk 设备

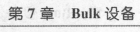

图 7-5 驱动安装完成

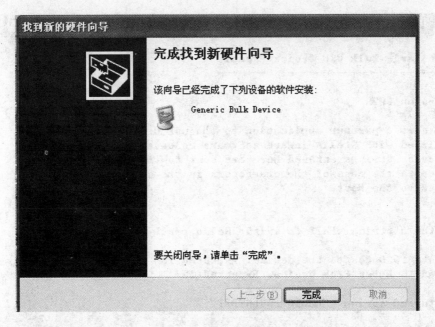

图 7-6 Bulk 设备

```
# include <initguid.h>
# include "lmusbdll.h"
# include "luminary_guids.h"
//***********************
// Buffer size definitions
//***********************
# define MAX_STRING_LEN 256
# define MAX_ENTRY_LEN 256
# define USB_BUFFER_LEN 1216
//***********************
// The build version number
//*****************************************************
# define BLDVER "6075"
//*****************************************************
// The number of bytes we read and write per transaction if in echo mode
//*****************************************************
# define ECHO_PACKET_SIZE 1216
//*****************************************************
// Buffer into which error messages are written
//*****************************************************
TCHAR g_pcErrorString[MAX_STRING_LEN];
//*****************************************************
// The number of bytes transfered in the last measurement interval
```

第7章 Bulk 设备

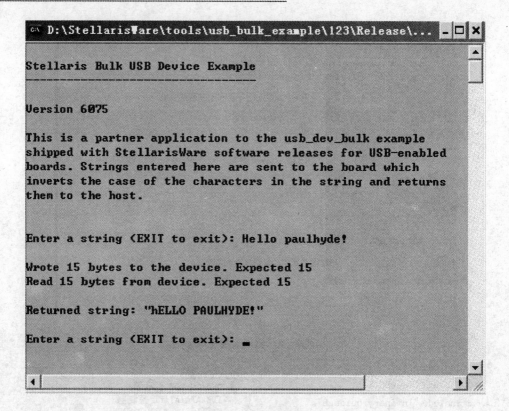

图 7-7 上位机通信

```
//************************************************************
ULONG g_ulByteCount = 0;
//************************************************************
// The total number of packets transfered
//************************************************************
ULONG g_ulPacketCount = 0;
//************************************************************
LPTSTR GetSystemErrorString(DWORD dwError)
{
    DWORD dwRetcode;
    // Ask Windows for the error message description
    dwRetcode = FormatMessage(FORMAT_MESSAGE_FROM_SYSTEM, "%0", dwError, 0,
                        g_pcErrorString, MAX_STRING_LEN, NULL);
    if(dwRetcode == 0)
    {
        return((LPTSTR)L"Unknown");
    }
    else
```

```c
    {
        // Remove the trailing "\n\r" if present
        if(dwRetcode >= 2)
        {
            if(g_pcErrorString[dwRetcode - 2] == '\r')
            {
                g_pcErrorString[dwRetcode - 2] = '\0';
            }
        }
        return(g_pcErrorString);
    }
}
//*************************************************************
// Print the throughput in terms of Kbps once per second
//*************************************************************
void UpdateThroughput(void)
{
    static ULONG ulStartTime = 0;
    static ULONG ulLast = 0;
    ULONG ulNow;
    ULONG ulElapsed;
    SYSTEMTIME sSysTime;
    // Get the current system time
    GetSystemTime(&sSysTime);
    ulNow = (((((sSysTime.wHour * 60) +
            sSysTime.wMinute) * 60) +
            sSysTime.wSecond) * 1000) + sSysTime.wMilliseconds;
    // If this is the first call, set the start time
    if(ulStartTime == 0)
    {
        ulStartTime = ulNow;
        ulLast = ulNow;
        return;
    }
    // How much time has elapsed since the last measurement
    ulElapsed = (ulNow > ulStartTime) ? (ulNow - ulStartTime) : (ulStartTime - ul-
            Now);
    // We dump a new measurement every second
    if(ulElapsed > 1000)
    {
        printf("\r%6dKbps Packets: %10d ", ((g_ulByteCount * 8) / ulElapsed), g_
            ulPacketCount);
```

```c
            g_ulByteCount = 0;
            ulStartTime = ulNow;
        }
    }
// ****************************************************************
// The main application entry function
// ****************************************************************
int main(int argc, char * argv[])
{
    BOOL bResult;
    BOOL bDriverInstalled;
    BOOL bEcho;
    char szBuffer[USB_BUFFER_LEN];
    ULONG ulWritten;
    ULONG ulRead;
    ULONG ulLength;
    DWORD dwError;
    LMUSB_HANDLE hUSB;
    // Are we operating in echo mode or not? The " - e" parameter tells the
    // app to echo everything it receives back to the device unchanged
    bEcho = ((argc > 1) && (argv[1][1] == 'e')) ? TRUE : FALSE;
    // Print a cheerful welcome.
    printf("\nStellaris Bulk USB Device Example\n");
    printf( "---------------------------------\n\n");
    printf("Version % s\n\n", BLDVER);
    if(! bEcho)
    {
        printf("This is a partner application to the usb_dev_bulk example\n");
        printf("shipped with StellarisWare software releases for USB - enabled\n");
        printf("boards. Strings entered here are sent to the board which\n");
        printf("inverts the case of the characters in the string and returns\n");
        printf("them to the host.\n\n");
    }
    else
    {
        printf("If run with the \" - e\" command line switch, this application\n");
        printf("echoes all data received on the bulk IN endpoint to the bulk\n");
        printf("OUT endpoint.  This feature may be helpful during development\n");
        printf("and debug of your own USB devices.  Note that this will not\n");
        printf("do anything exciting if run with the usb_dev_bulk example\n");
        printf("device attached since it expects the host to initiate transfers.\n\n");
    }
```

```c
// Find our USB device and prepare it for communication
hUSB = InitializeDevice(BULK_VID, BULK_PID,
                        (LPGUID)&(GUID_DEVINTERFACE_STELLARIS_BULK),
                        &bDriverInstalled);
if(hUSB)
{
    // Are we operating in echo mode or not? The "-e" parameter tells the
    // app to echo everything it receives back to the device unchanged
    if(bEcho)
    {
        printf("Running in echo mode. Press Ctrl + C to exit.\n\n"
            "Throughput:       0Kbps Packets:            0");
        while(1)
        {
            // Read a block of data from the device
            dwError = ReadUSBPacket(hUSB, szBuffer, USB_BUFFER_LEN, &ulRead,
                                    INFINITE, NULL);
            if(dwError != ERROR_SUCCESS)
            {
                // We failed to read from the device
                printf("\n\nError %d (%S) reading from bulk IN pipe.\n", dwError,
                    GetSystemErrorString(dwError));
                break;
            }
            else
            {
                // Update our byte and packet counters
                g_ulByteCount += ulRead;
                g_ulPacketCount++;
                // Write the data back out to the device
                bResult = WriteUSBPacket(hUSB, szBuffer, ulRead, &ulWritten);
                if(!bResult)
                {
                    // We failed to write the data for some reason
                    dwError = GetLastError();
                    printf("\n\nError %d (%S) writing to bulk OUT pipe.\n",
                        dwError,GetSystemErrorString(dwError));
                    break;
                }
                // Display the throughput
                UpdateThroughput();
            }
```

第 7 章 Bulk 设备

```c
            }
        }
        else
        {
            // We are running in normal mode. Keep sending and receiving
            // strings until the user indicates that it is time to exit
            while(1)
            {
                // The device was found and successfully configured. Now get a
                // string from
                // the user...
                do
                {
                    printf("\nEnter a string (EXIT to exit): ");
                    fgets(szBuffer, MAX_ENTRY_LEN, stdin);
                    printf("\n");
                    // How many characters were entered (including the trailing '\n')
                    ulLength = (ULONG)strlen(szBuffer);
                    if(ulLength <= 1)
                    {
                        printf("\nPlease enter some text.\n");
                        ulLength = 0;
                    }
                    else
                    {
                        // Get rid of the trailing '\n' if there is one there
                        if(szBuffer[ulLength - 1] == '\n')
                        {
                            szBuffer[ulLength - 1] = '\0';
                            ulLength--;
                        }
                    }
                }
                while(ulLength == 0);
                if(! (strcmp("EXIT", szBuffer)))
                {
                    printf("Exiting on user request.\n");
                    break;
                }
                // Write the user's string to the device
                bResult = WriteUSBPacket(hUSB, szBuffer, ulLength, &ulWritten);
                if(! bResult)
```

```c
            {
                dwError = GetLastError();
                printf("Error %d (%S) writing to bulk OUT pipe.\n", dwError,
                    GetSystemErrorString(dwError));
            }
            else
            {
                // We wrote data successfully so now read it back
                printf("Wrote %d bytes to the device. Expected %d\n",
                    ulWritten, ulLength);
                // We expect the same number of bytes as we just sent
                dwError = ReadUSBPacket(hUSB, szBuffer, ulWritten, &ulRead,
                                INFINITE, NULL);
                if(dwError != ERROR_SUCCESS)
                {
                    // We failed to read from the device
                    printf("Error %d (%S) reading from bulk IN pipe.\n", dwError,
                        GetSystemErrorString(dwError));
                }
                else
                {
                    szBuffer[ulRead] = '\0';
                    printf("Read %d bytes from device. Expected %d\n",
                        ulRead, ulWritten);
                    printf("\nReturned string: \"%s\"\n", szBuffer);
                }
            }
        }
    }
}
else
{
    // An error was reported while trying to connect to the device
    dwError = GetLastError();
    printf("\nUnable to initialize the Stellaris Bulk USB Device.\n");
    printf("Error code is %d (%S)\n\n", dwError, GetSystemErrorString(dwEr-
        ror));
    printf("Please make sure you have a Stellaris USB-enabled evaluation\n");
    printf("or development kit running the usb_dev_bulk example\n");
    printf("application connected to this system via the \"USB OTG\" or\n");
    printf("\"USB DEVICE\" connectors. Once the device is connected, run\n");
    printf("this application again.\n\n");
```

```c
        printf("\nPress \"Enter\" to exit: ");
        fgets(szBuffer, MAX_STRING_LEN, stdin);
        printf("\n");
        return(2);
    }
    TerminateDevice(hUSB);
    return(0);
}
```

Bulk 设备开发源码如下所示:

```c
#include "inc/hw_ints.h"
#include "inc/hw_memmap.h"
#include "inc/hw_types.h"
#include "inc/hw_sysctl.h"
#include "inc/hw_udma.h"
#include "inc/hw_gpio.h"
#include "driverlib/gpio.h"
#include "driverlib/interrupt.h"
#include "driverlib/sysctl.h"
#include "driverlib/usb.h"
#include "usblib/usblib.h"
#include "usblib/usb-ids.h"
#include "usblib/device/usbdevice.h"
#include "usblib/device/usbdbulk.h"
#include "uartstdio.h"
#include "ustdlib.h"
//每次传输数据大小
#define BULK_BUFFER_SIZE 256
unsigned long RxHandler(void * pvCBData, unsigned long ulEvent,
                       unsigned long ulMsgValue, void * pvMsgData);
unsigned long TxHandler(void * pvlCBData, unsigned long ulEvent,
                       unsigned long ulMsgValue, void * pvMsgData);
unsigned long EchoNewDataToHost(tUSBDBulkDevice * psDevice, unsigned char * pcData,
                       unsigned long ulNumBytes);
#define COMMAND_PACKET_RECEIVED 0x00000001
#define COMMAND_STATUS_UPDATE   0x00000002
volatile unsigned long g_ulFlags = 0;
char * g_pcStatus;
static volatile tBoolean g_bUSBConfigured = false;
volatile unsigned long g_ulTxCount = 0;
volatile unsigned long g_ulRxCount = 0;
const tUSBBuffer g_sRxBuffer;
```

```c
const tUSBBuffer g_sTxBuffer;
//*************************************************************
// 设备语言描述符
//*************************************************************
const unsigned char g_pLangDescriptor[] =
{
    4,
    USB_DTYPE_STRING,
    USBShort(USB_LANG_EN_US)
};
//*************************************************************
// 制造商 字符串 描述符
//*************************************************************
const unsigned char g_pManufacturerString[] =
{
    (17 + 1) * 2,
    USB_DTYPE_STRING,
    'T', 0, 'e', 0, 'x', 0, 'a', 0, 's', 0, ' ', 0, 'I', 0, 'n', 0, 's', 0,
    't', 0, 'r', 0, 'u', 0, 'm', 0, 'e', 0, 'n', 0, 't', 0, 's', 0,
};
//*************************************************************
//产品 字符串 描述符
//*************************************************************
const unsigned char g_pProductString[] =
{
    (19 + 1) * 2,
    USB_DTYPE_STRING,
    'G', 0, 'e', 0, 'n', 0, 'e', 0, 'r', 0, 'i', 0, 'c', 0, ' ', 0, 'B', 0,
    'u', 0, 'l', 0, 'k', 0, ' ', 0, 'D', 0, 'e', 0, 'v', 0, 'i', 0, 'c', 0,
    'e', 0
};
//*************************************************************
//产品 序列号 描述符
//*************************************************************
const unsigned char g_pSerialNumberString[] =
{
    (8 + 1) * 2,
    USB_DTYPE_STRING,
    '1', 0, '2', 0, '3', 0, '4', 0, '5', 0, '6', 0, '7', 0, '8', 0
};
//*************************************************************
// 设备接口字符串描述符
```

```c
//*****************************************************************
const unsigned char g_pDataInterfaceString[] =
{
    (19 + 1) * 2,
    USB_DTYPE_STRING,
    'B', 0, 'u', 0, 'l', 0, 'k', 0, ' ', 0, 'D', 0, 'a', 0, 't', 0,
    'a', 0, ' ', 0, 'I', 0, 'n', 0, 't', 0, 'e', 0, 'r', 0, 'f', 0,
    'a', 0, 'c', 0, 'e', 0
};
//*****************************************************************
// 设备配置字符串描述符
//*****************************************************************
const unsigned char g_pConfigString[] =
{
    (23 + 1) * 2,
    USB_DTYPE_STRING,
    'B', 0, 'u', 0, 'l', 0, 'k', 0, ' ', 0, 'D', 0, 'a', 0, 't', 0,
    'a', 0, ' ', 0, 'C', 0, 'o', 0, 'n', 0, 'f', 0, 'i', 0, 'g', 0,
    'u', 0, 'r', 0, 'a', 0, 't', 0, 'i', 0, 'o', 0, 'n', 0
};
//*****************************************************************
// 字符串描述符集合
//*****************************************************************
const unsigned char * const g_pStringDescriptors[] =
{
    g_pLangDescriptor,
    g_pManufacturerString,
    g_pProductString,
    g_pSerialNumberString,
    g_pDataInterfaceString,
    g_pConfigString
};
#define NUM_STRING_DESCRIPTORS (sizeof(g_pStringDescriptors) /       \
                                sizeof(unsigned char *))
//*****************************************************************
// 定义 Bulk 设备实例
//*****************************************************************
tBulkInstance g_sBulkInstance;
//*****************************************************************
// 定义 Bulk 设备
//*****************************************************************
const tUSBDBulkDevice g_sBulkDevice =
```

```
{
    0x1234,
    USB_PID_BULK,
    500,
    USB_CONF_ATTR_SELF_PWR,
    USBBufferEventCallback,
    (void *)&g_sRxBuffer,
    USBBufferEventCallback,
    (void *)&g_sTxBuffer,
    g_pStringDescriptors,
    NUM_STRING_DESCRIPTORS,
    &g_sBulkInstance
};
//*************************************************************
// 定义 Buffer
//*************************************************************
unsigned char g_pucUSBRxBuffer[BULK_BUFFER_SIZE];
unsigned char g_pucUSBTxBuffer[BULK_BUFFER_SIZE];
unsigned char g_pucTxBufferWorkspace[USB_BUFFER_WORKSPACE_SIZE];
unsigned char g_pucRxBufferWorkspace[USB_BUFFER_WORKSPACE_SIZE];
const tUSBBuffer g_sRxBuffer =
{
    false,                          // This is a receive buffer
    RxHandler,                      // pfnCallback
    (void *)&g_sBulkDevice,         // Callback data is our device pointer
    USBDBulkPacketRead,             // pfnTransfer
    USBDBulkRxPacketAvailable,      // pfnAvailable
    (void *)&g_sBulkDevice,         // pvHandle
    g_pucUSBRxBuffer,               // pcBuffer
    BULK_BUFFER_SIZE,               // ulBufferSize
    g_pucRxBufferWorkspace          // pvWorkspace
};
const tUSBBuffer g_sTxBuffer =
{
    true,                           // This is a transmit buffer
    TxHandler,                      // pfnCallback
    (void *)&g_sBulkDevice,         // Callback data is our device pointer
    USBDBulkPacketWrite,            // pfnTransfer
    USBDBulkTxPacketAvailable,      // pfnAvailable
    (void *)&g_sBulkDevice,         // pvHandle
    g_pucUSBTxBuffer,               // pcBuffer
    BULK_BUFFER_SIZE,               // ulBufferSize
```

```c
            g_pucTxBufferWorkspace              // pvWorkspace
};
//***************************************************************
//USB Bulk 设备类返回事件处理函数(Callback)
//***************************************************************
unsigned long   TxHandler(void * pvCBData, unsigned long ulEvent, unsigned long ulMs-
                    gValue,void * pvMsgData)
{
    //发送完成事件
    if(ulEvent == USB_EVENT_TX_COMPLETE)
    {
        g_ulTxCount += ulMsgValue;
    }
    return(0);
}
unsigned long RxHandler(void * pvCBData, unsigned long ulEvent,
                    unsigned long ulMsgValue, void * pvMsgData)
{
    // 接收事件
    switch(ulEvent)
    {
        //连接成功
        case USB_EVENT_CONNECTED:
        {
            GPIOPinWrite(GPIO_PORTF_BASE,0x40,0x40);
            g_bUSBConfigured = true;
            g_pcStatus = "Host connected.";
            g_ulFlags |= COMMAND_STATUS_UPDATE;
            // Flush our buffers
            USBBufferFlush(&g_sTxBuffer);
            USBBufferFlush(&g_sRxBuffer);
            break;
        }
        // 断开连接
        case USB_EVENT_DISCONNECTED:
        {
            GPIOPinWrite(GPIO_PORTF_BASE,0x40,0x00);
            g_bUSBConfigured = false;
            g_pcStatus = "Host disconnected.";
            g_ulFlags |= COMMAND_STATUS_UPDATE;
            break;
        }
```

```c
        // 有可能数据接收
        case USB_EVENT_RX_AVAILABLE:
        {
            tUSBDBulkDevice * psDevice;
            psDevice = (tUSBDBulkDevice *)pvCBData;
            // 把接收到的数据发送回去
            return(EchoNewDataToHost(psDevice, pvMsgData, ulMsgValue));
        }
        //挂起,唤醒
        case USB_EVENT_SUSPEND:
        case USB_EVENT_RESUME:break;
        default:break;
    }
    return(0);
}
//*************************************************************
//EchoNewDataToHost 函数
//*************************************************************
unsigned long EchoNewDataToHost(tUSBDBulkDevice * psDevice, unsigned char * pcData,
                                unsigned long ulNumBytes)
{
    unsigned long ulLoop, ulSpace, ulCount;
    unsigned long ulReadIndex;
    unsigned long ulWriteIndex;
    tUSBRingBufObject sTxRing;
    // 获取 Buffer 信息
    USBBufferInfoGet(&g_sTxBuffer, &sTxRing);
    // 有多少可能的空间
    ulSpace = USBBufferSpaceAvailable(&g_sTxBuffer);
    // 改变数据
    ulLoop = (ulSpace < ulNumBytes) ? ulSpace : ulNumBytes;
    ulCount = ulLoop;
    // 更新接收到的数据个数
    g_ulRxCount += ulNumBytes;
    ulReadIndex = (unsigned long)(pcData - g_pucUSBRxBuffer);
    ulWriteIndex = sTxRing.ulWriteIndex;
    while(ulLoop)
    {
        //更新接收的数据
        if((g_pucUSBRxBuffer[ulReadIndex] >= 'a') &&
           (g_pucUSBRxBuffer[ulReadIndex] <= 'z'))
        {
```

```c
            //转换
            g_pucUSBTxBuffer[ulWriteIndex] =
                (g_pucUSBRxBuffer[ulReadIndex] - 'a') + 'A';
        }
        else
        {
            //转换
            if((g_pucUSBRxBuffer[ulReadIndex] >= 'A') &&
                (g_pucUSBRxBuffer[ulReadIndex] <= 'Z'))
            {
                //转换
                g_pucUSBTxBuffer[ulWriteIndex] =
                    (g_pucUSBRxBuffer[ulReadIndex] - 'Z') + 'z';
            }
            else
            {
                //转换
                g_pucUSBTxBuffer[ulWriteIndex] = g_pucUSBRxBuffer[ulReadIndex];
            }
        }
        // 更新指针
        ulWriteIndex++;
        ulWriteIndex = (ulWriteIndex == BULK_BUFFER_SIZE) ? 0 : ulWriteIndex;

        ulReadIndex++;
        ulReadIndex = (ulReadIndex == BULK_BUFFER_SIZE) ? 0 : ulReadIndex;
        ulLoop--;
    }
    // 发送数据
    USBBufferDataWritten(&g_sTxBuffer, ulCount);
    return(ulCount);
}
//*************************************************************
// 应用主函数
//*************************************************************
int main(void)
{
    unsigned long ulTxCount = 0;
    unsigned long ulRxCount = 0;
    //char pcBuffer[16];
    //设置内核电压、主频 50 MHz
    SysCtlLDOSet(SYSCTL_LDO_2_75V);
```

```
SysCtlClockSet(SYSCTL_XTAL_8MHZ | SYSCTL_SYSDIV_4 | SYSCTL_USE_PLL | SYSCTL_OSC
            _MAIN );
SysCtlPeripheralEnable(SYSCTL_PERIPH_GPIOF);
GPIOPinTypeGPIOOutput(GPIO_PORTF_BASE,0xf0);
GPIOPinTypeGPIOInput(GPIO_PORTF_BASE,0x0f);
HWREG(GPIO_PORTF_BASE + GPIO_O_PUR) |= 0x0f;
// 初始化发送与接收 Buffer
USBBufferInit((tUSBBuffer *)&g_sTxBuffer);
USBBufferInit((tUSBBuffer *)&g_sRxBuffer);
// 初始化 Bulk 设备
USBDBulkInit(0, (tUSBDBulkDevice *)&g_sBulkDevice);
while(1)
{
    if(g_ulFlags & COMMAND_STATUS_UPDATE)
    {
        //清除更新标志
        g_ulFlags &= ~COMMAND_STATUS_UPDATE;
        GPIOPinWrite(GPIO_PORTF_BASE,0x30,0x30);
    }
    // 发送完成
    if(ulTxCount != g_ulTxCount)
    {
        ulTxCount = g_ulTxCount;
        GPIOPinWrite(GPIO_PORTF_BASE,0x10,0x10);
        //usnprintf(pcBuffer, 16, " %d ", ulTxCount);
    }
    // 接收完成
    if(ulRxCount != g_ulRxCount)
    {
        ulRxCount = g_ulRxCount;
        GPIOPinWrite(GPIO_PORTF_BASE,0x20,0x20);
    }
}
```

7.5 小 结

本章主要介绍 USB 使用 Bulk 与 PC 通信的开发，通过开发一个通信实例，让读者明白 Bulk 的通信原理、Bulk 通信 API 函数、Bulk 通信上位机开发等。Bulk 数据传输在便携式设备、数据采集、移动设备等方面有重要应用。

第 8 章

CDC 设备

CDC 是通信设备类,具有两个接口类,可以完成数据传输与控制,不同的 CDC 类使用的端点数量不一样。常用的 USB 转 RS232 就是典型的 CDC 类,本章主要介绍 USB 转 RS232 的开发流程与程序编写。

8.1 CDC 设备介绍

USB 的 CDC 类是 USB 通信设备类(Communication Device Class)的简称。CDC 类是 USB 组织定义的一类专门给各种通信设备(电信通信设备和中速网络通信设备)使用的 USB 子类。根据 CDC 类所针对通信设备的不同,CDC 类又分成以下不同的模型:USB 传统纯电话业务(POTS)模型,USB ISDN 模型和 USB 网络模型。通常一个 CDC 类又由两个接口子类组成通信接口类(Communication Interface Class)和数据接口类(Data Interface Class)。通信接口类对设备进行管理和控制,而数据接口类传送数据。这两个接口子类占有不同数量和类型的终端点(endpoints),不同 CDC 类模型,其所对应的接口的终端点需求也是不同的。

8.2 CDC 数据类型

usbdcdc.h 中已经定义好 CDC 设备类中使用的所有数据类型和函数,同时也会使用 Buffer 数据类型及 API 函数,第 7 章已介绍了 Buffer 数据类型及 API 函数,下面只介绍 CDC 设备类使用的数据类型。

```
typedef enum
{
    //CDC 状态没定义
    CDC_STATE_UNCONFIGURED,
    //空闲状态
    CDC_STATE_IDLE,
    //等待数据发送或者结束
    CDC_STATE_WAIT_DATA,
    //等待数据处理
```

```
    CDC_STATE_WAIT_CLIENT
} tCDCState;
```

tCDCState 定义 CDC 端点状态。定义在 usbdcdc.h 文件中。用于端点状态标记与控制,可以保证数据传输不相互冲突。

```
typedef struct
{
    //USB 基地址
    unsigned long ulUSBBase;
    //设备信息
    tDeviceInfo *psDevInfo;
    //配置信息
    tConfigDescriptor *psConfDescriptor;
    //CDC 接收端点状态
    volatile tCDCState eCDCRxState;
    //CDC 发送端点状态
    volatile tCDCState eCDCTxState;
    //CDC 请求状态
    volatile tCDCState eCDCRequestState;
    //CDC 中断状态
    volatile tCDCState eCDCInterruptState;
    //请求更新标志
    volatile unsigned char ucPendingRequest;
    //暂时结束
    unsigned short usBreakDuration;
    //控制
    unsigned short usControlLineState;
    //UART 状态
    unsigned short usSerialState;
    //标志位
    volatile unsigned short usDeferredOpFlags;
    //最后一次发送数据大小
    unsigned short usLastTxSize;
    //UART 控制参数
    tLineCoding sLineCoding;
    //接收数据
    volatile tBoolean bRxBlocked;
    //控制数据
    volatile tBoolean bControlBlocked;
    //连接是否成功
    volatile tBoolean bConnected;
    //控制端点
```

第 8 章 CDC 设备

```c
    unsigned char ucControlEndpoint;
    //Bulk IN 端点
    unsigned char ucBulkINEndpoint;
    //Bulk Out 端点
    unsigned char ucBulkOUTEndpoint;
    //接口控制
    unsigned char ucInterfaceControl;
    //接口数据
    unsigned char ucInterfaceData;
}
tCDCInstance;
```

tCDCInstance, CDC 设备类实例。定义了 CDC 设备类的 USB 基地址、设备信息、IN 端点和 OUT 端点等信息。

```c
typedef struct
{
    //VID
    unsigned short usVID;
    //PID
    unsigned short usPID;
    //最大耗电量
    unsigned short usMaxPowermA;
    //电源属性
    unsigned char ucPwrAttributes;
    //控制回调函数
    tUSBCallback pfnControlCallback;
    //控制回调函数的第一个参数
    void * pvControlCBData;
    //接收回调函数
    tUSBCallback pfnRxCallback;
    //接收回调函数的第一个参数
    void * pvRxCBData;
    //发送回调函数
    tUSBCallback pfnTxCallback;
    //发送回调函数的第一个参数
    void * pvTxCBData;
    //字符串描述符集合
    const unsigned char * const * ppStringDescriptors;
    //字符串描述符个数
    unsigned long ulNumStringDescriptors;
    //CDC 类实例
    tCDCSerInstance * psPrivateCDCSerData;
```

}
tUSBDCDCDevice;

tUSBDCDCDevice,CDC 设备类,定义了 VID、PID、电源属性和字符串描述符等,还包括了一个 CDC 设备类实例。其他设备描述符、配置信息通过 API 函数输入 tCDCSerInstance 定义的 CDC 设备实例中,如图 8-1 所示。

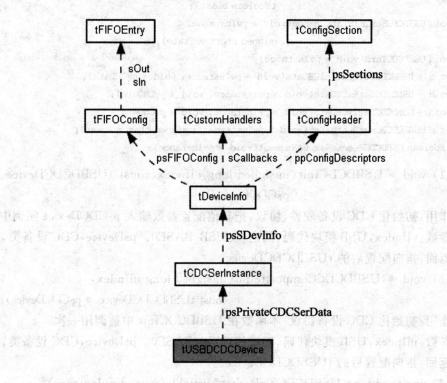

图 8-1 tUSBDCDCDevice 调用图

8.3 API 函数

在 CDC 设备类 API 库中定义了 13 个函数,用于完成 USB CDC 设备初始化、配置及数据处理、以及 11 个 Buffer 操作函数,下面为 usbdcdc.h 中定义的 API 函数。

```
void * USBDCDCInit(unsigned long ulIndex,
                   const tUSBDCDCDevice * psCDCDevice);
void * USBDCDCCompositeInit(unsigned long ulIndex,
                   const tUSBDCDCDevice * psCDCDevice);
unsigned long USBDCDCTxPacketAvailable(void * pvInstance);
unsigned long USBDCDCPacketWrite(void * pvInstance,
                   unsigned char * pcData,
```

```
                              unsigned long ulLength,
                              tBoolean bLast);
unsigned long USBDCDCRxPacketAvailable(void * pvInstance);
unsigned long USBDCDCPacketRead(void * pvInstance,
                              unsigned char * pcData,
                              unsigned long ulLength,
                              tBoolean bLast);
void USBDCDCSerialStateChange(void * pvInstance,
                              unsigned short usState);
void USBDCDCTerm(void * pvInstance);
void * USBDCDCSetControlCBData(void * pvInstance, void * pvCBData);
void * USBDCDCSetRxCBData(void * pvInstance, void * pvCBData);
void * USBDCDCSetTxCBData(void * pvInstance, void * pvCBData);
void USBDCDCPowerStatusSet(void * pvInstance, unsigned char ucPower);
tBoolean USBDCDCRemoteWakeupRequest(void * pvInstance);
```

(1) void * USBDCDCInit(unsigned long ulIndex, const tUSBDCDCDevice * psCDCDevice);

作用：初始化 CDC 设备硬件、协议，把其他配置参数输入 psCDCDevice 实例中。

参数：ulIndex，USB 模块代码，固定值：USB_BASE0。psDevice，CDC 设备类。

返回：指向配置后的 tUSBDCDCDevice。

(2) void * USBDCDCCompositeInit(unsigned long ulIndex,
                              const tUSBDCDCDevice * psCDCDevice);

作用：初始化 CDC 设备协议，本函数在 USBDCDCInit 中被调用一次。

参数：ulIndex，USB 模块代码，固定值：USB_BASE0。psDevice，CDC 设备类。

返回：指向配置后的 tUSBDCDCDevice。

(3) unsigned long USBDCDCTxPacketAvailable(void * pvInstance);

作用：获取可用发送数据的长度。

参数：pvInstance，tUSBDCDCDevice 设备指针。

返回：待发送数据包的字节数。

(4) unsigned long USBDCDCPacketWrite(void * pvInstance,
                              unsigned char * pcData,
                              unsigned long ulLength,
                              tBoolean bLast);

作用：通过 CDC 传输发送一个包数据，底层驱动，在 Buffer 中使用。

参数：pvInstance，tUSBDCDCDevice 设备指针。pcData，待写入的数据指针。ulLength，待写入数据的长度。bLast，是否传输结束包。

返回：成功发送长度，可能与 ulLength 长度不一样。

(5) unsigned long USBDCDCRxPacketAvailable(void * pvInstance);

作用:获取接收数据的长度。
参数:pvInstance,tUSBDCDCDevice 设备指针。
返回:可用接收数据个数,可读取的有效数据。

(6) unsigned long USBDCDCPacketRead(void * pvInstance,
                                    unsigned char * pcData,
                                    unsigned long ullLength,
                                    tBoolean bLast);

作用:通过 CDC 传输接收一个包数据,底层驱动,在 Buffer 中使用。
参数:pvInstance,tUSBDCDCDevice 设备指针。pcData,读出数据指针。ulLength,读出数据的长度。bLast,是否是结束包。
返回:成功接收长度,可能与 ullLength 长度不一样。

(7) void USBDCDCSerialStateChange(void * pvInstance,
                                   unsigned short usState);

作用:UART 收到数据后,调用此函数进行数据处理。
参数:pvInstance,tUSBDCDCDevice 设备指针。usState,UART 状态。
返回:无。

(8) void USBDCDCTerm(void * pvInstance);
作用:结束 CDC 设备。
参数:pvInstance,指向 tUSBDCDCDevice。
返回:无。

(9) void * USBDCDCSetControlCBData(void * pvInstance, void * pvCBData);
作用:改变控制回调函数的第一个参数。
参数:pvInstance,指向 tUSBDCDCDevice。pvCBData,用于替换的参数。
返回:旧参数指针。

(10) void * USBDCDCSetRxCBData(void * pvInstance, void * pvCBData);
作用:改变接收回调函数的第一个参数。
参数:pvInstance,指向 tUSBDCDCDevice。pvCBData,用于替换的参数。
返回:旧参数指针。

(11) void * USBDCDCSetTxCBData(void * pvInstance, void * pvCBData);
作用:改变发送回调函数的第一个参数。
参数:pvInstance,指向 tUSBDCDCDevice。pvCBData,用于替换的参数。
返回:旧参数指针。

(12) void USBDCDCPowerStatusSet(void * pvInstance, unsigned char
                                 ucPower);

作用:修改电源属性、状态。
参数:pvInstance,指向 tUSBDCDCDevice。ucPower,电源属性。

返回:无。

(13) tBoolean USBDCDCRemoteWakeupRequest(void * pvInstance);
作用:唤醒请求。
参数:pvInstance,指向 tUSBDCDCDevice。
返回:无。

在这些函数中 USBDCDCInit、USBDCDCPacketWrite、USBDCDCPacketRead、USBDCDCTxPacketAvailable、USBDCDCRxPacketAvailable 函数最重要并且使用最多,第一次使用 CDC 设备时,USBDCDCInit 用于初始化 CDC 设备的配置与控制。USBDCDCPacketRead、USBDCDCPacketWrite、USBDCDCTxPacketAvailable、USBDCDCRxPacketAvailable 为 CDC 传输数据的底层驱动函数用于驱动 Buffer。

在 CDC 类中也会使用到 11 个 Buffer 处理函数,用于 Buffer 数据的接收、发送及处理。在 CDC 传输中大量使用,使用方法在第 7 章已讲解。

8.4 CDC 设备开发

CDC 设备开发只需要 4 步就能完成。图 8-2 为 CDC 类开发流程图,其主要流程为 CDC 设备配置(主要是字符串描述符)、Callback 函数编写、USB 处理器初始化、数据处理。

下面以"USB 转 UART"实例说明使用 USB 库开发 USB CDC 类的过程。

(1) CDC 设备配置(主要是字符串描述符),按字符串描述符标准完成串描述符配置,进而完成 CDC 设备配置。

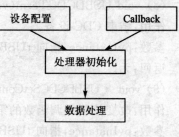

图 8-2 CDC 类开发流程图

```
# include "inc/hw_ints.h"
# include "inc/hw_memmap.h"
# include "inc/hw_types.h"
# include "inc/hw_uart.h"
# include "inc/hw_gpio.h"
# include "driverlib/debug.h"
# include "driverlib/gpio.h"
# include "driverlib/interrupt.h"
# include "driverlib/sysctl.h"
# include "driverlib/systick.h"
# include "driverlib/timer.h"
# include "driverlib/uart.h"
# include "driverlib/usb.h"
```

```c
#include "usblib/usblib.h"
#include "usblib/usbcdc.h"
#include "usblib/usb-ids.h"
#include "usblib/device/usbdevice.h"
#include "usblib/device/usbdcdc.h"

#define UART_BUFFER_SIZE 256
volatile unsigned long g_ulUARTTxCount = 0;
volatile unsigned long g_ulUARTRxCount = 0;

// UART 设置
#define USB_UART_BASE           UART0_BASE
#define USB_UART_PERIPH         SYSCTL_PERIPH_UART0
#define USB_UART_INT            INT_UART0
#define TX_GPIO_BASE            GPIO_PORTA_BASE
#define TX_GPIO_PERIPH          SYSCTL_PERIPH_GPIOA
#define TX_GPIO_PIN             GPIO_PIN_1
#define RX_GPIO_BASE            GPIO_PORTA_BASE
#define RX_GPIO_PERIPH          SYSCTL_PERIPH_GPIOA
#define RX_GPIO_PIN             GPIO_PIN_0
#define DEFAULT_BIT_RATE        115200
#define DEFAULT_UART_CONFIG     (UART_CONFIG_WLEN_8 | UART_CONFIG_PAR_NONE | \
                                 UART_CONFIG_STOP_ONE)
// 发送中断标志
static tBoolean g_bSendingBreak = false;
//系统时钟
volatile unsigned long g_ulSysTickCount = 0;
// g_ulFlags 使用的标志位
#define COMMAND_PACKET_RECEIVED  0x00000001
#define COMMAND_STATUS_UPDATE    0x00000002
//全局标志
volatile unsigned long g_ulFlags = 0;
//状态
char * g_pcStatus;
//全局 USB 配置标志
static volatile tBoolean g_bUSBConfigured = false;
unsigned long RxHandler(void * pvCBData, unsigned long ulEvent,
                        unsigned long ulMsgValue, void * pvMsgData);
unsigned long TxHandler(void * pvCBData, unsigned long ulEvent,
                        unsigned long ulMsgValue, void * pvMsgData);
unsigned long ControlHandler(void * pvCBData, unsigned long ulEvent,
                             unsigned long ulMsgValue, void * pvMsgData);
```

```c
void USBUARTPrimeTransmit(unsigned long ulBase);
void CheckForSerialStateChange(const tUSBDCDCDevice * psDevice, long lErrors);
void SetControlLineState(unsigned short usState);
tBoolean SetLineCoding(tLineCoding * psLineCoding);
void GetLineCoding(tLineCoding * psLineCoding);
void SendBreak(tBoolean bSend);

const tUSBBuffer g_sTxBuffer;
const tUSBBuffer g_sRxBuffer;
const tUSBDCDCDevice g_sCDCDevice;
unsigned char g_pucUSBTxBuffer[];
unsigned char g_pucUSBRxBuffer[];

//**************************************************************
// 设备语言描述符
//**************************************************************
const unsigned char g_pLangDescriptor[] =
{
    4,
    USB_DTYPE_STRING,
    USBShort(USB_LANG_EN_US)
};
//**************************************************************
// 制造商 字符串 描述符
//**************************************************************
const unsigned char g_pManufacturerString[] =
{
    (17 + 1) * 2,
    USB_DTYPE_STRING,
    'T', 0, 'e', 0, 'x', 0, 'a', 0, 's', 0, ' ', 0, 'I', 0, 'n', 0, 's', 0,
    't', 0, 'r', 0, 'u', 0, 'm', 0, 'e', 0, 'n', 0, 't', 0, 's', 0,
};
//**************************************************************
//产品 字符串 描述符
//**************************************************************
const unsigned char g_pProductString[] =
{
    2 + (16 * 2),
    USB_DTYPE_STRING,
    'V', 0, 'i', 0, 'r', 0, 't', 0, 'u', 0, 'a', 0, 'l', 0, ' ', 0,
    'C', 0, 'O', 0, 'M', 0, ' ', 0, 'P', 0, 'o', 0, 'r', 0, 't', 0
```

```c
};
//***************************************************************
//产品 序列号 描述符
//***************************************************************
const unsigned char g_pSerialNumberString[] =
{
    2 + (8 * 2),
    USB_DTYPE_STRING,
    '1', 0, '2', 0, '3', 0, '4', 0, '5', 0, '6', 0, '7', 0, '9', 0
};
//***************************************************************
// 设备接口字符串描述符
//***************************************************************
const unsigned char g_pControlInterfaceString[] =
{
    2 + (21 * 2),
    USB_DTYPE_STRING,
    'A', 0, 'C', 0, 'M', 0, ' ', 0, 'C', 0, 'o', 0, 'n', 0, 't', 0,
    'r', 0, 'o', 0, 'l', 0, ' ', 0, 'I', 0, 'n', 0, 't', 0, 'e', 0,
    'r', 0, 'f', 0, 'a', 0, 'c', 0, 'e', 0
};
//***************************************************************
// 设备配置字符串描述符
//***************************************************************
const unsigned char g_pConfigString[] =
{
    2 + (26 * 2),
    USB_DTYPE_STRING,
    'S', 0, 'e', 0, 'l', 0, 'f', 0, ' ', 0, 'P', 0, 'o', 0, 'w', 0,
    'e', 0, 'r', 0, 'e', 0, 'd', 0, ' ', 0, 'C', 0, 'o', 0, 'n', 0,
    'f', 0, 'i', 0, 'g', 0, 'u', 0, 'r', 0, 'a', 0, 't', 0, 'i', 0,
    'o', 0, 'n', 0
};
//***************************************************************
// 字符串描述符集合
//***************************************************************
const unsigned char * const g_pStringDescriptors[] =
{
    g_pLangDescriptor,
    g_pManufacturerString,
    g_pProductString,
    g_pSerialNumberString,
```

```c
    g_pControlInterfaceString,
    g_pConfigString
};
#define NUM_STRING_DESCRIPTORS (sizeof(g_pStringDescriptors) /    \
                                sizeof(unsigned char *))
//*************************************************************
// 定义 CDC 设备实例
//*************************************************************
tCDCSerInstance g_sCDCInstance;
//*************************************************************
// 定义 CDC 设备
//*************************************************************
const tUSBDCDCDevice g_sCDCDevice =
{
    0x1234,
    USB_PID_SERIAL,
    0,
    USB_CONF_ATTR_SELF_PWR,
    ControlHandler,
    (void *)&g_sCDCDevice,
    USBBufferEventCallback,
    (void *)&g_sRxBuffer,
    USBBufferEventCallback,
    (void *)&g_sTxBuffer,
    g_pStringDescriptors,
    NUM_STRING_DESCRIPTORS,
    &g_sCDCInstance
};
//*************************************************************
// 定义 Buffer
//*************************************************************
unsigned char g_pcUSBRxBuffer[UART_BUFFER_SIZE];
unsigned char g_pucRxBufferWorkspace[USB_BUFFER_WORKSPACE_SIZE];
unsigned char g_pcUSBTxBuffer[UART_BUFFER_SIZE];
unsigned char g_pucTxBufferWorkspace[USB_BUFFER_WORKSPACE_SIZE];
const tUSBBuffer g_sRxBuffer =
{
    false,                          // This is a receive buffer
    RxHandler,                      // pfnCallback
    (void *)&g_sCDCDevice,          // Callback data is our device pointer
    USBDCDCPacketRead,              // pfnTransfer
    USBDCDCRxPacketAvailable,       // pfnAvailable
```

```
    (void *)&g_sCDCDevice,          // pvHandle
    g_pcUSBRxBuffer,                // pcBuffer
    UART_BUFFER_SIZE,               // ulBufferSize
    g_pucRxBufferWorkspace          // pvWorkspace
};
const tUSBBuffer g_sTxBuffer =
{
    true,                           // This is a transmit buffer
    TxHandler,                      // pfnCallback
    (void *)&g_sCDCDevice,          // Callback data is our device pointer
    USBDCDCPacketWrite,             // pfnTransfer
    USBDCDCTxPacketAvailable,       // pfnAvailable
    (void *)&g_sCDCDevice,          // pvHandle
    g_pcUSBTxBuffer,                // pcBuffer
    UART_BUFFER_SIZE,               // ulBufferSize
    g_pucTxBufferWorkspace          // pvWorkspace
};
```

(2) 完成 Callback 函数。Callback 函数用于处理输出端点、输入端点数据事务。CDC 设备接收回调函数包含以下事务：USB_EVENT_RX_AVAILABLE、USB_EVENT_DATA_REMAINING、USB_EVENT_REQUEST_BUFFER；CDC 设备发送回调函数包含以下事务：USB_EVENT_TX_COMPLETE；CDC 设备控制回调函数包含了以下事务：USB_EVENT_CONNECTED、USB_EVENT_DISCONNECTED、USBD_CDC_EVENT_GET_LINE_CODING、USBD_CDC_EVENT_SET_LINE_CODING、USBD_CDC_EVENT_SET_CONTROL_LINE_STATE、USBD_CDC_EVENT_SEND_BREAK、USBD_CDC_EVENT_CLEAR_BREAK、USB_EVENT_SUSPEND、USB_EVENT_RESUME。如表 8-1 所列，为 CDC 事务，负责检测 USB 通信事件，并发送给 Callback。

表 8-1 CDC 事务

名 称	属 性	说 明
USB_EVENT_RX_AVAILABLE	接收	有数据可接收
USB_EVENT_DATA_REMAINING	接收	剩余数据
USB_EVENT_REQUEST_BUFFER	接收	请求 Buffer
USB_EVENT_TX_COMPLETE	发送	发送完成
USB_EVENT_RESUME	控制	唤醒
USB_EVENT_SUSPEND	控制	挂起
USBD_CDC_EVENT_CLEAR_BREAK	控制	清除 Break 信号

续表 8-1

名 称	属 性	说 明
USBD_CDC_EVENT_SEND_BREAK	控制	发送 Break 信号
USBD_CDC_EVENT_SET_CONTROL_LINE_STATE	控制	控制信号
USBD_CDC_EVENT_SET_LINE_CODING	控制	配置 UART 通信参数
USBD_CDC_EVENT_GET_LINE_CODING	控制	获取 UART 通信参数
USB_EVENT_DISCONNECTED	控制	断开
USB_EVENT_CONNECTED	控制	连接

根据以上事务编写 Callback 函数，如下所示。

```
//***************************************************************
//CDC 设备类控制回调函数
//***************************************************************
unsigned long  ControlHandler(void * pvCBData, unsigned long ulEvent,
                            unsigned long ulMsgValue, void * pvMsgData)
{
    unsigned long ulIntsOff;
    // 判断处理事务
    switch(ulEvent)
    {
        //连接成功
        case USB_EVENT_CONNECTED:
            g_bUSBConfigured = true;
            //清空 Buffer
            USBBufferFlush(&g_sTxBuffer);
            USBBufferFlush(&g_sRxBuffer);
            //更新状态
            ulIntsOff = IntMasterDisable();
            g_pcStatus = "Host connected.";
            g_ulFlags |= COMMAND_STATUS_UPDATE;
            if(!ulIntsOff)
            {
                IntMasterEnable();
            }
            break;
        //断开连接
        case USB_EVENT_DISCONNECTED:
            g_bUSBConfigured = false;
            ulIntsOff = IntMasterDisable();
            g_pcStatus = "Host disconnected.";
```

```c
            g_ulFlags |= COMMAND_STATUS_UPDATE;
            if(!ulIntsOff)
            {
                IntMasterEnable();
            }
            break;
        // 获取 UART 通信参数
        case USBD_CDC_EVENT_GET_LINE_CODING:
            GetLineCoding(pvMsgData);
            break;
        //设置 UART 通信参数
        case USBD_CDC_EVENT_SET_LINE_CODING:
            SetLineCoding(pvMsgData);
            break;
        // 设置 RS232 RTS 和 DTR
        case USBD_CDC_EVENT_SET_CONTROL_LINE_STATE:
            SetControlLineState((unsigned short)ulMsgValue);
            break;
        // 发送 Break 信号
        case USBD_CDC_EVENT_SEND_BREAK:
            SendBreak(true);
            break;
        // 清除 Break 信号
        case USBD_CDC_EVENT_CLEAR_BREAK:
            SendBreak(false);
            break;
        // 挂起与唤醒事务
        case USB_EVENT_SUSPEND:
        case USB_EVENT_RESUME:
            break;
        default:
            break;
    }
    return(0);
}


//*************************************************************
//CDC 设备类发送回调函数
//*************************************************************
unsigned long TxHandler(void * pvCBData, unsigned long ulEvent, unsigned long ulMsgValue,void * pvMsgData)
```

```c
    {
        switch(ulEvent)
        {
            //发送结束,在此不用处理数据
            case USB_EVENT_TX_COMPLETE:
                break;
            default:
                break;
        }
        return(0);
    }
//***************************************************************
//CDC设备类发送回调函数
//***************************************************************
unsigned long TxHandler(void * pvCBData, unsigned long ulEvent, unsigned long ulMsgValue,void * pvMsgData)
{
    switch(ulEvent)
    {
        //发送结束,在此不用处理数据
        case USB_EVENT_TX_COMPLETE:
            break;
        default:
            break;
    }
    return(0);
}
//***************************************************************
//CDC设备类接收回调函数
//***************************************************************
unsigned long RxHandler(void * pvCBData, unsigned long ulEvent, unsigned long ulMsgValue,void * pvMsgData)
{
    unsigned long ulCount;
    //判断事务类型
    switch(ulEvent)
    {
        //接收数据
        case USB_EVENT_RX_AVAILABLE:
        {
            //UART接收数据并能过USB发给主机
            USBUARTPrimeTransmit(USB_UART_BASE);
            UARTIntEnable(USB_UART_BASE, UART_INT_TX);
```

```c
            break;
        }
        // 检查剩余数据
        case USB_EVENT_DATA_REMAINING:
        {
            ulCount = UARTBusy(USB_UART_BASE) ? 1 : 0;
            return(ulCount);
        }
        //请求 Buffer
        case USB_EVENT_REQUEST_BUFFER:
        {
            return(0);
        }
        default:
            break;
    }

    return(0);
}
//**************************************************************
// 设置 RS232 RTS 和 DTR
//**************************************************************
void SetControlLineState(unsigned short usState)
{
    // 根据 MCU 引脚自行添加
}
//**************************************************************
// 设置 UART 通信参数
//**************************************************************
tBoolean SetLineCoding(tLineCoding * psLineCoding)
{
    unsigned long ulConfig;
    tBoolean bRetcode;
    bRetcode = true;
    //数据长度
    switch(psLineCoding->ucDatabits)
    {
        case 5:
        {
            ulConfig = UART_CONFIG_WLEN_5;
            break;
        }
```

```c
        case 6:
        {
            ulConfig = UART_CONFIG_WLEN_6;
            break;
        }
        case 7:
        {
            ulConfig = UART_CONFIG_WLEN_7;
            break;
        }
        case 8:
        {
            ulConfig = UART_CONFIG_WLEN_8;
            break;
        }
        default:
        {
            ulConfig = UART_CONFIG_WLEN_8;
            bRetcode = false;
            break;
        }
    }
    // 校验位
    switch(psLineCoding->ucParity)
    {
        case USB_CDC_PARITY_NONE:
        {
            ulConfig |= UART_CONFIG_PAR_NONE;
            break;
        }
        case USB_CDC_PARITY_ODD:
        {
            ulConfig |= UART_CONFIG_PAR_ODD;
            break;
        }
        case USB_CDC_PARITY_EVEN:
        {
            ulConfig |= UART_CONFIG_PAR_EVEN;
            break;
        }
        case USB_CDC_PARITY_MARK:
        {
```

```
            ulConfig |= UART_CONFIG_PAR_ONE;
            break;
        }
        case USB_CDC_PARITY_SPACE:
        {
            ulConfig |= UART_CONFIG_PAR_ZERO;
            break;
        }
        default:
        {
            ulConfig |= UART_CONFIG_PAR_NONE;
            bRetcode = false;
            break;
        }
    }
    //停止位
    switch(psLineCoding->ucStop)
    {
        case USB_CDC_STOP_BITS_1:
        {
            ulConfig |= UART_CONFIG_STOP_ONE;
            break;
        }
        case USB_CDC_STOP_BITS_2:
        {
            ulConfig |= UART_CONFIG_STOP_TWO;
            break;
        }
        default:
        {
            ulConfig = UART_CONFIG_STOP_ONE;
            bRetcode |= false;
            break;
        }
    }
    UARTConfigSetExpClk(USB_UART_BASE, SysCtlClockGet(), psLineCoding->ulRate,
                        ulConfig);
    return(bRetcode);
}

//****************************************************************
// 获取 UART 通信参数
```

```c
//****************************************************************
void GetLineCoding(tLineCoding * psLineCoding)
{
    unsigned long ulConfig;
    unsigned long ulRate;
    UARTConfigGetExpClk(USB_UART_BASE, SysCtlClockGet(), &ulRate,
                        &ulConfig);
    psLineCoding->ulRate = ulRate;
    //发送数据长度
    switch(ulConfig & UART_CONFIG_WLEN_MASK)
    {
        case UART_CONFIG_WLEN_8:
        {
            psLineCoding->ucDatabits = 8;
            break;
        }
        case UART_CONFIG_WLEN_7:
        {
            psLineCoding->ucDatabits = 7;
            break;
        }
        case UART_CONFIG_WLEN_6:
        {
            psLineCoding->ucDatabits = 6;
            break;
        }
        case UART_CONFIG_WLEN_5:
        {
            psLineCoding->ucDatabits = 5;
            break;
        }
    }
    //校验位
    switch(ulConfig & UART_CONFIG_PAR_MASK)
    {
        case UART_CONFIG_PAR_NONE:
        {
            psLineCoding->ucParity = USB_CDC_PARITY_NONE;
            break;
        }
        case UART_CONFIG_PAR_ODD:
        {
```

```c
            psLineCoding->ucParity = USB_CDC_PARITY_ODD;
            break;
        }
        case UART_CONFIG_PAR_EVEN:
        {
            psLineCoding->ucParity = USB_CDC_PARITY_EVEN;
            break;
        }
        case UART_CONFIG_PAR_ONE:
        {
            psLineCoding->ucParity = USB_CDC_PARITY_MARK;
            break;
        }
        case UART_CONFIG_PAR_ZERO:
        {
            psLineCoding->ucParity = USB_CDC_PARITY_SPACE;
            break;
        }
    }
    //停止位
    switch(ulConfig & UART_CONFIG_STOP_MASK)
    {
        case UART_CONFIG_STOP_ONE:
        {
            psLineCoding->ucStop = USB_CDC_STOP_BITS_1;
            break;
        }
        case UART_CONFIG_STOP_TWO:
        {
            psLineCoding->ucStop = USB_CDC_STOP_BITS_2;
            break;
        }
    }
}
// ***************************************************************
// UART 发送 Break 信号
// ***************************************************************
void SendBreak(tBoolean bSend)
{
    if(! bSend)
    {
        UARTBreakCtl(USB_UART_BASE, false);
```

```
            g_bSendingBreak = false;
        }
        else
        {
            UARTBreakCtl(USB_UART_BASE, true);
            g_bSendingBreak = true;
        }
    }
```

(3) 系统初始化,配置内核电压、系统主频、使能端口、LED 控制等,本例中使用 4 个 LED 进行指示数据传输。在这个例子中,CDC 传输接收的数据发送给主机。LED 与按键电路图如图 5-6 所示。

系统初始化程序如下所示。

```
    unsigned long ulTxCount;
    unsigned long ulRxCount;
    //char pcBuffer[16];
    //设置内核电压、主频 50 MHz
    SysCtlLDOSet(SYSCTL_LDO_2_75V);
    SysCtlClockSet(SYSCTL_XTAL_8MHZ | SYSCTL_SYSDIV_4 | SYSCTL_USE_PLL | SYSCTL_OSC_
                MAIN );
    SysCtlPeripheralEnable(SYSCTL_PERIPH_GPIOF);
    GPIOPinTypeGPIOOutput(GPIO_PORTF_BASE,0xf0);
    GPIOPinTypeGPIOInput(GPIO_PORTF_BASE,0x0f);
    HWREG(GPIO_PORTF_BASE + GPIO_O_PUR) |= 0x0f;

    g_bUSBConfigured = false;
    //UART 配置
    SysCtlPeripheralEnable(USB_UART_PERIPH);
    SysCtlPeripheralEnable(TX_GPIO_PERIPH);
    SysCtlPeripheralEnable(RX_GPIO_PERIPH);
    GPIOPinTypeUART(TX_GPIO_BASE, TX_GPIO_PIN);
    GPIOPinTypeUART(RX_GPIO_BASE, RX_GPIO_PIN);
    UARTConfigSetExpClk(USB_UART_BASE, SysCtlClockGet(), DEFAULT_BIT_RATE,
                        DEFAULT_UART_CONFIG);
    UARTFIFOLevelSet(USB_UART_BASE, UART_FIFO_TX4_8, UART_FIFO_RX4_8);
    //配置和使能 UART 中断
    UARTIntClear(USB_UART_BASE, UARTIntStatus(USB_UART_BASE, false));
    UARTIntEnable(USB_UART_BASE, (UART_INT_OE | UART_INT_BE | UART_INT_PE |
                UART_INT_FE | UART_INT_RT | UART_INT_TX | UART_INT_RX));
```

```
//初始化发送与接收 Buffer
USBBufferInit((tUSBBuffer *)&g_sTxBuffer);
USBBufferInit((tUSBBuffer *)&g_sRxBuffer);
//初始化 CDC 设备
USBDCDCInit(0, (tUSBDCDCDevice *)&g_sCDCDevice);
ulRxCount = 0;
ulTxCount = 0;
IntEnable(USB_UART_INT);
```

(4) 数据处理。主要使用 11 个 Buffer 处理函数,用于 Buffer 数据的接收、发送及处理。在 CDC 传输中大量使用。

```
//*************************************************************
//CheckForSerialStateChange 在接收到串行数据后,处理标志
//*************************************************************
void CheckForSerialStateChange(const tUSBDCDCDevice * psDevice, long lErrors)
{
    unsigned short usSerialState;
    // 设置 TXCARRIER (DSR)和 RXCARRIER (DCD)位
    usSerialState = USB_CDC_SERIAL_STATE_TXCARRIER |
                    USB_CDC_SERIAL_STATE_RXCARRIER;
    // 判断是什么标志
    if(lErrors)
    {
        if(lErrors & UART_DR_OE)
        {
            usSerialState |= USB_CDC_SERIAL_STATE_OVERRUN;
        }
        if(lErrors & UART_DR_PE)
        {
            usSerialState |= USB_CDC_SERIAL_STATE_PARITY;
        }
        if(lErrors & UART_DR_FE)
        {
            usSerialState |= USB_CDC_SERIAL_STATE_FRAMING;
        }
        if(lErrors & UART_DR_BE)
        {
            usSerialState |= USB_CDC_SERIAL_STATE_BREAK;
        }
    }
    // 改变状态
    USBDCDCSerialStateChange((void *)psDevice, usSerialState);
}
```

第8章 CDC 设备

```c
}
//*****************************************************************
//从 UART 中读取数据,并放在 CDC 设备 Buffer 中发送给 USB 主机
//*****************************************************************
long ReadUARTData(void)
{
    long lChar, lErrors;
    unsigned char ucChar;
    unsigned long ulSpace;
    lErrors = 0;
    //检查有多少可用空间
    ulSpace = USBBufferSpaceAvailable((tUSBBuffer *)&g_sTxBuffer);
    //从 UART 中读取数据并写入到 CDC 设备类的 Buffer 中发送
    while(ulSpace && UARTCharsAvail(USB_UART_BASE))
    {
        //读一个字节
        lChar = UARTCharGetNonBlocking(USB_UART_BASE);
        //是不是控制或错误标志
        if(!(lChar & ~0xFF))
        {
            ucChar = (unsigned char)(lChar & 0xFF);
            USBBufferWrite((tUSBBuffer *)&g_sTxBuffer,
                           (unsigned char *)&ucChar, 1);
            ulSpace--;
        }
        else
        {
            lErrors |= lChar;
        }
        g_ulUARTRxCount++;
    }
    return(lErrors);
}
//*****************************************************************
// 从 Buffer 中读取数据,通过 UART 发送
//*****************************************************************
void USBUARTPrimeTransmit(unsigned long ulBase)
{
    unsigned long ulRead;
    unsigned char ucChar;
    if(g_bSendingBreak)
```

```
    {
        return;
    }
    //检查 UART 中的可用空间
    while(UARTSpaceAvail(ulBase))
    {
        //从 Buffer 中读取一个字节
        ulRead = USBBufferRead((tUSBBuffer *)&g_sRxBuffer, &ucChar, 1);
        if(ulRead)
        {
            //放在 UART TXFIFO 中发送
            UARTCharPutNonBlocking(ulBase, ucChar);
            g_ulUARTTxCount ++ ;
        }
        else
        {
            return;
        }
    }
}

//**************************************************************
// UART 中断处理函数
//**************************************************************
void USBUARTIntHandler(void)
{
    unsigned long ulInts;
    long lErrors;
    //获取中断标志并清除
    ulInts = UARTIntStatus(USB_UART_BASE, true);
    UARTIntClear(USB_UART_BASE, ulInts);
    //发送中断
    if(ulInts & UART_INT_TX)
    {
        //从 USB 中获取数据并通过 UART 发送
        USBUARTPrimeTransmit(USB_UART_BASE);
        // If the output buffer is empty, turn off the transmit interrupt
        if(! USBBufferDataAvailable(&g_sRxBuffer))
        {
            UARTIntDisable(USB_UART_BASE, UART_INT_TX);
        }
    }
```

```c
        // 接收中断
        if(ulInts & (UART_INT_RX | UART_INT_RT))
        {
            //从 UART 中读取数据并通过 Buffer 发送给 USB 主机
            lErrors = ReadUARTData();
            //检查是否有控制信号或者错误信号
            CheckForSerialStateChange(&g_sCDCDevice, lErrors);
        }
    }
    while(1)
    {
        if(g_ulFlags & COMMAND_STATUS_UPDATE)
        {
            //有新数据到达,并清除数据更新标志
            IntMasterDisable();
            g_ulFlags &= ~COMMAND_STATUS_UPDATE;
            IntMasterEnable();

            GPIOPinWrite(GPIO_PORTF_BASE,0x30,0x30);
        }
        //发送完成
        if(ulTxCount != g_ulUARTTxCount)
        {
            ulTxCount = g_ulUARTTxCount;
            GPIOPinWrite(GPIO_PORTF_BASE,0x10,0x10);
            //usnprintf(pcBuffer, 16, " %d ", ulTxCount);
        }
        //接收完成
        if(ulRxCount != g_ulUARTTxCount)
        {
            ulRxCount = g_ulUARTRxCount;
            GPIOPinWrite(GPIO_PORTF_BASE,0x20,0x20);
        }
    }
```

使用上面 4 步就可完成 CDC 设备的开发。CDC 设备开发时要加入两个库函数,分别是 usblib.lib 和 DriverLib.lib,在启动代码中加入 USB0DeviceIntHandler 中断服务函数和 USBUARTIntHandler 中断服务程序。UART→RS232 的 USB 设备开发完成后,在 Windows XP 下运行效果如图 8-3 所示,Virtual COM Port 正在枚举,可以在计算机右下脚观察到图中的提示。

第 8 章 CDC 设备

在枚举过程中可以看出,在计算机右下脚可以看到 Virtual COM Port 字样,表示正在进行枚举,并手动安装驱动,如图 8-4 所示。枚举成功后,在"设备管理器"的"端口"中看到 DESCRIPTION_0 设备,如图 8-5 所示,为提示 Virtual COM Port 驱动安装图,根据计算机提供的信

图 8-3 Virtual COM Port 枚举

息,使用 TI 提供的 CDC 类驱动程序,按提示逐步安装驱动。到此为止 CDC 设备就可以正常使用了。

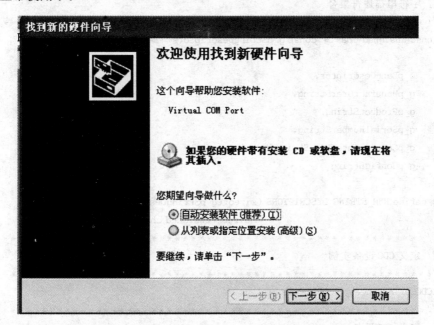

图 8-4 Virtual COM Port 驱动安装

图 8-5 枚举成功

CDC 设备需要配合上位机使用,上位机将待发送的字符串先发送到 USB→UART 的 CDC 类设备中,该设备将 USB 信号转换为 UART 的 TTL 电平信号,并通过设备的 UART 引脚发送给其他 UART 设备。如图 8-6 所示,串口调试助手通过"USB→UART 设备"的 USB 口发送一组字符串,在其 UART 端口可以接收到发送的数据。

CDC 设备 USB→UART 开发源码较多,下面只列出一部分,如下所示:

第 8 章 CDC 设备

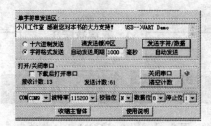
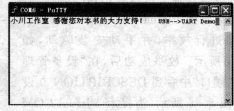

图 8-6　USB 转 UART 工作过程

```
//*************************************************************
// 字符串描述符集合
//*************************************************************
const unsigned char * const g_pStringDescriptors[] =
{
    g_pLangDescriptor,
    g_pManufacturerString,
    g_pProductString,
    g_pSerialNumberString,
    g_pControlInterfaceString,
    g_pConfigString
};
#define NUM_STRING_DESCRIPTORS (sizeof(g_pStringDescriptors) /      \
                                sizeof(unsigned char *))
//*************************************************************
// 定义 CDC 设备实例
//*************************************************************
tCDCSerInstance g_sCDCInstance;
//*************************************************************
// 定义 CDC 设备
//*************************************************************
const tUSBDCDCDevice g_sCDCDevice =
{
    0x1234,
    USB_PID_SERIAL,
    0,
    USB_CONF_ATTR_SELF_PWR,
    ControlHandler,
    (void *)&g_sCDCDevice,
    USBBufferEventCallback,
    (void *)&g_sRxBuffer,
    USBBufferEventCallback,
    (void *)&g_sTxBuffer,
```

```c
    g_pStringDescriptors,
    NUM_STRING_DESCRIPTORS,
    &g_sCDCInstance
};
//*****************************************************************
// 定义 Buffer
//*****************************************************************
unsigned char g_pcUSBRxBuffer[UART_BUFFER_SIZE];
unsigned char g_pucRxBufferWorkspace[USB_BUFFER_WORKSPACE_SIZE];
unsigned char g_pcUSBTxBuffer[UART_BUFFER_SIZE];
unsigned char g_pucTxBufferWorkspace[USB_BUFFER_WORKSPACE_SIZE];
const tUSBBuffer g_sRxBuffer =
{
    false,                            // This is a receive buffer
    RxHandler,                        // pfnCallback
    (void *)&g_sCDCDevice,            // Callback data is our device pointer
    USBDCDCPacketRead,                // pfnTransfer
    USBDCDCRxPacketAvailable,         // pfnAvailable
    (void *)&g_sCDCDevice,            // pvHandle
    g_pcUSBRxBuffer,                  // pcBuffer
    UART_BUFFER_SIZE,                 // ulBufferSize
    g_pucRxBufferWorkspace            // pvWorkspace
};
const tUSBBuffer g_sTxBuffer =
{
    true,                             // This is a transmit buffer
    TxHandler,                        // pfnCallback
    (void *)&g_sCDCDevice,            // Callback data is our device pointer
    USBDCDCPacketWrite,               // pfnTransfer
    USBDCDCTxPacketAvailable,         // pfnAvailable
    (void *)&g_sCDCDevice,            // pvHandle
    g_pcUSBTxBuffer,                  // pcBuffer
    UART_BUFFER_SIZE,                 // ulBufferSize
    g_pucTxBufferWorkspace            // pvWorkspace
};

//*****************************************************************
//CheckForSerialStateChange 在接收到串行数据后,处理标志
//*****************************************************************
void CheckForSerialStateChange(const tUSBDCDCDevice *psDevice, long lErrors)
{
```

```c
        unsigned short usSerialState;
        // 设置 TXCARRIER (DSR)和 RXCARRIER (DCD)位
        usSerialState = USB_CDC_SERIAL_STATE_TXCARRIER |
                        USB_CDC_SERIAL_STATE_RXCARRIER;
        // 判断是什么标志
        if(lErrors)
        {
            if(lErrors & UART_DR_OE)
            {
                usSerialState |= USB_CDC_SERIAL_STATE_OVERRUN;
            }
            if(lErrors & UART_DR_PE)
            {
                usSerialState |= USB_CDC_SERIAL_STATE_PARITY;
            }
            if(lErrors & UART_DR_FE)
            {
                usSerialState |= USB_CDC_SERIAL_STATE_FRAMING;
            }
            if(lErrors & UART_DR_BE)
            {
                usSerialState |= USB_CDC_SERIAL_STATE_BREAK;
            }
            // 改变状态
            USBDCDCSerialStateChange((void *)psDevice, usSerialState);
        }
}
//*************************************************************
//从 UART 中读取数据,并放在 CDC 设备 Buffer 中发送给 USB 主机
//*************************************************************
long ReadUARTData(void)
{
    long lChar, lErrors;
    unsigned char ucChar;
    unsigned long ulSpace;
    lErrors = 0;
    //检查有多少可用空间
    ulSpace = USBBufferSpaceAvailable((tUSBBuffer *)&g_sTxBuffer);
    //从 UART 中读取数据并写入到 CDC 设备类的 Buffer 中发送
    while(ulSpace && UARTCharsAvail(USB_UART_BASE))
    {
        //读一个字节
```

```c
        lChar = UARTCharGetNonBlocking(USB_UART_BASE);
        //是不是控制或错误标志
        if(!(lChar & ~0xFF))
        {
            ucChar = (unsigned char)(lChar & 0xFF);
            USBBufferWrite((tUSBBuffer *)&g_sTxBuffer,
                           (unsigned char *)&ucChar, 1);
            ulSpace--;
        }
        else
        {
            lErrors |= lChar;
        }
        g_ulUARTRxCount++;
    }
    return(lErrors);
}
//*****************************************************************
// 从 Buffer 中读取数据,通过 UART 发送
//*****************************************************************
void USBUARTPrimeTransmit(unsigned long ulBase)
{
    unsigned long ulRead;
    unsigned char ucChar;
    if(g_bSendingBreak)
    {
        return;
    }
    //检查 UART 中的可用空间
    while(UARTSpaceAvail(ulBase))
    {
        //从 Buffer 中读取一个字节
        ulRead = USBBufferRead((tUSBBuffer *)&g_sRxBuffer, &ucChar, 1);
        if(ulRead)
        {
            // 放在 UART TXFIFO 中发送
            UARTCharPutNonBlocking(ulBase, ucChar);
            g_ulUARTTxCount++;
        }
        else
        {
            return;
```

第 8 章 CDC 设备

```
        }
    }
}
//*************************************************************
// 设置 RS232 RTS 和 DTR
//*************************************************************
void SetControlLineState(unsigned short usState)
{
    // 根据 MCU 引脚自行添加
}
//*************************************************************
// 设置 UART 通信参数
//*************************************************************
tBoolean SetLineCoding(tLineCoding * psLineCoding)
{
    unsigned long ulConfig;
    tBoolean bRetcode;
    bRetcode = true;
    // 数据长度
    switch(psLineCoding->ucDatabits)
    {
        case 5:
        {
            ulConfig = UART_CONFIG_WLEN_5;
            break;
        }
        case 6:
        {
            ulConfig = UART_CONFIG_WLEN_6;
            break;
        }
        case 7:
        {
            ulConfig = UART_CONFIG_WLEN_7;
            break;
        }
        case 8:
        {
            ulConfig = UART_CONFIG_WLEN_8;
            break;
        }
        default:
```

```c
            {
                ulConfig = UART_CONFIG_WLEN_8;
                bRetcode = false;
                break;
            }
        }
        // 校验位
        switch(psLineCoding->ucParity)
        {
            case USB_CDC_PARITY_NONE:
            {
                ulConfig |= UART_CONFIG_PAR_NONE;
                break;
            }
            case USB_CDC_PARITY_ODD:
            {
                ulConfig |= UART_CONFIG_PAR_ODD;
                break;
            }
            case USB_CDC_PARITY_EVEN:
            {
                ulConfig |= UART_CONFIG_PAR_EVEN;
                break;
            }
            case USB_CDC_PARITY_MARK:
            {
                ulConfig |= UART_CONFIG_PAR_ONE;
                break;
            }
            case USB_CDC_PARITY_SPACE:
            {
                ulConfig |= UART_CONFIG_PAR_ZERO;
                break;
            }
            default:
            {
                ulConfig |= UART_CONFIG_PAR_NONE;
                bRetcode = false;
                break;
            }
        }
        //停止位
```

```c
        switch(psLineCoding->ucStop)
        {
            case USB_CDC_STOP_BITS_1:
            {
                ulConfig |= UART_CONFIG_STOP_ONE;
                break;
            }
            case USB_CDC_STOP_BITS_2:
            {
                ulConfig |= UART_CONFIG_STOP_TWO;
                break;
            }
            default:
            {
                ulConfig = UART_CONFIG_STOP_ONE;
                bRetcode |= false;
                break;
            }
        }
        UARTConfigSetExpClk(USB_UART_BASE, SysCtlClockGet(), psLineCoding->ulRate,
                      ulConfig);
        return(bRetcode);
}

//****************************************************************
// 获取 UART 通信参数
//****************************************************************
void GetLineCoding(tLineCoding * psLineCoding)
{
    unsigned long ulConfig;
    unsigned long ulRate;
    UARTConfigGetExpClk(USB_UART_BASE, SysCtlClockGet(), &ulRate,
                      &ulConfig);
    psLineCoding->ulRate = ulRate;
    //发送数据长度
    switch(ulConfig & UART_CONFIG_WLEN_MASK)
    {
        case UART_CONFIG_WLEN_8:
        {
            psLineCoding->ucDatabits = 8;
            break;
        }
```

```c
        case UART_CONFIG_WLEN_7:
        {
            psLineCoding->ucDatabits = 7;
            break;
        }
        case UART_CONFIG_WLEN_6:
        {
            psLineCoding->ucDatabits = 6;
            break;
        }
        case UART_CONFIG_WLEN_5:
        {
            psLineCoding->ucDatabits = 5;
            break;
        }
    }
    // 校验位
    switch(ulConfig & UART_CONFIG_PAR_MASK)
    {
        case UART_CONFIG_PAR_NONE:
        {
            psLineCoding->ucParity = USB_CDC_PARITY_NONE;
            break;
        }
        case UART_CONFIG_PAR_ODD:
        {
            psLineCoding->ucParity = USB_CDC_PARITY_ODD;
            break;
        }
        case UART_CONFIG_PAR_EVEN:
        {
            psLineCoding->ucParity = USB_CDC_PARITY_EVEN;
            break;
        }
        case UART_CONFIG_PAR_ONE:
        {
            psLineCoding->ucParity = USB_CDC_PARITY_MARK;
            break;
        }
        case UART_CONFIG_PAR_ZERO:
        {
            psLineCoding->ucParity = USB_CDC_PARITY_SPACE;
```

```c
            break;
        }
    }
    //停止位
    switch(ulConfig & UART_CONFIG_STOP_MASK)
    {
        case UART_CONFIG_STOP_ONE:
        {
            psLineCoding->ucStop = USB_CDC_STOP_BITS_1;
            break;
        }
        case UART_CONFIG_STOP_TWO:
        {
            psLineCoding->ucStop = USB_CDC_STOP_BITS_2;
            break;
        }
    }
}
//****************************************************************
// UART 发送 Break 信号
//****************************************************************
void SendBreak(tBoolean bSend)
{
    if(! bSend)
    {
        UARTBreakCtl(USB_UART_BASE, false);
        g_bSendingBreak = false;
    }
    else
    {
        UARTBreakCtl(USB_UART_BASE, true);
        g_bSendingBreak = true;
    }
}
//****************************************************************
//CDC 设备类控制回调函数
//****************************************************************
unsigned long   ControlHandler(void * pvCBData, unsigned long ulEvent,
                               unsigned long ulMsgValue, void * pvMsgData)
{
    unsigned long ulIntsOff;
    // 判断处理事务
```

```c
switch(ulEvent)
{
    //连接成功
    case USB_EVENT_CONNECTED:
        g_bUSBConfigured = true;
        //清空 Buffer
        USBBufferFlush(&g_sTxBuffer);
        USBBufferFlush(&g_sRxBuffer);
        // 更新状态
        ulIntsOff = IntMasterDisable();
        g_pcStatus = "Host connected.";
        g_ulFlags |= COMMAND_STATUS_UPDATE;
        if(! ulIntsOff)
        {
            IntMasterEnable();
        }
        break;
    //断开连接
    case USB_EVENT_DISCONNECTED:
        g_bUSBConfigured = false;
        ulIntsOff = IntMasterDisable();
        g_pcStatus = "Host disconnected.";
        g_ulFlags |= COMMAND_STATUS_UPDATE;
        if(! ulIntsOff)
        {
            IntMasterEnable();
        }
        break;
    // 获取 UART 通信参数
    case USBD_CDC_EVENT_GET_LINE_CODING:
        GetLineCoding(pvMsgData);
        break;
    //设置 UART 通信参数
    case USBD_CDC_EVENT_SET_LINE_CODING:
        SetLineCoding(pvMsgData);
        break;
    // 设置 RS232 RTS 和 DTR
    case USBD_CDC_EVENT_SET_CONTROL_LINE_STATE:
        SetControlLineState((unsigned short)ulMsgValue);
        break;
    // 发送 Break 信号
    case USBD_CDC_EVENT_SEND_BREAK:
```

```
                SendBreak(true);
                break;
            // 清除 Break 信号
            case USBD_CDC_EVENT_CLEAR_BREAK:
                SendBreak(false);
                break;
            // 挂起与唤醒事务
            case USB_EVENT_SUSPEND:
            case USB_EVENT_RESUME:
                break;
            default:
                break;
        }
        return(0);
    }
    //****************************************************************
    //CDC 设备类发送回调函数
    //****************************************************************
    unsigned long TxHandler(void * pvCBData, unsigned long ulEvent, unsigned long ulMs-
                    gValue,void * pvMsgData)
    {
        switch(ulEvent)
        {
            //发送结束,在此不用处理数据
            case USB_EVENT_TX_COMPLETE:
                break;
            default:
                break;
        }
        return(0);
    }
    //****************************************************************
    //CDC 设备类接收回调函数
    //****************************************************************
    unsigned long RxHandler(void * pvCBData, unsigned long ulEvent, unsigned long ulMs-
                    gValue,void * pvMsgData)
    {
        unsigned long ulCount;
        //判断事务类型
        switch(ulEvent)
        {
            //接收数据
```

```c
        case USB_EVENT_RX_AVAILABLE:
        {
            //UART 接收数据并能过 USB 发给主机
            USBUARTPrimeTransmit(USB_UART_BASE);
            UARTIntEnable(USB_UART_BASE, UART_INT_TX);
            break;
        }
        // 检查剩余数据
        case USB_EVENT_DATA_REMAINING:
        {
            ulCount = UARTBusy(USB_UART_BASE) ? 1 : 0;
            return(ulCount);
        }
        //请求 Buffer
        case USB_EVENT_REQUEST_BUFFER:
        {
            return(0);
        }
        default:
            break;
    }

    return(0);
}

//**************************************************************
// UART 中断处理函数
//**************************************************************
void USBUARTIntHandler(void)
{
    unsigned long ulInts;
    long lErrors;
    //获取中断标志并清除
    ulInts = UARTIntStatus(USB_UART_BASE, true);
    UARTIntClear(USB_UART_BASE, ulInts);
    // 发送中断
    if(ulInts & UART_INT_TX)
    {
        // 从 USB 中获取数据并能过 UART 发送
        USBUARTPrimeTransmit(USB_UART_BASE);
        // If the output buffer is empty, turn off the transmit interrupt
        if(! USBBufferDataAvailable(&g_sRxBuffer))
```

```c
            {
                UARTIntDisable(USB_UART_BASE, UART_INT_TX);
            }
        }
        // 接收中断
        if(ulInts & (UART_INT_RX | UART_INT_RT))
        {
            //从 UART 中读取数据并通过 Buffer 发送给 USB 主机
            lErrors = ReadUARTData();
            //检查是否有控制信号或者错误信号
            CheckForSerialStateChange(&g_sCDCDevice, lErrors);
        }
}
//****************************************************************
// 应用主函数
//****************************************************************
int main(void)
{
    unsigned long ulTxCount;
    unsigned long ulRxCount;
    //char pcBuffer[16];
    //设置内核电压、主频 50 MHz
    SysCtlLDOSet(SYSCTL_LDO_2_75V);
    SysCtlClockSet(SYSCTL_XTAL_8MHZ | SYSCTL_SYSDIV_4 | SYSCTL_USE_PLL  | SYSCTL_OSC
                    _MAIN );
    SysCtlPeripheralEnable(SYSCTL_PERIPH_GPIOF);
    GPIOPinTypeGPIOOutput(GPIO_PORTF_BASE,0xf0);
    GPIOPinTypeGPIOInput(GPIO_PORTF_BASE,0x0f);
    HWREG(GPIO_PORTF_BASE + GPIO_O_PUR) |= 0x0f;

    g_bUSBConfigured = false;
    //UART 配置
    SysCtlPeripheralEnable(USB_UART_PERIPH);
    SysCtlPeripheralEnable(TX_GPIO_PERIPH);
    SysCtlPeripheralEnable(RX_GPIO_PERIPH);
    GPIOPinTypeUART(TX_GPIO_BASE, TX_GPIO_PIN);
    GPIOPinTypeUART(RX_GPIO_BASE, RX_GPIO_PIN);
    UARTConfigSetExpClk(USB_UART_BASE, SysCtlClockGet(), DEFAULT_BIT_RATE,
                        DEFAULT_UART_CONFIG);
    UARTFIFOLevelSet(USB_UART_BASE, UART_FIFO_TX4_8, UART_FIFO_RX4_8);
    // 配置和使能 UART 中断
    UARTIntClear(USB_UART_BASE, UARTIntStatus(USB_UART_BASE, false));
```

```c
UARTIntEnable(USB_UART_BASE, (UART_INT_OE | UART_INT_BE | UART_INT_PE |
             UART_INT_FE | UART_INT_RT | UART_INT_TX | UART_INT_RX));

// 初始化发送与接收 Buffer
USBBufferInit((tUSBBuffer *)&g_sTxBuffer);
USBBufferInit((tUSBBuffer *)&g_sRxBuffer);
// 初始化 CDC 设备
USBDCDCInit(0, (tUSBDCDCDevice *)&g_sCDCDevice);

ulRxCount = 0;
ulTxCount = 0;
IntEnable(USB_UART_INT);
while(1)
{
    if(g_ulFlags & COMMAND_STATUS_UPDATE)
    {
        //有新数据到达,并清除数据更新标志
        IntMasterDisable();
        g_ulFlags &= ~COMMAND_STATUS_UPDATE;
        IntMasterEnable();

        GPIOPinWrite(GPIO_PORTF_BASE,0x30,0x30);
    }
    // 发送完成
    if(ulTxCount != g_ulUARTTxCount)
    {
        ulTxCount = g_ulUARTTxCount;
        GPIOPinWrite(GPIO_PORTF_BASE,0x10,0x10);
        //usnprintf(pcBuffer, 16, " %d ", ulTxCount);
    }
    // 接收完成
    if(ulRxCount != g_ulUARTTxCount)
    {
        ulRxCount = g_ulUARTRxCount;
        GPIOPinWrite(GPIO_PORTF_BASE,0x20,0x20);
    }
}
}
```

8.5 小 结

本章主要通过 USB 转 RS232 实例开发,介绍了 USB CDC 类、CDC 通信原理以及开发流程。本章介绍的 USB→RS232 没有硬件控制功能,读者可以自行添加代码,实现全功能 USB→RS232 设备开发。

第 9 章 Mass Storage 设备

目前，随着便携式设备的大力发展，需要更多、容量更大的便携式存储设备，当然首选 USB 系统，在 USB 系统中，定义了一个专用的大容量存储设备类——Mass Storage Device Class，此类是专为便携式存储设备而定义的，包括 USB CD-ROM 和 U 盘等。本章主要介绍 Mass Storage 类的开发过程。

9.1 Mass Storage 设备介绍

USB 的 Mass Storage 类是 USB 大容量存储设备类（Mass Storage Device Class），专门用于大容量存储设备，比如 U 盘、移动硬盘、USB CD-ROM、读卡器等，在日常生活中经常用到。USB Mass Storage 设备开发相对简单。

9.2 MSC 数据类型

usbdmsc.h 中已经定义好 MSC 设备类中使用的所有数据类型和函数。下面介绍 MSC 设备类使用的数据类型。

```
typedef enum
{
    //设备加入
    USBDMSC_MEDIA_PRESENT,
    //设备移出
    USBDMSC_MEDIA_NOTPRESENT,
    //设备未知
    USBDMSC_MEDIA_UNKNOWN
}
tUSBDMSCMediaStatus;
```

tUSBDMSCMediaStatus，定义存储设备状态，定义在 usbdmsc.h 文件中。USB-DMSCMediaChange() 函数用于修改此状态。

第9章 Mass Storage 设备

```c
typedef struct
{
    //Open 函数指针,用户自己完成该函数的编写
    void *(* Open)(unsigned long ulDrive);
    //Close 函数指针,用户自己完成该函数的编写
    void (* Close)(void * pvDrive);
    // BlockRead 函数指针,用于读取存储设备,用户自己完成该函数的编写
    unsigned long (* BlockRead)(void * pvDrive, unsigned char * pucData,
                        unsigned long ulSector,
                        unsigned long ulNumBlocks);
    // BlockWrite 函数指针,用于写入存储设备,用户自己完成该函数的编写
    unsigned long (* BlockWrite)(void * pvDrive, unsigned char * pucData,
                        unsigned long ulSector,
                        unsigned long ulNumBlocks);
    // 读取当前扇区数
    unsigned long (* NumBlocks)(void * pvDrive);
}
tMSCDMedia;
```

tMSCDMedia,存储设备底层操作驱动。用于 MSC 设备对存储设备的操作。

```c
typedef struct
{
    unsigned long ulUSBBase;
    tDeviceInfo * psDevInfo;
    tConfigDescriptor * psConfDescriptor;
    unsigned char ucErrorCode;
    unsigned char ucSenseKey;
    unsigned short usAddSenseCode;
    void * pvMedia;
    volatile tBoolean bConnected;
    unsigned long ulFlags;
    tUSBDMSCMediaStatus eMediaStatus;
    unsigned long pulBuffer[DEVICE_BLOCK_SIZE>>2];
    unsigned long ulBytesToTransfer;
    unsigned long ulCurrentLBA;
    unsigned char ucINEndpoint;
    unsigned char ucINDMA;
    unsigned char ucOUTEndpoint;
    unsigned char ucOUTDMA;
    unsigned char ucInterface;
    unsigned char ucSCSIState;
}
tMSCInstance;
```

tMSCInstance，MSC 设备类实例。定义了 MSC 设备类的 USB 基地址、设备信息、IN 端点、OUT 端点等信息。

```
typedef struct
{
    unsigned short usVID;
    unsigned short usPID;
    unsigned char pucVendor[8];
    unsigned char pucProduct[16];
    unsigned char pucVersion[4];
    unsigned short usMaxPowermA;
    unsigned char ucPwrAttributes;
    const unsigned char * const ppStringDescriptors;
    unsigned long ulNumStringDescriptors;
    tMSCDMedia sMediaFunctions;
    tUSBCallback pfnEventCallback;
    tMSCInstance * psPrivateData;
}
tUSBDMSCDevice;
```

tUSBDMSCDevice，MSC 设备类，定义了 VID、PID、电源属性、字符串描述符等，还包括一个 MSC 设备类实例。其他设备描述符、配置信息通过 API 函数输入 tMSCInstance 定义的 MSC 设备实例中，图 9-1 为 tUSBDMSCDevice 调用图。

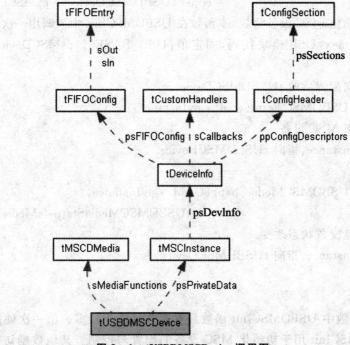

图 9-1 tUSBDMSCDevice 调用图

9.3 API 函数

在 MSC 设备类 API 库中定义了 4 个函数，完成 USB MSC 设备初始化、配置及数据处理。下面为 usbdMSC.h 中定义的 API 函数。

```
void * USBDMSCInit(unsigned long ulIndex,
                   const tUSBDMSCDevice * psMSCDevice);
void * USBDMSCCompositeInit(unsigned long ulIndex,
                   const tUSBDMSCDevice * psMSCDevice);
void USBDMSCTerm(void * pvInstance);
void USBDMSCMediaChange(void * pvInstance,
                   tUSBDMSCMediaStatus eMediaStatus);
```

(1) void * USBDMSCInit(unsigned long ulIndex,
                   const tUSBDMSCDevice * psMSCDevice);

作用：初始化 MSC 设备硬件、协议，把其他配置参数输入 psMSCDevice 实例中。

参数：ulIndex，USB 模块代码，固定值：USB_BASE0。psMSCDevice，MSC 设备类。

返回：指向配置后的 tUSBDMSCDevice。

(2) void * USBDMSCCompositeInit(unsigned long ulIndex,
                   const tUSBDMSCDevice * psMSCDevice);

作用：初始化 MSC 设备协议，本函数在 USBDMSCInit 中被调用一次。

参数：ulIndex，USB 模块代码，固定值：USB_BASE0。psMSCDevice，MSC 设备类。

返回：指向配置后的 tUSBDMSCDevice。

(3) void USBDMSCTerm(void * pvInstance);

作用：结束 MSC 设备。

参数：pvInstance，指向 tUSBDMSCDevice。

返回：无。

(4) void USBDMSCMediaChange(void * pvInstance,
                   tUSBDMSCMediaStatus eMediaStatus);

作用：存储设备状态改变。

参数：pvInstance，指向 tUSBDMSCDevice。

返回：无。

在这些函数中 USBDMSCInit 函数最重要并且使用最多。第一次使用 MSC 设备时，USBDMSCInit 用于初始化 MSC 设备的配置与控制。其他数据访问、控制处理由中断直接调用 tMSCDMedia 定义的 5 个底层驱动函数完成。

9.4 MSC 设备开发

MSC 设备开发只需要 5 步就能完成。MSC 开发流程如图 9-2 所示,包括 MSC 设备配置(主要是字符串描述符)、Callback 函数编写、存储设备底层驱动编写、USB 处理器初始化、数据处理步骤。

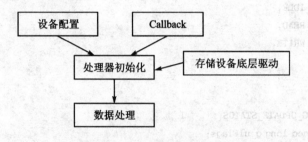

图 9-2 MSC 开发流程图

下面以"USB 转 UART"的实例说明使用 USB 库开发 USB MSC 类的过程。

(1) MSC 设备配置(主要是字符串描述符),按字符串描述符标准完成串描述符配置,进而完成 MSC 设备配置。

```
# include "inc/hw_ints.h"
# include "inc/hw_memmap.h"
# include "inc/hw_gpio.h"
# include "inc/hw_types.h"
# include "inc/hw_ints.h"
# include "driverlib/sysctl.h"
# include "driverlib/gpio.h"
# include "driverlib/interrupt.h"
# include "driverlib/rom.h"
# include "driverlib/systick.h"
# include "driverlib/usb.h"
# include "driverlib/udma.h"
# include "usblib/usblib.h"
# include "usblib/usb-ids.h"
# include "usblib/device/usbdevice.h"
# include "usblib/device/usbdmsc.h"
# include "diskio.h"
# include "usbdsdcard.h"

//声明函数原型
unsigned long USBDMSCEventCallback(void * pvCBData, unsigned long ulEvent,
```

```c
                                        unsigned long ulMsgParam,
                                        void * pvMsgData);
const tUSBDMSCDevice g_sMSCDevice;
//msc 状态
volatile enum
{
    MSC_DEV_DISCONNECTED,
    MSC_DEV_CONNECTED,
    MSC_DEV_IDLE,
    MSC_DEV_READ,
    MSC_DEV_WRITE,
}
g_eMSCState;
//全局标志
#define FLAG_UPDATE_STATUS      1
static unsigned long g_ulFlags;
//DMA
tDMAControlTable sDMAControlTable[64] __attribute__ ((aligned(1024)));

//*************************************************************
// 语言描述符
//*************************************************************
const unsigned char g_pLangDescriptor[] =
{
    4,
    USB_DTYPE_STRING,
    USBShort(USB_LANG_EN_US)
};
//*************************************************************
// 制造商 字符串 描述符
//*************************************************************
const unsigned char g_pManufacturerString[] =
{
    (17 + 1) * 2,
    USB_DTYPE_STRING,
    'T', 0, 'e', 0, 'x', 0, 'a', 0, 's', 0, ' ', 0, 'I', 0, 'n', 0, 's', 0,
    't', 0, 'r', 0, 'u', 0, 'm', 0, 'e', 0, 'n', 0, 't', 0, 's', 0,
};
//*************************************************************
//产品 字符串 描述符
//*************************************************************
const unsigned char g_pProductString[] =
```

```c
{
    (19 + 1) * 2,
    USB_DTYPE_STRING,
    'M', 0, 'a', 0, 's', 0, 's', 0, ' ', 0, 'S', 0, 't', 0, 'o', 0, 'r', 0,
    'a', 0, 'g', 0, 'e', 0, ' ', 0, 'D', 0, 'e', 0, 'v', 0, 'i', 0, 'c', 0,
    'e', 0
};
//*************************************************************
// 产品 序列号 描述符
//*************************************************************
const unsigned char g_pSerialNumberString[] =
{
    (8 + 1) * 2,
    USB_DTYPE_STRING,
    '1', 0, '2', 0, '3', 0, '4', 0, '5', 0, '6', 0, '7', 0, '8', 0
};


//*************************************************************
// 设备接口字符串描述符
//*************************************************************
const unsigned char g_pDataInterfaceString[] =
{
    (19 + 1) * 2,
    USB_DTYPE_STRING,
    'B', 0, 'u', 0, 'l', 0, 'k', 0, ' ', 0, 'D', 0, 'a', 0, 't', 0,
    'a', 0, ' ', 0, 'I', 0, 'n', 0, 't', 0, 'e', 0, 'r', 0, 'f', 0,
    'a', 0, 'c', 0, 'e', 0
};
//*************************************************************
// 设备配置字符串描述符
//*************************************************************
const unsigned char g_pConfigString[] =
{
    (23 + 1) * 2,
    USB_DTYPE_STRING,
    'B', 0, 'u', 0, 'l', 0, 'k', 0, ' ', 0, 'D', 0, 'a', 0, 't', 0,
    'a', 0, ' ', 0, 'C', 0, 'o', 0, 'n', 0, 'f', 0, 'i', 0, 'g', 0,
    'u', 0, 'r', 0, 'a', 0, 't', 0, 'i', 0, 'o', 0, 'n', 0
};
//*************************************************************
// 字符串描述符集合
//*************************************************************
```

第9章 Mass Storage 设备

```c
const unsigned char * const g_pStringDescriptors[] =
{
    g_pLangDescriptor,
    g_pManufacturerString,
    g_pProductString,
    g_pSerialNumberString,
    g_pDataInterfaceString,
    g_pConfigString
};
#define NUM_STRING_DESCRIPTORS (sizeof(g_pStringDescriptors) /    \
                                sizeof(unsigned char *))
//****************************************************************
//MSC 实例,配置并为设备信息提供空间
//****************************************************************
tMSCInstance g_sMSCInstance;
//****************************************************************
//设备配置
//****************************************************************
const tUSBDMSCDevice g_sMSCDevice =
{
    USB_VID_STELLARIS,
    USB_PID_MSC,
    "TI         ",
    "Mass Storage    ",
    "1.00",
    500,
    USB_CONF_ATTR_SELF_PWR,
    g_pStringDescriptors,
    NUM_STRING_DESCRIPTORS,
    {
        USBDMSCStorageOpen,
        USBDMSCStorageClose,
        USBDMSCStorageRead,
        USBDMSCStorageWrite,
        USBDMSCStorageNumBlocks
    },
    USBDMSCEventCallback,
    &g_sMSCInstance
};
#define MSC_BUFFER_SIZE 512
```

(2) 完成 Callback 函数。Callback 函数用于返回数据处理状态。

```
// *************************************************************
//Callback 函数
// *************************************************************
unsigned long USBDMSCEventCallback(void * pvCBData, unsigned long ulEvent,
                            unsigned long ulMsgParam, void * pvMsgData)
{
    switch(ulEvent)
    {
        // 正在写数据到存储设备
        case USBD_MSC_EVENT_WRITING:
        {
            if(g_eMSCState != MSC_DEV_WRITE)
            {
                g_eMSCState = MSC_DEV_WRITE;
                g_ulFlags |= FLAG_UPDATE_STATUS;
            }
            break;
        }
        //读取数据
        case USBD_MSC_EVENT_READING:
        {
            if(g_eMSCState != MSC_DEV_READ)
            {
                g_eMSCState = MSC_DEV_READ;
                g_ulFlags |= FLAG_UPDATE_STATUS;
            }

            break;
        }
        //空闲
        case USBD_MSC_EVENT_IDLE:
        default:
        {
            break;
        }
    }
    return(0);
}
```

第 9 章 Mass Storage 设备

(3) 完成接口函数编写，接口函数主要完成磁盘打开、扇区读写、关闭等。

```c
#define SDCARD_PRESENT          0x00000001
#define SDCARD_IN_USE           0x00000002

struct
{
    unsigned long ulFlags;
}
g_sDriveInformation;

//*************************************************************
//打开存储设备
//*************************************************************
void * USBDMSCStorageOpen(unsigned long ulDrive)
{
    unsigned char ucPower;
    unsigned long ulTemp;
    // 检查是否在使用
    if(g_sDriveInformation.ulFlags & SDCARD_IN_USE)
    {
        return(0);
    }
    // 初始化存储设备
    ulTemp = disk_initialize(0);
    if(ulTemp == RES_OK)
    {
        //打开电源
        ucPower = 1;
        disk_ioctl(0, CTRL_POWER, &ucPower);
        //设置标志
        g_sDriveInformation.ulFlags = SDCARD_PRESENT | SDCARD_IN_USE;
    }
    else if(ulTemp == STA_NODISK)
    {
        // 没有存储设备
        g_sDriveInformation.ulFlags = SDCARD_IN_USE;
    }
    else
    {
        return(0);
    }
```

```c
    return((void *)&g_sDriveInformation);
}
//*************************************************************
// 关闭存储设备
//*************************************************************
void USBDMSCStorageClose(void * pvDrive)
{
    unsigned char ucPower;
    g_sDriveInformation.ulFlags = 0;
    ucPower = 0;
    disk_ioctl(0, CTRL_POWER, &ucPower);
}
//*************************************************************
//读取扇区数据
//*************************************************************
unsigned long USBDMSCStorageRead(void * pvDrive,
                                 unsigned char * pucData,
                                 unsigned long ulSector,
                                 unsigned long ulNumBlocks)
{
    if(disk_read (0, pucData, ulSector, ulNumBlocks) == RES_OK)
    {
        return(ulNumBlocks * 512);
    }
    return(0);
}
//*************************************************************
//写数据到扇区
//*************************************************************
unsigned long USBDMSCStorageWrite(void * pvDrive,
                                  unsigned char * pucData,
                                  unsigned long ulSector,
                                  unsigned long ulNumBlocks)
{
    if(disk_write(0, pucData, ulSector, ulNumBlocks) == RES_OK)
    {
        return(ulNumBlocks * 512);
    }
    return(0);
}
//*************************************************************
// 获取当前扇区
```

```c
//*************************************************************
unsigned long USBDMSCStorageNumBlocks(void * pvDrive)
{
    unsigned long ulSectorCount;
    disk_ioctl(0, GET_SECTOR_COUNT, &ulSectorCount);
    return(ulSectorCount);
}
#define USBDMSC_IDLE              0x00000000
#define USBDMSC_NOT_PRESENT       0x00000001
//*************************************************************
// 存储设备当前状态
//*************************************************************
unsigned long USBDMSCStorageStatus(void * pvDrive);
```

(4) 系统初始化,配置内核电压、系统主频、使能端口等。

```c
//系统初始化
SysCtlLDOSet(SYSCTL_LDO_2_75V);
SysCtlClockSet(SYSCTL_XTAL_8MHZ | SYSCTL_SYSDIV_4 | SYSCTL_USE_PLL   | SYSCTL_OSC_
               MAIN );
SysCtlPeripheralEnable(SYSCTL_PERIPH_GPIOF);
GPIOPinTypeGPIOOutput(GPIO_PORTF_BASE,0xf0);
GPIOPinTypeGPIOInput(GPIO_PORTF_BASE,0x0f);
HWREG(GPIO_PORTF_BASE + GPIO_O_PUR) |= 0x0f;
// ucDMA 配置
 SysCtlPeripheralEnable(SYSCTL_PERIPH_UDMA);
 SysCtlDelay(10);
 uDMAControlBaseSet(&sDMAControlTable[0]);
 uDMAEnable();
 g_ulFlags = 0;
 g_eMSCState = MSC_DEV_IDLE;
 //msc 设备初始化
 USBDMSCInit(0, (tUSBDMSCDevice * )&g_sMSCDevice);
 //初始化存储设备
 disk_initialize(0);
```

(5) 状态处理,其他控制。

```c
    while(1)
    {
        switch(g_eMSCState)
        {
            case MSC_DEV_READ:
            {
```

第 9 章　Mass Storage 设备

```
            if(g_ulFlags & FLAG_UPDATE_STATUS)
            {
                g_ulFlags &= ~FLAG_UPDATE_STATUS;
            }
            break;
        }
        case MSC_DEV_WRITE:
        {
            if(g_ulFlags & FLAG_UPDATE_STATUS)
            {
                g_ulFlags &= ~FLAG_UPDATE_STATUS;
            }
            break;
        }
        case MSC_DEV_IDLE:
        default:
        {
            break;
        }
    }
}
```

使用上面 5 步就能完成 MSC 设备的开发。MSC 设备开发时要加入两个库函数，分别是 usblib.lib 和 DriverLib.lib，在启动代码中加入 USB0DeviceIntHandler 中断服务函数。以上 MSC 设备开发完成，在 Windows XP 下的运行效果如图 9-3 所示。

在枚举过程中，在计算机右下脚可以看到 Mass Storage Device 字样，表示电脑 USB 控制器正在枚举 MSC 设备，自动搜索可用驱动并安装，如果驱动安装失败，请手动安装 MSC 设备驱动程序。枚举成功后，在"设备管理器"的"通用串行总线控制器"中看到 USB Mass Storage Device 设备，如图 9-4 所示。现在 MSC 设备就可以正式使用了。

图 9-3　MSC 设备正在枚举

MSC 设备可以在"我的电脑"中"有可移动存储的设备"中找到，如图 9-5 所示，在"我的电脑"中有两个 U 盘，其中一个为笔者笔记本电脑上自带的 MMC，另一个为设计的 MSC 设备(U 盘)。

MSC 设备开发源码较多，下面只列出一部分，如下所示：

第9章 Mass Storage 设备

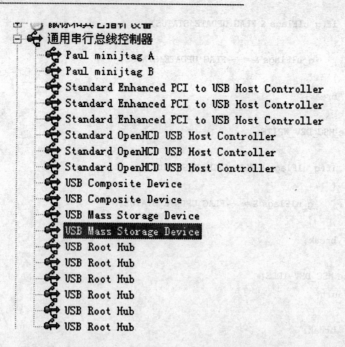

图 9-4 MSC 设备识别完成

图 9-5 识别可移动磁盘

```
# include "inc/hw_ints.h"
# include "inc/hw_memmap.h"
# include "inc/hw_gpio.h"
# include "inc/hw_types.h"
# include "inc/hw_ints.h"
# include "driverlib/sysctl.h"
# include "driverlib/gpio.h"
# include "driverlib/interrupt.h"
# include "driverlib/rom.h"
# include "driverlib/systick.h"
# include "driverlib/usb.h"
# include "driverlib/udma.h"
# include "usblib/usblib.h"
```

```c
#include "usblib/usb-ids.h"
#include "usblib/device/usbdevice.h"
#include "usblib/device/usbdmsc.h"
#include "diskio.h"
#include "usbdsdcard.h"

//声明函数原型
unsigned long USBDMSCEventCallback(void * pvCBData, unsigned long ulEvent,
                                   unsigned long ulMsgParam,
                                   void * pvMsgData);
const tUSBDMSCDevice g_sMSCDevice;
//msc 状态
volatile enum
{
    MSC_DEV_DISCONNECTED,
    MSC_DEV_CONNECTED,
    MSC_DEV_IDLE,
    MSC_DEV_READ,
    MSC_DEV_WRITE,
}
g_eMSCState;
//全局标志
#define FLAG_UPDATE_STATUS      1
static unsigned long g_ulFlags;
//DMA
tDMAControlTable sDMAControlTable[64] __attribute__ ((aligned(1024)));

// ***************************************************************
// 语言描述符
// ***************************************************************
const unsigned char g_pLangDescriptor[] =
{
    4,
    USB_DTYPE_STRING,
    USBShort(USB_LANG_EN_US)
};
// ***************************************************************
// 制造商 字符串 描述符
// ***************************************************************
const unsigned char g_pManufacturerString[] =
{
    (17 + 1) * 2,
```

```c
    USB_DTYPE_STRING,
    'T', 0, 'e', 0, 'x', 0, 'a', 0, 's', 0, ' ', 0, 'I', 0, 'n', 0, 's', 0,
    't', 0, 'r', 0, 'u', 0, 'm', 0, 'e', 0, 'n', 0, 't', 0, 's', 0,
};
//*************************************************************
//产品 字符串 描述符
//*************************************************************
const unsigned char g_pProductString[] =
{
    (19 + 1) * 2,
    USB_DTYPE_STRING,
    'M', 0, 'a', 0, 's', 0, 's', 0, ' ', 0, 'S', 0, 't', 0, 'o', 0, 'r', 0,
    'a', 0, 'g', 0, 'e', 0, ' ', 0, 'D', 0, 'e', 0, 'v', 0, 'i', 0, 'c', 0,
    'e', 0
};
//*************************************************************
// 产品 序列号 描述符
//*************************************************************
const unsigned char g_pSerialNumberString[] =
{
    (8 + 1) * 2,
    USB_DTYPE_STRING,
    '1', 0, '2', 0, '3', 0, '4', 0, '5', 0, '6', 0, '7', 0, '8', 0
};

//*************************************************************
// 设备接口字符串描述符
//*************************************************************
const unsigned char g_pDataInterfaceString[] =
{
    (19 + 1) * 2,
    USB_DTYPE_STRING,
    'B', 0, 'u', 0, 'l', 0, 'k', 0, ' ', 0, 'D', 0, 'a', 0, 't', 0,
    'a', 0, ' ', 0, 'I', 0, 'n', 0, 't', 0, 'e', 0, 'r', 0, 'f', 0,
    'a', 0, 'c', 0, 'e', 0
};
//*************************************************************
// 设备配置字符串描述符
//*************************************************************
const unsigned char g_pConfigString[] =
{
    (23 + 1) * 2,
```

```c
    USB_DTYPE_STRING,
    'B', 0, 'u', 0, 'l', 0, 'k', 0, ' ', 0, 'D', 0, 'a', 0, 't', 0,
    'a', 0, ' ', 0, 'C', 0, 'o', 0, 'n', 0, 'f', 0, 'i', 0, 'g', 0,
    'u', 0, 'r', 0, 'a', 0, 't', 0, 'i', 0, 'o', 0, 'n', 0
};
//*************************************************************
// 字符串描述符集合
//*************************************************************
const unsigned char * const g_pStringDescriptors[] =
{
    g_pLangDescriptor,
    g_pManufacturerString,
    g_pProductString,
    g_pSerialNumberString,
    g_pDataInterfaceString,
    g_pConfigString
};
#define NUM_STRING_DESCRIPTORS (sizeof(g_pStringDescriptors) /      \
                                sizeof(unsigned char *))
//*************************************************************
//MSC 实例,配置并为设备信息提供空间
//*************************************************************
tMSCInstance g_sMSCInstance;
//*************************************************************
//设备配置
//*************************************************************
const tUSBDMSCDevice g_sMSCDevice =
{
    USB_VID_STELLARIS,
    USB_PID_MSC,
    "TI              ",
    "Mass Storage    ",
    "1.00",
    500,
    USB_CONF_ATTR_SELF_PWR,
    g_pStringDescriptors,
    NUM_STRING_DESCRIPTORS,
    {
        USBDMSCStorageOpen,
        USBDMSCStorageClose,
        USBDMSCStorageRead,
        USBDMSCStorageWrite,
```

```c
                    USBDMSCStorageNumBlocks
    },
    USBDMSCEventCallback,
    &g_sMSCInstance
};
#define MSC_BUFFER_SIZE 512
//****************************************************************
//Callback 函数
//****************************************************************
unsigned long USBDMSCEventCallback(void * pvCBData, unsigned long ulEvent,
                            unsigned long ulMsgParam, void * pvMsgData)
{
    switch(ulEvent)
    {
        // 正在写数据到存储设备
        case USBD_MSC_EVENT_WRITING:
        {
            if(g_eMSCState != MSC_DEV_WRITE)
            {
                g_eMSCState = MSC_DEV_WRITE;
                g_ulFlags |= FLAG_UPDATE_STATUS;
            }
            break;
        }
        //读取数据
        case USBD_MSC_EVENT_READING:
        {
            if(g_eMSCState != MSC_DEV_READ)
            {
                g_eMSCState = MSC_DEV_READ;
                g_ulFlags |= FLAG_UPDATE_STATUS;
            }

            break;
        }
        //空闲
        case USBD_MSC_EVENT_IDLE:
        default:
        {
            break;
        }
    }
```

```c
    return(0);
}

//**************************************************************
//主函数
//**************************************************************
int main(void)
{
    //系统初始化
    SysCtlLDOSet(SYSCTL_LDO_2_75V);
    SysCtlClockSet(SYSCTL_XTAL_8MHZ | SYSCTL_SYSDIV_4 | SYSCTL_USE_PLL  | SYSCTL_OSC
                _MAIN );
    SysCtlPeripheralEnable(SYSCTL_PERIPH_GPIOF);
    GPIOPinTypeGPIOOutput(GPIO_PORTF_BASE,0xf0);
    GPIOPinTypeGPIOInput(GPIO_PORTF_BASE,0x0f);
    HWREG(GPIO_PORTF_BASE + GPIO_O_PUR) |= 0x0f;
    // ucDMA 配置
    SysCtlPeripheralEnable(SYSCTL_PERIPH_UDMA);
    SysCtlDelay(10);
    uDMAControlBaseSet(&sDMAControlTable[0]);
    uDMAEnable();
    g_ulFlags = 0;
    g_eMSCState = MSC_DEV_IDLE;
    //msc 设备初始化
    USBDMSCInit(0, (tUSBDMSCDevice * )&g_sMSCDevice);
    //初始化存储设备
    disk_initialize(0);
    while(1)
    {
        switch(g_eMSCState)
        {
            case MSC_DEV_READ:
            {
                if(g_ulFlags & FLAG_UPDATE_STATUS)
                {
                    g_ulFlags &= ~FLAG_UPDATE_STATUS;
                }
                break;
            }
            case MSC_DEV_WRITE:
            {
                if(g_ulFlags & FLAG_UPDATE_STATUS)
```

```
                {
                    g_ulFlags &= ~FLAG_UPDATE_STATUS;
                }
                break;
            }
            case MSC_DEV_IDLE:
            default:
            {
                break;
            }
        }
    }
}
```

9.5 小　结

　　本章主要介绍大容量存储设备的开发,通过本章的学习,可以了解 U 盘的开发过程及其开发难点。其他大容量存储设备开发与 U 盘开发类似,其他大容量存储设备读者可以自己尝试完成其设计。注意,由于 TI 提供的 USB 处理器现只支持低速与全速,其数据传输速率相对较低。

第 10 章

Composite 设备

第 5~9 章已经介绍了 USB 常用设备的开发,但是只能开发较单一的 USB 设备,如果想开发两个或两个以上功能的 USB 设备,怎么办呢? 本章介绍复合设备,可以实现多个基本功能的复合 USB 系统的开发,以满足更高级、更复杂的 USB 系统设计。

10.1 Composite 设备介绍

USB 的 Composite 类是 USB 复合设备类,一个 USB 设备具有多种设备功能,比如一个 USB 设备同时具有鼠标和键盘功能。单一的 USB 设备开发相对简单,但在很多时候使用的 USB 设备具有多种功能。Composite 类可以满足这种要求。

10.2 Composite 数据类型

usbdcomp.h 中已经定义好 Composite 设备类中使用的所有数据类型和函数。下面介绍 Composite 设备类使用的数据类型。

```
typedef struct
{
    const tDeviceInfo * pDeviceInfo;
    const tConfigHeader * psConfigHeader;
    unsigned char ucIfaceOffset;
} tUSBDCompositeEntry;
```

tUSBDCompositeEntry,定义 Composite 设备信息。定义在 usbdcomp.h 文件中。

```
typedef struct
{
    unsigned long ulUSBBase;
    tDeviceInfo * psDevInfo;
    tConfigDescriptor sConfigDescriptor;
```

```
    tDeviceDescriptor sDeviceDescriptor;
    tConfigHeader sCompConfigHeader;
    tConfigSection psCompSections[2];
    tConfigSection * ppsCompSections[2];
    unsigned long ulDataSize;
    unsigned char * pucData;
}
tCompositeInstance;
```

tCompositeInstance,设备类实例。定义了 Composite 设备类的 USB 基地址、设备信息、IN 端点、OUT 端点等信息。

```
typedef struct
{
    const tDeviceInfo * psDevice;
    void * pvInstance;
}
tCompositeEntry;
```

tCompositeEntry,Composite 各设备的设备信息。

```
typedef struct
{
    unsigned short usVID;
    unsigned short usPID;
    unsigned short usMaxPowermA;
    unsigned char ucPwrAttributes;
    tUSBCallback pfnCallback;
    const unsigned char * const * ppStringDescriptors;
    unsigned long ulNumStringDescriptors;
    unsigned long ulNumDevices;
    tCompositeEntry * psDevices;
    tCompositeInstance * psPrivateData;
}
tUSBDCompositeDevice;
```

tUSBDCompositeDevice,Composite 设备类,定义了 VID、PID、电源属性、字符串描述符等,还包括了 Composite 设备类实例。其他设备描述符、配置信息通过 API 函数输入 tCompositeInstance 定义的 Composite 设备实例中。

10.3 API 函数

在 Composite 设备类 API 库中定义了 2 个函数,完成 USB Composite 设备初始化、配置及数据处理。下面为 usbdcomp.h 中定义的 API 函数。

```
void * USBDCompositeInit(unsigned long ulIndex,
                tUSBDCompositeDevice * psCompDevice,
                unsigned long ulSize,
                unsigned char * pucData);
void USBDCompositeTerm(void * pvInstance);
```

(1) void * USBDCompositeInit(unsigned long ulIndex,
　　　　　　　　tUSBDCompositeDevice * psCompDevice,
　　　　　　　　unsigned long ulSize,
　　　　　　　　unsigned char * pucData);

作用：初始化 Composite 设备硬件、协议，把其他配置参数填入 psCompDevice 实例中。

参数：ulIndex，USB 模块代码，固定值：USB_BASE0。psMSCDevice，MSC 设备类。

返回：指向配置后的 tUSBDCompositeDevice。

(2) void USBDCompositeTerm(void * pvInstance);

作用：结束 Composite 设备。

参数：pvInstance，指向 tUSBDCompositeDevice。

返回：无。

在这些函数中 USBDCompositeInit 函数最重要，用于处理各子设备信息，保存所有子设备配置及其他数据。完成配置后，子设备操作相互独立，可参考第 5～9 章 USB 设备介绍。

10.4 Composite 设备开发

Composite 设备开发只需要 3 步就能完成：各子设备配置、完善接口函数；Composite 设备配置、协调；各子设备数据处理。图 10-1 为 Composite 设备开发流程图。

下面以"电子教鞭"实例说明使用 USB 库开发 USB Composite 设备的过程，电子教鞭有两个重要功能，即 U 盘功能和控制功能。所以要做两个子类：大容量存储类与键盘类。通过大容量存储类完成 U 盘数据存储，以备教师复制课件；通过键盘的 F5、up、down、ESC 等键完成幻灯片播放控制。

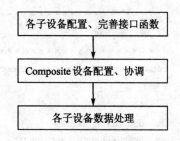

图 10-1　Composite 设备开发流程图

第 10 章 Composite 设备

(1) 各子设备配置、完善接口函数。

```c
#define DESCRIPTOR_DATA_SIZE      (COMPOSITE_DHID_SIZE + COMPOSITE_DMSC_SIZE)
unsigned char g_pucDescriptorData[DESCRIPTOR_DATA_SIZE];

//声明函数原型
unsigned long USBDMSCEventCallback(void * pvCBData, unsigned long ulEvent,
                                   unsigned long ulMsgParam,
                                   void * pvMsgData);
//声明函数原型
unsigned long KeyboardHandler(void * pvCBData,
                              unsigned long ulEvent,
                              unsigned long ulMsgData,
                              void * pvMsgData);
unsigned long EventHandler(void * pvCBData, unsigned long ulEvent,
                           unsigned long ulMsgData, void * pvMsgData);
const tUSBDMSCDevice g_sMSCDevice;
//msc 状态
volatile enum
{
    MSC_DEV_DISCONNECTED,
    MSC_DEV_CONNECTED,
    MSC_DEV_IDLE,
    MSC_DEV_READ,
    MSC_DEV_WRITE,
}
g_eMSCState;
//全局标志
#define FLAG_UPDATE_STATUS       1
static unsigned long g_ulFlags;
//DMA
tDMAControlTable sDMAControlTable[64] __attribute__ ((aligned(1024)));

//*************************************************************
// 语言描述符
//*************************************************************
const unsigned char g_pLangDescriptor[] =
{
    4,
    USB_DTYPE_STRING,
    USBShort(USB_LANG_EN_US)
};
```

```c
//*************************************************************
// 制造商 字符串 描述符
//*************************************************************
const unsigned char g_pManufacturerString[] =
{
    (17 + 1) * 2,
    USB_DTYPE_STRING,
    'T', 0, 'e', 0, 'x', 0, 'a', 0, 's', 0, ' ', 0, 'I', 0, 'n', 0, 's', 0,
    't', 0, 'r', 0, 'u', 0, 'm', 0, 'e', 0, 'n', 0, 't', 0, 's', 0,
};
//*************************************************************
//产品 字符串 描述符
//*************************************************************
const unsigned char g_pProductString[] =
{
    (19 + 1) * 2,
    USB_DTYPE_STRING,
    'M', 0, 'a', 0, 's', 0, 's', 0, ' ', 0, 'S', 0, 't', 0, 'o', 0, 'r', 0,
    'a', 0, 'g', 0, 'e', 0, ' ', 0, 'D', 0, 'e', 0, 'v', 0, 'i', 0, 'c', 0,
    'e', 0
};
//*************************************************************
// 产品 序列号 描述符
//*************************************************************
const unsigned char g_pSerialNumberString[] =
{
    (8 + 1) * 2,
    USB_DTYPE_STRING,
    '1', 0, '2', 0, '3', 0, '4', 0, '5', 0, '6', 0, '7', 0, '8', 0
};
//*************************************************************
// 字符串描述符集合
//*************************************************************
const unsigned char * const g_pStringDescriptors[] =
{
    g_pLangDescriptor,
    g_pManufacturerString,
    g_pProductString,
    g_pSerialNumberString,
};
#define NUM_STRING_DESCRIPTORS (sizeof(g_pStringDescriptors) /      \
                                sizeof(unsigned char *))
```

第10章 Composite 设备

```c
//***************************************************************
//MSC 实例,配置并为设备信息提供空间
//***************************************************************
tMSCInstance g_sMSCInstance;
//***************************************************************
//msc 设备配置
//***************************************************************
const tUSBDMSCDevice g_sMSCDevice =
{
    USB_VID_STELLARIS,
    USB_PID_MSC,
    "TI            ",
    "Mass Storage       ",
    "1.00",
    200,
    USB_CONF_ATTR_SELF_PWR,
    g_pStringDescriptors,
    NUM_STRING_DESCRIPTORS,
    {
        USBDMSCStorageOpen,
        USBDMSCStorageClose,
        USBDMSCStorageRead,
        USBDMSCStorageWrite,
        USBDMSCStorageNumBlocks
    },
    USBDMSCEventCallback,
    &g_sMSCInstance
};
#define MSC_BUFFER_SIZE 512
//***************************************************************
//键盘实例,配置并为设备信息提供空间
//***************************************************************
tHIDKeyboardInstance g_KeyboardInstance;
//***************************************************************
//键盘设备配置
//***************************************************************
const tUSBDHIDKeyboardDevice g_sKeyboardDevice =
{
    USB_VID_STELLARIS,
    USB_VID_STELLARIS,
    200,
    USB_CONF_ATTR_SELF_PWR | USB_CONF_ATTR_RWAKE,
```

```c
    KeyboardHandler,
    (void *)&g_sKeyboardDevice,
    0,
    0,
    &g_KeyboardInstance
};

//*************************************************************
//Callback 函数
//*************************************************************
unsigned long USBDMSCEventCallback(void * pvCBData, unsigned long ulEvent,
                        unsigned long ulMsgParam, void * pvMsgData)
{
    switch(ulEvent)
    {
        // 正在写数据到存储设备
        case USBD_MSC_EVENT_WRITING:
        {
            break;
        }
        //读取数据
        case USBD_MSC_EVENT_READING:
        {
            GPIOPinWrite(GPIO_PORTF_BASE,0x10,0x10);
            break;
        }
        //空闲
        case USBD_MSC_EVENT_IDLE:
        default:
        {
            GPIOPinWrite(GPIO_PORTF_BASE,0x10,0x00);
            break;
        }
    }
    return(0);
}
//*************************************************************
//键盘 Callback 函数
//*************************************************************
unsigned long KeyboardHandler(void * pvCBData, unsigned long ulEvent,
                    unsigned long ulMsgData, void * pvMsgData)
{
```

第10章 Composite 设备

```
        switch (ulEvent)
        {
            case USB_EVENT_CONNECTED:
            {
                GPIOPinWrite(GPIO_PORTF_BASE,0x20,0x20);
                break;
            }
            case USB_EVENT_DISCONNECTED:
            {
                GPIOPinWrite(GPIO_PORTF_BASE,0x20,0x00);
                break;
            }
            case USB_EVENT_TX_COMPLETE:
            {
                break;
            }
            case USB_EVENT_SUSPEND:
            {
                break;
            }
            case USB_EVENT_RESUME:
            {
                break;
            }
            case USBD_HID_KEYB_EVENT_SET_LEDS:
            {
                break;
            }
            default:
            {
                break;
            }
        }
        return (0);
    }
```

(2) 完成 Composite 设备配置、协调。

```
//**************************************************************
//复合设备配置
//**************************************************************
tCompositeEntry g_psCompDevices[] =
{
```

```c
    {
        &g_sMSCDeviceInfo,
        (void *)&g_sMSCDeviceInfo
    },
    {
        &g_sHIDDeviceInfo,
        (void *)&g_sHIDDeviceInfo
    }
};
#define NUM_DEVICES             (sizeof(g_psCompDevices)/sizeof(tCompositeEntry))
tCompositeInstance g_CompInstance;
unsigned long xxx[10];
tUSBDCompositeDevice g_sCompDevice =
{
    USB_VID_STELLARIS,
    0x0123,
    500,
    USB_CONF_ATTR_BUS_PWR,
    EventHandler,
    g_pStringDescriptors,
    NUM_STRING_DESCRIPTORS,
    2,
    g_psCompDevices,
    xxx,
    &g_CompInstance
};
//****************************************************************
//复合设备 Callback 函数
//****************************************************************
unsigned long EventHandler(void * pvCBData, unsigned long ulEvent, unsigned long ulMsgData,void * pvMsgData)
{
    unsigned long ulNewEvent;
    ulNewEvent = 1;
    switch(ulEvent)
    {
        case USB_EVENT_CONNECTED:
        {
            break;
        }
        case USB_EVENT_DISCONNECTED:
        {
```

第 10 章 Composite 设备

```
            break;
        }
        case USB_EVENT_SUSPEND:
        {
            break;
        }
        case USB_EVENT_RESUME:
        {
            break;
        }

        default:
        {
            ulNewEvent = 0;
            break;
        }
    }
    if(ulNewEvent)
    {
    }
    return(0);
}
```

(3) 各子设备数据处理,主要是按键处理,U 盘功能自动调用底层驱动自动完成。

```
//系统初始化
SysCtlLDOSet(SYSCTL_LDO_2_75V);
SysCtlClockSet(SYSCTL_XTAL_8MHZ | SYSCTL_SYSDIV_8 | SYSCTL_USE_PLL     | SYSCTL_
              OSC_MAIN );
SysCtlPeripheralEnable(SYSCTL_PERIPH_GPIOF);
GPIOPinTypeGPIOOutput(GPIO_PORTF_BASE,0xf0);
GPIOPinTypeGPIOInput(GPIO_PORTF_BASE,0x0f);
HWREG(GPIO_PORTF_BASE + GPIO_O_PUR) |= 0x0f;
// ucDMA 配置
SysCtlPeripheralEnable(SYSCTL_PERIPH_UDMA);
SysCtlDelay(10);
uDMAControlBaseSet(&sDMAControlTable[0]);
uDMAEnable();

g_ulFlags = 0;
g_eMSCState = MSC_DEV_IDLE;
//复合设备初始化
```

```c
    g_sCompDevice.psDevices[0].pvInstance =
        USBDMSCCompositeInit(0, &g_sMSCDevice);
    g_sCompDevice.psDevices[1].pvInstance =
        USBDHIDKeyboardInit(0, &g_sKeyboardDevice);
    USBDCompositeInit(0, &g_sCompDevice, DESCRIPTOR_DATA_SIZE,
                      g_pucDescriptorData);
    //初始化存储设备
    disk_initialize(0);
    while(1)
    {
        USBDHIDKeyboardKeyStateChange((void *)&g_sKeyboardDevice, HID_KEYB_CAPS_LOCK,
                    HID_KEYB_USAGE_A,
                    (GPIOPinRead(GPIO_PORTF_BASE, 0x0f) & GPIO_PIN_0)
                    ? false : true);
        USBDHIDKeyboardKeyStateChange((void *)&g_sKeyboardDevice, 0,
                    HID_KEYB_USAGE_DOWN_ARROW,
                    (GPIOPinRead(GPIO_PORTF_BASE, 0x0f) & GPIO_PIN_1)
                    ? false : true);
        USBDHIDKeyboardKeyStateChange((void *)&g_sKeyboardDevice, 0,
                    HID_KEYB_USAGE_UP_ARROW,
                    (GPIOPinRead(GPIO_PORTF_BASE, 0x0f) & GPIO_PIN_2)
                    ? false : true);
        USBDHIDKeyboardKeyStateChange((void *)&g_sKeyboardDevice, 0,
                    HID_KEYB_USAGE_ESCAPE,
                    (GPIOPinRead(GPIO_PORTF_BASE, 0x0f) & GPIO_PIN_3)
                    ? false : true);
        SysCtlDelay(SysCtlClockGet()/3000);
    }
```

使用上面3步就完成了Composite设备的开发。Composite设备开发时要加入两个库函数，分别为usblib.lib和DriverLib.lib，在启动代码中加入USB0DeviceIntHandler中断服务函数。以上Composite设备开发完成，在Windows XP下运行效果如图10-2所示，为该复合设备正在枚举提示图。

图10-2 复合设备枚举

在计算机中可以发现多了USB MSC设备和HID设备，同时还多了一个Composite设备。

Composite设备开发源码较多，下面只列出一部分，如下所示。

第10章 Composite 设备

```c
#include "inc/hw_ints.h"
#include "inc/hw_memmap.h"
#include "inc/hw_gpio.h"
#include "inc/hw_types.h"
#include "inc/hw_ints.h"
#include "driverlib/sysctl.h"
#include "driverlib/gpio.h"
#include "driverlib/interrupt.h"
#include "driverlib/rom.h"
#include "driverlib/systick.h"
#include "driverlib/usb.h"
#include "driverlib/udma.h"
#include "usblib/usblib.h"
#include "usblib/usb-ids.h"
#include "usblib/device/usbdevice.h"
#include "usblib/device/usbdmsc.h"
#include "diskio.h"
#include "usbdsdcard.h"
#include "usblib/usblib.h"
#include "usblib/usbhid.h"
#include "usblib/device/usbdhid.h"
#include "usblib/device/usbdcomp.h"
#include "usblib/device/usbdhidkeyb.h"

#define DESCRIPTOR_DATA_SIZE     (COMPOSITE_DHID_SIZE + COMPOSITE_DMSC_SIZE)
unsigned char g_pucDescriptorData[DESCRIPTOR_DATA_SIZE];

//声明函数原型
unsigned long USBDMSCEventCallback(void * pvCBData, unsigned long ulEvent,
                                   unsigned long ulMsgParam,
                                   void * pvMsgData);
//声明函数原型
unsigned long KeyboardHandler(void * pvCBData,
                              unsigned long ulEvent,
                              unsigned long ulMsgData,
                              void * pvMsgData);
unsigned long EventHandler(void * pvCBData, unsigned long ulEvent,
                           unsigned long ulMsgData, void * pvMsgData);
const tUSBDMSCDevice g_sMSCDevice;
//msc 状态
volatile enum
{
```

```c
    MSC_DEV_DISCONNECTED,
    MSC_DEV_CONNECTED,
    MSC_DEV_IDLE,
    MSC_DEV_READ,
    MSC_DEV_WRITE,
}
g_eMSCState;
//全局标志
#define FLAG_UPDATE_STATUS     1
static unsigned long g_ulFlags;
//DMA
tDMAControlTable sDMAControlTable[64] __attribute__ ((aligned(1024)));

//*************************************************************
// 语言描述符
//*************************************************************
const unsigned char g_pLangDescriptor[] =
{
    4,
    USB_DTYPE_STRING,
    USBShort(USB_LANG_EN_US)
};
//*************************************************************
// 制造商 字符串 描述符
//*************************************************************
const unsigned char g_pManufacturerString[] =
{
    (17 + 1) * 2,
    USB_DTYPE_STRING,
    'T', 0, 'e', 0, 'x', 0, 'a', 0, 's', 0, ' ', 0, 'I', 0, 'n', 0, 's', 0,
    't', 0, 'r', 0, 'u', 0, 'm', 0, 'e', 0, 'n', 0, 't', 0, 's', 0,
};
//*************************************************************
//产品 字符串 描述符
//*************************************************************
const unsigned char g_pProductString[] =
{
    (19 + 1) * 2,
    USB_DTYPE_STRING,
    'M', 0, 'a', 0, 's', 0, 's', 0, ' ', 0, 'S', 0, 't', 0, 'o', 0, 'r', 0,
    'a', 0, 'g', 0, 'e', 0, ' ', 0, 'D', 0, 'e', 0, 'v', 0, 'i', 0, 'c', 0,
    'e', 0
};
```

第 10 章 Composite 设备

```c
//*************************************************************
// 产品 序列号 描述符
//*************************************************************
const unsigned char g_pSerialNumberString[] =
{
    (8 + 1) * 2,
    USB_DTYPE_STRING,
    '1', 0, '2', 0, '3', 0, '4', 0, '5', 0, '6', 0, '7', 0, '8', 0
};
//*************************************************************
// 字符串描述符集合
//*************************************************************
const unsigned char * const g_pStringDescriptors[] =
{
    g_pLangDescriptor,
    g_pManufacturerString,
    g_pProductString,
    g_pSerialNumberString,
};
#define NUM_STRING_DESCRIPTORS (sizeof(g_pStringDescriptors) / \
                                sizeof(unsigned char *))
//*************************************************************
//MSC 实例,配置并为设备信息提供空间
//*************************************************************
tMSCInstance g_sMSCInstance;
//*************************************************************
//msc 设备配置
//*************************************************************
const tUSBDMSCDevice g_sMSCDevice =
{
    USB_VID_STELLARIS,
    USB_PID_MSC,
    "TI            ",
    "Mass Storage    ",
    "1.00",
    200,
    USB_CONF_ATTR_SELF_PWR,
    g_pStringDescriptors,
    NUM_STRING_DESCRIPTORS,
    {
        USBDMSCStorageOpen,
        USBDMSCStorageClose,
```

```c
        USBDMSCStorageRead,
        USBDMSCStorageWrite,
        USBDMSCStorageNumBlocks
    },
    USBDMSCEventCallback,
    &g_sMSCInstance
};
#define MSC_BUFFER_SIZE 512
//**************************************************************
//键盘实例,配置并为设备信息提供空间
//**************************************************************
tHIDKeyboardInstance g_KeyboardInstance;
//**************************************************************
//键盘设备配置
//**************************************************************
const tUSBDHIDKeyboardDevice g_sKeyboardDevice =
{
    USB_VID_STELLARIS,
    USB_VID_STELLARIS,
    200,
    USB_CONF_ATTR_SELF_PWR | USB_CONF_ATTR_RWAKE,
    KeyboardHandler,
    (void *)&g_sKeyboardDevice,
    0,
    0,
    &g_KeyboardInstance
};
//**************************************************************
//复合设备配置
//**************************************************************
tCompositeEntry g_psCompDevices[] =
{
    {
        &g_sMSCDeviceInfo,
        (void    *)&g_sMSCDeviceInfo
    },
    {
        &g_sHIDDeviceInfo,
        (void    *)&g_sHIDDeviceInfo
    }
};
#define NUM_DEVICES                (sizeof(g_psCompDevices)/sizeof(tCompositeEntry))
```

第10章 Composite 设备

```c
tCompositeInstance g_CompInstance;
unsigned long xxx[10];
tUSBDCompositeDevice g_sCompDevice =
{
    USB_VID_STELLARIS,
    0x0124,
    500,
    USB_CONF_ATTR_BUS_PWR,
    EventHandler,
    g_pStringDescriptors,
    NUM_STRING_DESCRIPTORS,
    2,
    g_psCompDevices,
    xxx,
    &g_CompInstance
};

//***************************************************************
//Callback 函数
//***************************************************************
unsigned long USBDMSCEventCallback(void * pvCBData, unsigned long ulEvent,
                          unsigned long ulMsgParam, void * pvMsgData)
{
    switch(ulEvent)
    {
        // 正在写数据到存储设备
        case USBD_MSC_EVENT_WRITING:
        {
            break;
        }
        //读取数据
        case USBD_MSC_EVENT_READING:
        {
            GPIOPinWrite(GPIO_PORTF_BASE,0x10,0x10);
            break;
        }
        //空闲
        case USBD_MSC_EVENT_IDLE:
        default:
        {
            GPIOPinWrite(GPIO_PORTF_BASE,0x10,0x00);
            break;
```

```c
        }
    }
    return(0);
}
//**************************************************************
//键盘 Callback 函数
//**************************************************************
unsigned long KeyboardHandler(void * pvCBData, unsigned long ulEvent,
                              unsigned long ulMsgData, void * pvMsgData)
{
    switch (ulEvent)
    {
        case USB_EVENT_CONNECTED:
        {
            GPIOPinWrite(GPIO_PORTF_BASE,0x20,0x20);
            break;
        }
        case USB_EVENT_DISCONNECTED:
        {
            GPIOPinWrite(GPIO_PORTF_BASE,0x20,0x00);
            break;
        }
        case USB_EVENT_TX_COMPLETE:
        {
            break;
        }
        case USB_EVENT_SUSPEND:
        {
            break;
        }
        case USB_EVENT_RESUME:
        {
            break;
        }
        case USBD_HID_KEYB_EVENT_SET_LEDS:
        {
            break;
        }
        default:
        {
            break;
        }
```

第 10 章 Composite 设备

```c
        }
        return (0);
}
// **************************************************************
//复合设备 Callback 函数
// **************************************************************
unsigned long EventHandler(void * pvCBData, unsigned long ulEvent, unsigned long ulMs-
                           gData,void * pvMsgData)
{
    unsigned long ulNewEvent;
    ulNewEvent = 1;
    switch(ulEvent)
    {
        case USB_EVENT_CONNECTED:
        {
            break;
        }
        case USB_EVENT_DISCONNECTED:
        {
            break;
        }
        case USB_EVENT_SUSPEND:
        {
            break;
        }
        case USB_EVENT_RESUME:
        {
            break;
        }

        default:
        {
            ulNewEvent = 0;
            break;
        }
    }
    if(ulNewEvent)
    {
    }
    return(0);
}
// **************************************************************
```

```c
//主函数
//**************************************************************
int main(void)
{
    //系统初始化
    SysCtlLDOSet(SYSCTL_LDO_2_75V);
    SysCtlClockSet(SYSCTL_XTAL_8MHZ | SYSCTL_SYSDIV_8 | SYSCTL_USE_PLL   | SYSCTL_
                OSC_MAIN );
    SysCtlPeripheralEnable(SYSCTL_PERIPH_GPIOF);
    GPIOPinTypeGPIOOutput(GPIO_PORTF_BASE,0xf0);
    GPIOPinTypeGPIOInput(GPIO_PORTF_BASE,0x0f);
    HWREG(GPIO_PORTF_BASE + GPIO_O_PUR) | = 0x0f;
    // ucDMA 配置
    SysCtlPeripheralEnable(SYSCTL_PERIPH_UDMA);
    SysCtlDelay(10);
    uDMAControlBaseSet(&sDMAControlTable[0]);
    uDMAEnable();

    g_ulFlags = 0;
    g_eMSCState = MSC_DEV_IDLE;
    //复合设备初始化
    g_sCompDevice.psDevices[0].pvInstance =
        USBDMSCCompositeInit(0, &g_sMSCDevice);
    g_sCompDevice.psDevices[1].pvInstance =
        USBDHIDKeyboardInit(0, &g_sKeyboardDevice);
    USBDCompositeInit(0, &g_sCompDevice, DESCRIPTOR_DATA_SIZE,
                    g_pucDescriptorData);
    //初始化存储设备
    disk_initialize(0);
    while(1)
    {
        USBDHIDKeyboardKeyStateChange((void * )&g_sKeyboardDevice, HID_KEYB_CAPS_LOCK,
                            HID_KEYB_USAGE_A,
                            (GPIOPinRead(GPIO_PORTF_BASE, 0x0f) & GPIO_PIN_0)
                            ? false : true);
        USBDHIDKeyboardKeyStateChange((void * )&g_sKeyboardDevice, 0,
                            HID_KEYB_USAGE_DOWN_ARROW,
                            (GPIOPinRead(GPIO_PORTF_BASE, 0x0f) & GPIO_PIN_1)
                            ? false : true);
        USBDHIDKeyboardKeyStateChange((void * )&g_sKeyboardDevice, 0,
                            HID_KEYB_USAGE_UP_ARROW,
                            (GPIOPinRead(GPIO_PORTF_BASE, 0x0f) & GPIO_PIN_2)
```

```
                                   ? false : true);
        USBDHIDKeyboardKeyStateChange((void *)&g_sKeyboardDevice, 0,
                                      HID_KEYB_USAGE_ESCAPE,
                                      (GPIOPinRead(GPIO_PORTF_BASE, 0x0f) & GPIO_PIN_3)
                                      ? false : true);
        SysCtlDelay(SysCtlClockGet()/3000);
    }
}
```

10.5 小 结

本章介绍 USB 复合设备开发，通常 USB 设备不仅仅只有一个功能，如果需要多个功能、多个设备同时兼有时，复合设备能满足其要求。本章通过一个简单的电子教鞭实例，简单介绍了复合设备的开发流程。

第 11 章

USB 主机开发

第 5～10 章介绍了常用 USB 设备的开发,本章介绍使用 Stellaris 的 USB 处理器进行 USB 主机开发。USB 主机开发相对于 USB 设备开发较简单。USB 主机只有一个,只是驱动不一样而已,USB 主机只需识别 USB 设备,并能进行数据传输、控制就行。

11.1 USB 主机开发介绍

Luminary Micro Stellaris 的 USB 处理器具有主机控制功能,支持全速与低速。使用官方提供的 USB HOST 库可以轻松地开发 USB 主机。本章主要使用 USB 库编程为例讲解,涵盖整个 USB 主机开发。

Stellaris 所提供的 USB 处理器中,LM3S3XXX 与 LM3S5XXX 的 A0 版本有一个 bug。图 11-1 为 PB0 与 PB1 硬件连接图。在 A0 版本中,USB 处理器工作在主机和设备模式下时,PB0 与 PB1 不能当 GPIO 使用,因为在此版本中,PB0 与 PB1 为主机与设备提供电平信号。当 USB 处理器工作在主机模式下时,PB0 应该连接到低电平;工作在设备模式下时,PB0 应该连接到高电平。同时,PB0 引脚与电压信号之间应该连接一个电阻,其典型值为 10 Ω。PB1 必须连接到 5 V(4.75～5.25 V)。如果 USB 处理器不是 A0 版本,PB0 与 PB1 可以用作 GPIO 功能。

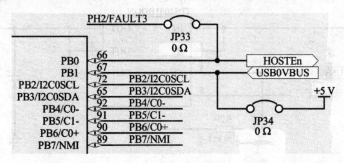

图 11-1 PB0 与 PB1 硬件连接图

图 11-2 为 LM3S3XXX 与 LM3S5XXX 系列的 USB 功能引脚连接图,USB 处理器工作在主机模式下时,USB 功能引脚的连接与设备模式下一样。①

第 11 章 USB 主机开发

USB0RBIAS,连接方式固定,为 USB 模拟电路内部必须的 9.1 kΩ 电阻(1‰精度),普通贴片即能满足需求;②USB0DP 和 USB0DM,USB0 的双向差分数据引脚,连接方式固定,分别连接到 USB 规范中的 D+ 和 D- 中,在使用时,请特别注意 D+ 和 D- 的连接方式。USB0EPEN,USB 主机电源输出使能引脚,主机输出电源使能信号,高电平有效,用于使能外部电源。如图 11-3 所示,通过 USB 处理器的 USB0EPEN 引脚输出高电平使能 VBUS 电源。USB0PFLT,主机模式下的外部电源异常输入引脚,指示外部电源的错误状态,低电平有效。如图 11-3 所示,当 TPS2051 的 VBUS 电源输出电流小于 1 mA 或者大于 500 mA 时,从 OCn 引脚输出低电平,开漏输出,通过 USB 处理器的 USB0PFLT 上拉,读取 OCn 的电平变化,并产生中断。注意,USB0EPEN 和 USB0PFLT 主要工作在主机模式下,为设备提供 VBUS 电源,当然这两个引脚可以不使用,可直接通过主板 5 V 提供 VBUS 电源,这会缺少 USB 电源欠压过流保护,但电路简单。开发人员可以根据自己的设计需要,自行剪裁硬件,但是 USB0RBIAS、USB0DP、USB0DM 必不可少,如果是 A0 版本处理器,工作在主机和设备模式下时,PB0 与 PB1 不能当 GPIO 使用,PB1 接 5 V;主机模式下,PB0 接地;设备模式下,PB0 接 5 V。

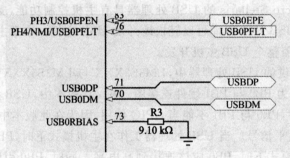

图 11-2 USB 功能引脚连接图

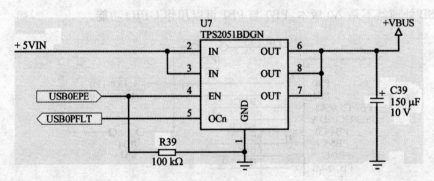

图 11-3 VBUS 控制电路

对 LM3S3XXX、LM3S5XXX、LM3S9XXX 系列处理器,有一部分具有 USB OTG 功能,与传统不具有 USB OTG 功能的处理器比较,PB0 作为 USB0ID,此信号用于检测 USB ID 信号的状态。此时 USBPHY 将在内部启用一个上拉电阻，通过外部元件（USB 连接器）检测 USB 控制器的初始状态（即电缆的 A 侧设置下拉电阻，B 侧设置上拉电阻）。该类 Stellaris 支持 OTG 标准的会话请求协议（SRP）和主机协商协议（HNP），提供完整的 OTG 协商。会话请求协议（SRP）允许连接在 USB 线缆 B 端的 B 类设备向连接在 A 端的 A 类设备发出请求，通知 A 类设备打开 VBUS 电源。主机协商协议（HNP）用于在初始会话请求协议提供供电后，决定 USB 线缆哪端的设备作为 USB Host 主控制器。

当该设备连接到非 OTG 外设或设备时，控制器可以检测出线缆终端所使用的设备类型，并提供一个寄存器来指示该控制器是用作主机控制器还是用作设备控制器。以上动作都是由 USB 控制器自动处理的。基于这种自动探测机制，系统使用 A 类/B 类连接器取代 A/B 类连接器，可支持完整的会话请求协议和主机协商协议。另外，USB 控制器还提供对连接到非 OTG 的外设或主机控制器的支持。可将 USB 控制器设置为专用的主机或设备模式，此时，USB0VBUS 和 USB0ID 引脚可被设置作为 GPIO 使用。但当 USB 控制器用作自供电设备时，必须将 GPIO 输入引脚或模拟比较器输入引脚连接到 VBUS 引脚上，并配置为在 VBUS 掉电时会产生中断。该中断用于禁用 USB0DP 信号上的上拉电阻。所以具有 USB OTG 功能的处理器工作在 A 设备模式下时，USB0VBUS 和 USB0ID 引脚可通过软件设置为通用 IO 端口（GPIO）功能；如果使用 USB OTG 去模拟 USB 主机功能时，USB0VBUS 和 USB0ID 不能用作通用 IO 端口（GPIO）功能，只能归 USB 使用，并且 USB0ID 引脚必须接地。

注意：对于具有 USB OTG 功能的 USB 处理器来说，也有一个重要的 bug 要注意，根据官方提供的数据手册可以看出，只要处理器不工作在 USB OTG 模式下，USB0VBUS 和 USB0ID 引脚可配置为 GPIO 使用，但在 B1 版本中（其他版本也有）无论 USB 处理器工作在什么模式，USB0ID 必须归 USB 使用，不能用于 GPIO 功能，如果工作在主机模式，USB0ID 连接低电平（地）；工作在设备模式，USB0ID 连接高电平（5 V）。同时 USB0VBUS 必须提供 5 V 输入。这个 bug 是怎么产生的呢？由于 USB 控制状态寄存器（USBGPCS）的 DEVMODOTG 工作不正常，造成无法通过寄存器来实现 USB 工作模式的切换，必须通过外面的 USB0ID 引脚来实现，相当于使用 USB OTG 来让 USB 处理器工作在主机或者设备模式下。

11.2 USB 主机开发过程

USB 主机开发相对简单，使用 USB 库编程时更加简单，USB 主机开发流程如图 11-4 所示，为 USB 主机开发的简单流程图。USB 主机开发时，首先确定主机要读取的设备类型，如果在 USB 库中没有该类设备的驱动，则由开发人员按照 USB 库中的设备驱动编写方式自行编写；编程时，首先配置 USB 处理器的内核；再根据情况编写、修改驱动；注册设备驱动，一个函数就能实现；通过 USB 库函数来运行主机。

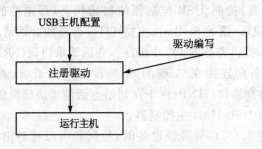

图 11-4 USB 主机开发流程图

USB 主机配置包括了中断控制、外设资源控制等一系列与处理器运行相关的模块、资源初始化与配置。在 USB 库中已经包含了常用设备驱动，比如 HID、MSC、Audio 等一系列驱动。在具有这些驱动的情况下，注册驱动，让 USB 处理器可以方便地控制 USB 设备。最后运行主机，枚举设备、进行数据传输并实现其功能。

11.2.1 主机配置

在使用 USB 主机之前必须配置好内部相关参数，并且主频最好不要低于 30 MHz，不然会出现不可预料的后果。根据外围 VBUS 电路，使能对应引脚，打开 VBUS 电源等。

（1）使用 SysCtlClockSet 设置 USB 处理器主频，最好不要低于 30 MHz，在本书中都是设置为 50 MHz。

```
SysCtlClockSet(SYSCTL_SYSDIV_4 | SYSCTL_USE_PLL | SYSCTL_OSC_MAIN |
               SYSCTL_XTAL_8MHZ);
```

（2）通过 SysCtlPeripheralEnable 使能 USB 模块时钟，在使用 USB 模块时，必须先打开 USB 资源时钟，由于具有 USB 功能的 LM3S 处理器目前都只有一个 USB 模块 USB0，所以打开 USB 模块时钟的函数方式也固定，参数也固定，如下所示：

```
SysCtlPeripheralEnable(SYSCTL_PERIPH_USB0);
```

（3）使能串口，方便调试。在嵌入式系统开发时，经常使用 UART 调试程序，输入控制命令、输出调试信息。在此，也使用 UART 调试 USB 主机，为了减轻开发人员的负担，处理器厂商的程序员已经把 UART 调试常用的命令与函数编写得非常完善并免费提供源代码，调用也很简单。打开 UART0，配置串口参数。

```
SysCtlPeripheralEnable(SYSCTL_PERIPH_UART0);
SysCtlPeripheralEnable(SYSCTL_PERIPH_GPIOA);
GPIOPinTypeUART(GPIO_PORTA_BASE, GPIO_PIN_0 | GPIO_PIN_1);
UARTStdioInit(0);
```

（4）为了让 USB 处理器能同时连接设备和主机，可以通过 FSUSB11MTCX 作 USB 扩展器，让处理器可能同时插入 USB 设备与 USB 主机，当然在某一个时刻只能工作在主机或者设备模式下，通过 USB 处理器的 GPIO 口控制 FSUSB11MTCX 的 S1 和 S2，让其选择接通主机与设备，如图 11-5 所示，当工作在主机模式下时，将与 S1 和 S2 连接的引脚置低。同时，打开 VBUS，给插入设备供电。

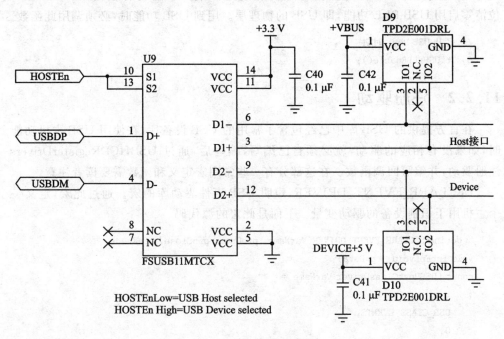

图 11-5 USB 接口扩展

根据图 11-5 所示，假设与 S1 和 S2 连接的是 PH2 引脚，那么选通主机连接的程序如下：

```
#define USB_MUX_GPIO_PERIPH     SYSCTL_PERIPH_GPIOH
#define USB_MUX_GPIO_BASE       GPIO_PORTH_BASE
#define USB_MUX_GPIO_PIN        GPIO_PIN_2
#define USB_MUX_SEL_DEVICE      USB_MUX_GPIO_PIN
#define USB_MUX_SEL_HOST        0
SysCtlPeripheralEnable(USB_MUX_GPIO_PERIPH);
// 切换为主机模式，外部 FSUSB11MTCX 切换
GPIOPinTypeGPIOOutput(USB_MUX_GPIO_BASE, USB_MUX_GPIO_PIN);
GPIOPinWrite(USB_MUX_GPIO_BASE, USB_MUX_GPIO_PIN, USB_MUX_SEL_HOST);
```

根据图 11-3，打开 VBUS 与异常检测，配置 USB0EPEN 和 USB0PFLT 为 USB 所用，在主机模式下通过 USB 处理器自动控制 VBUS 电源。

```
// VBUS 电源控制端口配置
```

```
GPIOPinTypeUSBDigital(GPIO_PORTH_BASE, GPIO_PIN_3 | GPIO_PIN_4);
SysCtlPeripheralEnable(SYSCTL_PERIPH_GPIOB);
```

(5) 打开 USB 的 PLL，为 USB 模块工作提供时钟源，当然只要使用 USB 功能，就必须打开 USB 模块的 PLL，不打开 PLL，USB 就不能正常工作。SysCtlUSBPLLEnable，USB PLL 在上电复位期间默认为禁用，将 RCC2 寄存器的 USBPWRDN 位清零启用 USB PLL 功能，即 USB 的物理层。用到 USB 功能时，必须调用此函数。

```
// 打开 USB Phy 时钟
SysCtlUSBPLLEnable();
```

11.2.2 注册驱动

在官方提供的 USB 库中已经包含了常用的 USB 设备库，在使用 USB 主机功能时，如果没有相应的驱动，就必须自己编写，编写后，通过 USBHCDRegisterDrivers 注册驱动，并编写回调函数。在这部分有一些常用宏定义和函数需要读者注意。

DECLARE_EVENT_DRIVER，声明 USB 事件驱动库的宏。通过此宏，定义一个主机用于识别设备的驱动变量。下面是此宏的源代码。

```
#define DECLARE_EVENT_DRIVER(VarName, pfnOpen, pfnClose, pfnEvent)
void IntFn(void * pvData);
const tUSBHostClassDriver VarName =
{
    USB_CLASS_EVENTS,
    0,
    0,
    pfnEvent
}
```

从此宏的源代码中可以看出：VarName 为变量的名称；pfnOpen 为开放调用这个驱动程序回调，这个值目前是保留的，应设置为 0；pfnClose 关闭这个驱动调用回调，这个值目前是保留的，应设置为 0；pfnEvent 将用于各种 USB 事件调用回调。第 1 个参数是变量的实际名称，用来声明本宏。第 2 个和第 3 个参数是保留的，并且未使用，应当设置为 0。最后一个参数是回调函数，实际被指定为一个函数指针的类型：void (* pfnEvent)(void * pvData)；pfnEvent 是一个 void 类型的函数指针，指向 tEventInfo 函数结构体。事件一旦发生就导致 pfnEvent 函数被调用。

tUSBHostClassDriver，这个结构定义了一个 USB 主机类驱动程序接口，一旦 USB 设备枚举就解析找到一个 USB 类驱动程序。此结构体定义如下：

```
typedef struct
{
    unsigned long ulInterfaceClass;
```

```
    void * ( * pfnOpen)(tUSBHostDevice * pDevice);
    void ( * pfnClose)(void * pvInstance);
    void ( * pfnIntHandler)(void * pvInstance);
}
tUSBHostClassDriver;
```

其中 ulInterfaceClass 为当前这个设备类驱动程序支持的接口类;(* pfnOpen)(tUSBHostDevice * pDevice)为函数指针,指向当这个结构体被调用时执行的函数;(* pfnClose)(void * pvInstance),指向本结构体所定义的设备断开连接时调用的函数;(* pfnIntHandler)(void * pvInstance),当端点收到设备传送的数据并产生中断时,指向中断处理回调函数。所以,DECLARE_EVENT_DRIVER 的最后一个参数必须由开发人员编写,其函数格式应该为 void fun(void * par)所定义的类型,同时这个函数名 fun 应该与 DECLARE_EVENT_DRIVER 的最后一个参数名字一致。

例如:编写一个 USB 处理键盘数据的回调函数,并注册驱动,程序代码如下。

```
// 声明 USB 事件驱动接口
DECLARE_EVENT_DRIVER(g_sUSBEventDriver, 0, 0, USBHCDEvents);
//事件处理函数
void  USBHCDEvents(void * pvData)
{
    tEventInfo * pEventInfo;
    // 指向事件参数
    pEventInfo = (tEventInfo * )pvData;
    switch(pEventInfo->ulEvent)
    {
        // 检测键盘
        case USB_EVENT_CONNECTED:
        {
            UARTprintf("Unknown Device Connected\n");
            //未知设备
            g_eUSBState = STATE_UNKNOWN_DEVICE;
            break;
        }
        // 键盘拔出
        case USB_EVENT_DISCONNECTED:
        {
            UARTprintf("Unknown Device Disconnected\n");
            // 键盘移除
            g_eUSBState = STATE_NO_DEVICE;
            break;
        }
```

```
            case USB_EVENT_POWER_FAULT:
            {
                UARTprintf("Power Fault\n");
                //电源异常
                g_eUSBState = STATE_POWER_FAULT;
                break;
            }
            default:
            {
                break;
            }
        }
    }
```

上面实例是 USB 主机读取 USB 键盘数据的回调函数,并注册驱动,可以看出,DECLARE_EVENT_DRIVER(g_sUSBEventDriver,0,0,USBHCDEvents)中的第 2 个和第 3 个参数为 0,最后一个参数 USBHCDEvents 为回调函数名称。pvData,为回调函数的入口参数,其类型为 tEventInfo,用于记录事件信息。目前支持以下事件:USB_EVENT_CONNECTED,USB_EVENT_DISCONNECTED 和 USB_EVENT_POWER_FAULT。tEventInfo 结构体定义如下:

```
typedef struct
{
    unsigned long ulEvent;
    unsigned long ulInstance;
}
tEventInfo;
```

tEventInfo 的 ulEvent 用于记录 USB 事件类型,以位的状态实现事件记录,当前支持 3 个事件,所以在编写回调函数时只需处理 3 个状态。ulInstance,暂时没有使用。

在一个 USB 主机系统中,可能有一个、两个、甚至更多个设备需要控制,那么会出现多个设备驱动,在注册驱动时,通过一个数组存放驱动名称与回调函数:tUSBHostClassDriver。

下面是 HID 设备驱动代码:

```
const tUSBHostClassDriver g_USBHIDClassDriver =
{
    USB_CLASS_HID,
    HIDDriverOpen,
    HIDDriverClose,
    0
```

第 11 章 USB 主机开发

tUSBHostClassDriver 可以定义回调函数,也可以定义设备驱动。对于某一种设备驱动来说,须具备两个 tUSBHostClassDriver 定义,一个用于定义驱动;一个用于回调,负责数据事件处理。在注册驱动时,需把两个包含在同一个数组里,完成驱动注册,下面是一个 HID 设备的驱动数组:

```
// 注册 HID 类驱动
static tUSBHostClassDriver const * const g_ppHostClassDrivers[] =
{
    &g_USBHIDClassDriver
    ,&g_sUSBEventDriver
};
static const unsigned long g_ulNumHostClassDrivers =
    sizeof(g_ppHostClassDrivers) / sizeof(tUSBHostClassDriver *);
```

有了设备驱动的数组后,还需使用函数 USBHCDRegisterDrivers 进行驱动注册,把驱动数组里的驱动全部注册到主机中,用于设备枚举。USBHCDRegisterDrivers 函数原型如下:

```
void USBHCDRegisterDrivers(unsigned long ulIndex,
                          const tUSBHostClassDriver * const * ppHClassDrvs,
                          unsigned long ulNumDrivers)
{
    // 保存所有驱动到 g_sUSBHCD 中
    g_sUSBHCD.pClassDrivers = ppHClassDrvs;
    // 保存驱动个数
    g_sUSBHCD.ulNumClassDrivers = ulNumDrivers;
}
```

这个函数用于初始化、注册一个 HCD(Host Contoller Driver)类设备列表,该列表包含多个驱动数组。ulIndex 指定使用哪个 USB 控制器,在 LM3S 处理器中,目前只有一个 USB 模块,所以,第一个参数为 0;ppHClassDrvs,一个被主机类设备支持的数组,包含所有设备驱动;ulNumDrivers 是驱动的数量。本函数将设置支持的驱动类,应该在 USBHCDInit() 函数前调用。

例如:注册一个驱动到 USB0 模块,代码如下所示。

```
// 注册驱动
USBHCDRegisterDrivers(0, g_ppHostClassDrivers, g_ulNumHostClassDrivers);
```

USBHCDPowerConfigInit,用于设置电源脚和电源故障配置。ulIndex USB,控制器。ulPwrConfig,应用程序所使用的电源配置,以下值中的其中一个需要被选择来设置电源故障灵敏度:

- USBHCD_FAULT_LOW：一个外部电源故障被检测到引脚驱动低。
- USBHCD_FAULT_HIGH：一个外部电源故障被检测到引脚驱动高。

以下值中的其中一个需要被选择来设置电源故障检测：
- USBHCD_FAULT_VBUS_NONE：不自动检测电源故障。
- USBHCD_FAULT_VBUS_TRI：故障时自动设置 USBnEPEN 脚为三态门。
- USBHCD_FAULT_VBUS_DIS：没有电源故障时自动驱动 USBnEPEN 脚。

以下值中的其中一个需要被选择来设置电源使能脚等级和来源：
- USBHCD_VBUS_MANUAL：电源控制完全由应用程序管理。
- USBHCD_VBUS_AUTO_LOW：USBEPEN 自动驱动为低。
- USBHCD_VBUS_AUTO_HIGH：USBEPEN 自动驱动为高。

如果 USBHCD_VBUS_MANUAL 被使用，那么应用程序需要提供一个事件驱动来接收 USB_EVENT_POWER_ENABLE 和 USB_EVENT_POWER_DISABLE 事件，以及使能和禁止 VBUS。USBHCDPowerConfigInit 的源码如下：

```
void USBHCDPowerConfigInit(unsigned long ulIndex, unsigned long ulPwrConfig)
{
    ASSERT(ulIndex == 0);
    g_ulPowerConfig = ulPwrConfig;
}
```

本函数必须在 HCDInit() 函数之前调用，以使电源配置能在电源使能前设置好，保证设备供电正常。参数 ulPwrConfig 为电源故障灵敏度参数，为 USB 处理器配置电源故障检测和电源使能脚等级和来源。

11.2.3 运行主机

USB 设备对应的驱动程序安装成功后，USB 主机便可通过 USB 驱动接口函数访问该 USB 设备，在这个阶段需要用到两个函数：USBHCDInit 和 USBHCDMain。USBHCDInit 用于初始化 USB 主机的控制程序，USBHCDMain 模拟实时操作系统，定义调用，用于检测 USB 通信。

USBHCDInit，用于初始化 HCD。ulIndex USB，模块编号；pvPool 指向一个缓存池；ulPoolSize 表示缓存池的大小。本函数将初始化一个 USB 主机控制，开始枚举以及与设备进行通信，本函数应该在应用程序开始时调用一次，本调用将启动一个 USB 主机控制器，并且连接任何的设备将立即开始枚举序列。USBHCDInit 源代码如下：

```
void USBHCDInit(unsigned long ulIndex, void * pvPool, unsigned long ulPoolSize)
{
    int iDriver;
    // 确保 ulPoolSize 定义的内存空间能够存放下读取的描述符
```

```c
ASSERT(ulPoolSize >= sizeof(tConfigDescriptor));
// 确保工作在设备模式下不调用这个函数
ASSERT(g_eUSBMode != USB_MODE_DEVICE);
// 如果没有设备 USB 处理器工作模式,就默认为主机模式
if(g_eUSBMode == USB_MODE_NONE)
{
    g_eUSBMode = USB_MODE_HOST;
}
//如果工作在主机模式下时,更新一下硬件,在 OTG 模式下不更新
if(g_eUSBMode == USB_MODE_HOST)
{
    SysCtlPeripheralEnable(SYSCTL_PERIPH_USB0);
    SysCtlUSBPLLEnable();
}
// 通过内部函数初始化 USB 控制器,准备枚举
USBHCDInitInternal(ulIndex, pvPool, ulPoolSize);
// 默认情况下没有事件驱动
g_sUSBHCD.iEventDriver = -1;
// 扫描驱动列表中的事件驱动,用于枚举
for(iDriver = 0; iDriver < g_sUSBHCD.ulNumClassDrivers; iDriver++)
{
    if(g_sUSBHCD.pClassDrivers[iDriver]->ulInterfaceClass ==
       USB_CLASS_EVENTS)
    {
        //如果有事件驱动,把事件驱动提取出来
        g_sUSBHCD.iEventDriver = iDriver;
    }
}
// 设置 ms 定时
g_ulTickms = SysCtlClockGet() / 3000;
// 检查是否要使用 DMA
if(CLASS_IS_DUSTDEVIL && REVISION_IS_A0)
{
    g_bUseDMAWA = 1;
}
}
```

通过 USBHCDInit 可以看出,USBHCDInit 主要负责初始化 HCD,根据需求更新硬件,通过 USBHCDInitInternal 初始化 USB 控制器,完成通道、DMA、定时器、端点、中断、电源初始化与配置,为枚举做准备。

USBHCDMain,主机控制设备的主程序,必须在应用程序中周期性地调用,使用时钟节拍机制来处理主机控制设备接口而无需使用一个实时多任务操作系统

第 11 章 USB 主机开发

(RTOS),所有时间边界代码(在操作系统中需要开中断和关中断的代码区)处理全部在一个中断上下文中,但是所有堵塞操作不在此函数中进行,从而允许这些堵塞操作函数完成等待而不挂起其他的中断。在使用 USB 主机时,需要为 USBHCDMain 提供基础时钟,来完成多任务操作系统的模拟,所以要循环调用此函数。USBHCD-Main 的源代码如下:

```c
void USBHCDMain(void)
{
    unsigned long ulIntState;
    tUSBHDeviceState eOldState;
    // 保存设备状态
    eOldState = g_sUSBHCD.eDeviceState[0];
    // 判断中断事件
    if(g_ulUSBHIntEvents)
    {
        // 关中断
        ulIntState = IntMasterDisable();
        if(g_ulUSBHIntEvents & INT_EVENT_POWER_FAULT)
        {
            // 电源故障
            if((g_sUSBHCD.iEventDriver != -1) &&
               (g_sUSBHCD.pClassDrivers[g_sUSBHCD.iEventDriver]->pfnIntHandler))
            {
                // 发送电源故障事件
                g_sUSBHCD.EventInfo.ulEvent = USB_EVENT_POWER_FAULT;
                //调用事件驱动
                g_sUSBHCD.pClassDrivers[g_sUSBHCD.iEventDriver]->pfnIntHandler(
                    &g_sUSBHCD.EventInfo);
            }
            g_sUSBHCD.eDeviceState[0] = HCD_POWER_FAULT;
        }
        else if(g_ulUSBHIntEvents & INT_EVENT_VBUS_ERR)
        {
            //VBUS 异常
            g_sUSBHCD.eDeviceState[0] = HCD_VBUS_ERROR;
        }
        else
        {
            // 检查是否有设备连接
            if(g_ulUSBHIntEvents & INT_EVENT_CONNECT)
            {
```

```c
            g_sUSBHCD.eDeviceState[0] = HCD_DEV_RESET;
        }
        else
        {
            // 检查是否有设备断开连接
            if(g_ulUSBHIntEvents & INT_EVENT_DISCONNECT)
            {
                g_sUSBHCD.eDeviceState[0] = HCD_DEV_DISCONNECTED;
            }
        }
    }
    //清除事件标志
    g_ulUSBHIntEvents = 0;
    // 打开总中断
    if(! ulIntState)
    {
        IntMasterEnable();
    }
}
//根据设备状态,做相应处理
switch(g_sUSBHCD.eDeviceState[0])
{
    //电源异常
    case HCD_POWER_FAULT:
    {
        break;
    }
    // VBUS 异常
    case HCD_VBUS_ERROR:
    {
    //关中断
        IntDisable(INT_USB0);
        if((eOldState != HCD_IDLE) && (eOldState != HCD_POWER_FAULT))
        {
            //断开连接
            USBHCDDeviceDisconnected(0);
        }
        //复位 USB 模块控制器
        SysCtlPeripheralReset(SYSCTL_PERIPH_USB0);
        //延时 100 ms
        SysCtlDelay(g_ulTickms * 100);
        //重新初始化 HCD
```

```c
            USBHCDInitInternal(0, g_sUSBHCD.pvPool, g_sUSBHCD.ulPoolSize);
            break;
        }
        case HCD_DEV_RESET:
        {
            // 触发 HCD 复位
            USBHCDReset(0);
            //设备连接
            g_sUSBHCD.eDeviceState[0] = HCD_DEV_CONNECTED;
            break;
        }
        //设备已经连接,开始枚举
        case HCD_DEV_CONNECTED:
        {
            // 在进行数据处理前,获取设备描述符
            if(g_sUSBHCD.USBDevice[0].DeviceDescriptor.bLength == 0)
            {
                // 发送设备描述符请求
                if(USBHCDGetDeviceDescriptor(0, &g_sUSBHCD.USBDevice[0]) == 0)
                {
                    // 如果不能读取设备描述符,设置为未知设备
                    g_sUSBHCD.eDeviceState[0] = HCD_DEV_ERROR;
                    //发送未知连接
                    SendUnknownConnect(0, 1);
                }
            }
            // 如果获取到设备描述符,现在设置设备地址
            else if(g_sUSBHCD.USBDevice[0].ulAddress == 0)
            {
                // 发送设置设备地址命令
                USBHCDSetAddress(1);
                // 保存设备地址
                g_sUSBHCD.USBDevice[0].ulAddress = 1;
                // 设备地址设置完成状态
                g_sUSBHCD.eDeviceState[0] = HCD_DEV_ADDRESSED;
            }
            break;
        }
        case HCD_DEV_ADDRESSED:
        {
            // 检查配置描述符
            if (g_sUSBHCD.USBDevice[0].pConfigDescriptor == 0)
```

```c
        {
            // 获取设备描述符
            if(USBHCDGetConfigDescriptor(0, &g_sUSBHCD.USBDevice[0]) == 0)
            {
                // 如果不能读取设备描述符,设置为未知设备
                g_sUSBHCD.eDeviceState[0] = HCD_DEV_ERROR;
                // 发送未知连接
                SendUnknownConnect(0, 1);
            }
            // 地址与设备描述符都已经获取,现在进行配置
            else
            {
                // 第一次进行设备配置
                USBHCDSetConfig(0, (unsigned long)&g_sUSBHCD.USBDevice[0], 1);
                // 配置状态
                g_sUSBHCD.eDeviceState[0] = HCD_DEV_CONFIGURED;
                // 打开驱动 0
                g_iUSBHActiveDriver = USBHCDOpenDriver(0, 0);
            }
            break;
        }
        // 获取设备请求
        case HCD_DEV_REQUEST:
        {
            g_sUSBHCD.eDeviceState[0] = HCD_DEV_CONNECTED;
            break;
        }

        // 获取字符串描述符
        case HCD_DEV_GETSTRINGS:
        {
            break;
        }
        //设备断开连接
        case HCD_DEV_DISCONNECTED:
        {
            // 断开 USB0
            USBHCDDeviceDisconnected(0);
            // 返回空闲状态
            g_sUSBHCD.eDeviceState[0] = HCD_IDLE;
            break;
        }
```

```
            // 连接与枚举完成,可以跳出本函数
            case HCD_DEV_CONFIGURED:
            {
                break;
            }
            //设备故障,等待移除
            case HCD_DEV_ERROR:
            default:
            {
                break;
            }
        }
    }
```

从 USBHCDMain 源代码中可以看出,此函数主要完成主机事件处理、枚举设备等,本函数首先检测电源、VBUS 异常,并发出响应;再通过 g_sUSBHCD.eDeviceState[0]记录的设备状态进行事件处理,HCD_DEV_RESET 后就开始获取设备描述符;如果获取成功,主机发送设备地址,让设备自动设置主机分配的地址;地址设置完成时,再通过设备的新地址与 USB 主机进行通信,再次获取设备描述符、字符串描述符,从源代码中可以看出,本函数并没有处理字符串描述符。USBHCDMain 调用函数图如图 11-6 所示。为了实时处理 USB 数据,在此 USB 库中简单地模拟了实时操作系统,并可以及时处理相关数据,USBHCDMain 必须周期性地调用。

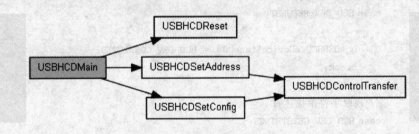

图 11-6　USBHCDMain 函数调用图

11.3　主机开发实例

前面两节介绍了 USB 主机开发的基础知识,本节主要以 USB 主机控制鼠标、键盘、U 盘、USB 音频设备为例,深入详细地介绍 USB 主机的开发过程。

11.3.1　鼠　标

本书的第 5 章介绍了 HID 设备开发,主要以鼠标与键盘为例进行了深入分析与讲解,因为鼠标与键盘的广泛使用,此 USB 类设备开发成为 USB 开发的一个热点,

第 11 章　USB 主机开发

同时为了适应嵌入式开发模式,要对 HID 设备类进行控制,必须开发对应 HID 设备的 USB 主机。

鼠标和键盘是日常生活中最常用的便携式设备之一,之前 ps2 接口的鼠标和键盘非常流行,但后来随着 USB 技术的发展,具有 USB 接口的鼠标和键盘遍地开花、发展迅速,并占有了 90% 的市场,鼠标和键盘生产厂商也充分利用 USB 的灵活特性,在此基础上研发出无线鼠标键盘,大量生产,并占有相当大的市场。所以 USB HID 设备与主机开发的重要性不言而喻,同时 USB 的 HID 设备离不开对应的 USB 主机支持,因此 USB HID 主机开发具有相当重要的地位。

本节主要介绍支持 USB 鼠标的 USB 主机开发,在 USB 库中已经编写好 USB 鼠标驱动,供 USB 主机使用,相关函数包含在 usbhhidmouse.h 头文件中。主机包含 3 个功能函数,实现对鼠标的数据读取、操作:USBHMouseOpen、USBHMouseClose、USBHMouseInit。在 USB 库中,默认报告为 4 个字节。如果有必要,可以打开 USBlib 源码,在 usbhhidmouse.c 中修改"#define USBHMS_REPORT_SIZE   4"为鼠标报告的实际字节数,由 USB 库实现的 USB 鼠标默认为 4 个字节。

```
unsigned long USBHMouseOpen(tUSBCallback pfnCallback,
                             unsigned char * pucBuffer,
                             unsigned long ulBufferSize);
unsigned long USBHMouseClose(unsigned long ulInstance);
unsigned long USBHMouseInit(unsigned long ulInstance);
```

(1) unsigned long USBHMouseOpen(tUSBCallback pfnCallback,
　　　　　　　　　　　　　　　　unsigned char * pucBuffer,
　　　　　　　　　　　　　　　　unsigned long ulBufferSize);

此函数用于打开一个鼠标实例。pfnCallback,回调函数,一但发生一个新的鼠标事件时会调用此回调函数;* pucBuffer,USB 主机与 USB 鼠标数据传输的数据缓存。该缓存的大小至少能容纳一个普通的设备描述符报告,一般定义为 128 即可满足需求;ulSize,USB 主机与 USB 鼠标数据传输的数据缓存大小。本函数主要用于打开一个鼠标实例。本函数返回的值用于区别其他所有 USB 鼠标的"句柄",一个常量对应一个鼠标实例,如果没有鼠标实例,则返回 0。

USBHMouseOpen 的函数代码如下:

```
unsigned long USBHMouseOpen(tUSBCallback pfnCallback, unsigned char * pucBuffer,
                             unsigned long ulSize)
{
    // 外部鼠标回调函数传递给全局鼠标结构体的回调函数, g_sUSBHMouse.pfnCallback
    g_sUSBHMouse.pfnCallback = pfnCallback;
    // 打开 USBHID 实例,用于初始化 HID 设备层
```

深入浅出 USB 系统开发——基于 ARM Cortex-M3

```
        g_sUSBHMouse.ulMouseInstance = USBHHIDOpen(USBH_HID_DEV_MOUSE,
                                                    USBHMouseCallback,
                                                    (unsigned long)&g_sUSBHMouse);
        // 传递数据缓存指针与大小
        g_sUSBHMouse.pucHeap = pucBuffer;
        g_sUSBHMouse.ulHeapSize = ulSize;
        return((unsigned long)&g_sUSBHMouse);
    }
```

通过 USBHMouseOpen 的函数代码可以看出,此函数通过 USBHHIDOpen 初始化 USBHID 驱动层,并返回鼠标实例,同时 USBHMouseOpen 函数完成鼠标实例 g_sUSBHMouse 的初始化,其中 g_sUSBHMouse 主要保存与鼠标实例相关的参数与函数,包含鼠标实例的所有数据。

例如:打开一个鼠标实例,程序代码如下所示。

```
// USB 鼠标接口需要的内存空间
#define MOUSE_MEMORY_SIZE        128
unsigned char g_pucBuffer[MOUSE_MEMORY_SIZE];
// 声明 USB 驱动,并产生 g_sUSBEventDriver
DECLARE_EVENT_DRIVER(g_sUSBEventDriver, 0, 0, USBHCDEvents);
//注册 g_USBHIDClassDriver HID 设备类,同时也注册一个 g_sUSBEventDriver 事件驱动
static tUSBHostClassDriver const * const g_ppHostClassDrivers[] =
{
    &g_USBHIDClassDriver
    ,&g_sUSBEventDriver
};
//驱动个数
//************************************************************
static const unsigned long g_ulNumHostClassDrivers =
    sizeof(g_ppHostClassDrivers) / sizeof(tUSBHostClassDriver *);


//鼠标实例,可以理解为句柄
//************************************************************
static unsigned long g_ulMouseInstance;
// 注册驱动
USBHCDRegisterDrivers(0, g_ppHostClassDrivers, g_ulNumHostClassDrivers);
UARTprintf("Host Mouse Application\n");
//打开鼠标驱动实例
g_ulMouseInstance =
        USBHMouseOpen(MouseCallback, g_pucBuffer, MOUSE_MEMORY_SIZE);
UARTprintf("USBHMouseOpen is OK! \n");
```

MOUSE_MEMORY_SIZE,定义 128 个字节,USB 主机与 USB 鼠标数据传输的数据缓存大小,该缓存的大小至少能容纳一个普通的设备描述符报告,所以定义 128 个字节完全足够;g_pucBuffer[MOUSE_MEMORY_SIZE]定义缓存,用于存放设备描述符报告;DECLARE_EVENT_DRIVER(g_sUSBEventDriver, 0, 0, USBHCDEvents),声明 USBHCDEvents 的事件驱动;g_ppHostClassDrivers,是设备驱动与事件驱动的集合,用于驱动注册时使用;g_ulNumHostClassDrivers,记录设备驱动与事件驱动个数;g_ulMouseInstance,鼠标实例,可以理解为句柄,是一个 32 位的无符号数;USBHCDRegisterDrivers,注册事件驱动与设备驱动;USBHMouseOpen,打开鼠标驱动,并返回一个驱动标号(句柄)g_ulMouseInstance。

(2) unsigned long USBHMouseClose(unsigned long ulInstance);

本函数用于关闭一个已打开的鼠标实例。参数 ulInstance,鼠标实例的标识值,应该为应用程序调用 USBHMouseOpen()函数时返回的值。本函数返回 0。

例如:关闭一个已经打开的鼠标实例,程序代码如下所示。

```
// USB 鼠标接口需要的内存空间
#define MOUSE_MEMORY_SIZE        128
unsigned char g_pucBuffer[MOUSE_MEMORY_SIZE];
// 声明 USB 驱动,并产生 g_sUSBEventDriver
DECLARE_EVENT_DRIVER(g_sUSBEventDriver, 0, 0, USBHCDEvents);
//注册 g_USBHIDClassDriver HID 设备类,同时也注册一个 g_sUSBEventDriver 事件驱动
static tUSBHostClassDriver const * const g_ppHostClassDrivers[] =
{
    &g_USBHIDClassDriver
    ,&g_sUSBEventDriver
};
//驱动个数
//**************************************************************
static const unsigned long g_ulNumHostClassDrivers =
    sizeof(g_ppHostClassDrivers) / sizeof(tUSBHostClassDriver *);

//鼠标实例,可以理解为句柄
//**************************************************************
static unsigned long g_ulMouseInstance;
// 注册驱动
USBHCDRegisterDrivers(0, g_ppHostClassDrivers, g_ulNumHostClassDrivers);
//打开鼠标驱动实例
g_ulMouseInstance =
        USBHMouseOpen(MouseCallback, g_pucBuffer, MOUSE_MEMORY_SIZE);
........
```

```
//关闭 g_ulMouseInstance 这个鼠标实例
USBHMouseClose(g_ulMouseInstance);
```

(3) unsigned long USBHMouseInit(unsigned long ulInstance);

本函数用于当检测到有一个鼠标实例时,初始化一个鼠标接口,然后向 USB 主机控制器报告此事件。ulInstance,鼠标实例的标识值,应该为应用程序调用 USBHMouseOpen()函数时返回的值,在调用函数 USBHMouseOpen()后接收到 USB_EVENT_CONNECTED 事件时,本函数必须调用,并且每次检测到 USB_EVENT_CONNECTED 事件发生时都要调用一次,不管怎么样,该函数只能在回调函数之外调用。本函数如果返回非零值表明发生了一个错误事件。

USBHMouseInit 的函数代码如下:

```
unsigned long USBHMouseInit(unsigned long ulInstance)
{
    tUSBHMouse * pUSBHMouse;
    // 获取鼠标实例的所有数据
    pUSBHMouse = (tUSBHMouse *)ulInstance;
    // 设置一个 HID 设备空闲
    USBHHIDSetIdle(pUSBHMouse->ulMouseInstance, 0, 0);
    // 获取一个给定设备实例的报告描述符,并存放于描述符的数据缓存空间内
    USBHHIDGetReportDescriptor(pUSBHMouse->ulMouseInstance,
                               pUSBHMouse->pucHeap,
                               pUSBHMouse->ulHeapSize);
    // 函数用于清除或设置一个设备的引导协议状态
    USBHHIDSetProtocol(pUSBHMouse->ulMouseInstance, 1);
    return(0);
}
```

通过 USBHMouseInit 的函数代码可看出,此函数调用 USBHHIDSetIdle 使 HID 主机处于空闲状态;通过 USBHHIDGetReportDescriptor 获取报告描述符;USBHHIDSetProtocol,设置设备的引导协议状态。从而完成鼠标实例的初始化。

例如:初始化鼠标实例,程序代码如下。

```
// 鼠标的回调函数
unsigned long MouseCallback(void * pvCBData, unsigned long ulEvent, unsigned long
                            ulMsgParam, void * pvMsgData)
{
    switch(ulEvent)
    {
        case USB_EVENT_CONNECTED:
        {
```

```
            UARTprintf("Mouse Connected\n");
            eUSBState = STATE_MOUSE_INIT;
            break;
        }
        case USB_EVENT_DISCONNECTED:
        {
            UARTprintf("Mouse Disconnected\n");
            eUSBState = STATE_NO_DEVICE;
            break;
        }
        ......
    return(0);
}
    while(1)
    {
        switch(eUSBState)
        {
            case STATE_MOUSE_INIT:
            {
                //初始化鼠标
                UARTprintf("init the mouse! \n");
                USBHMouseInit(g_ulMouseInstance);
                eUSBState = STATE_MOUSE_CONNECTED;
                break;
            }
            case STATE_MOUSE_CONNECTED:
            {
                UARTprintf("STATE_MOUSE_CONNECTED! \n");
                break;
            }
            ......
        }
        //循环调用USB主机控制单元
        USBHCDMain();
    }
```

USBHMouseInit 用于初始化一个鼠标实例，必须在检测到有 USB_EVENT_CONNECTED 事件时调用此函数，即使是 MouseCallback 内检测 USB_EVENT_

CONNECTED 事件,也不能直接在 MouseCallback 鼠标回调函数内调用,应该通过修改标志位,在 MouseCallback 之外调用 USBHMouseInit。例如上面的代码中,MouseCallback 函数检测到 USB_EVENT_CONNECTED 事件时,修改 eUSBState 标志位为 STATE_MOUSE_INIT,在主函数中,检测到 eUSBState 标志位为 STATE_MOUSE_INIT 时,调用 USBHMouseInit,初始化鼠标实例,之后修改 eUSBState 标志位为 STATE_MOUSE_CONNECTED,表示鼠标连接成功,可以使用。注意,每次 MouseCallback 检测到有 USB_EVENT_CONNECTED 事件发生时,都要调用 USBHMouseInit 对鼠标进行初始化。

通过以上 3 个函数,就可以完成 USB 鼠标的初始化与控制,那么 USB 鼠标数据(X、Y、按键)怎么传递给应用程序呢?

在调用 USBHMouseOpen 打开鼠标驱动实例时,其第一个参数 MouseCallback,用于 USB 鼠标事件回调,所有鼠标事件与数据都通过此函数传递。此函数的函数原型如下:

unsigned long MouseCallback(void * pvCBData, unsigned long ulEvent, unsigned long ulMsgParam, void * pvMsgData);

其中,pvCBData 为回调函数数据指针,此数据由内部函数给出,通常值为 0;ulMsgParam 为一个事件的具体参数,通过它检测状态:

USBH_EVENT_HID_MS_PRESS   //按键按下
USBH_EVENT_HID_MS_REL   //按键放开
USBH_EVENT_HID_MS_X   //X 值发生变化
USBH_EVENT_HID_MS_Y   //Y 值发生变化

pvMsgData,一个事件特定的数据指针。USBH_EVENT_HID_MS_PRESS 事件发生时,pvMsgData 提供按下按键的值;USBH_EVENT_HID_MS_REL 事件发生时,提供哪些按键松开;USBH_EVENT_HID_MS_X 事件发生时,提供当前 X 值;USBH_EVENT_HID_MS_Y 事件发生时,提供当前 Y 值。这个回调函数用于设备类驱动和主机管理。

本小结主要讲解 USB 鼠标的控制和数据读取。本小节的 USB 鼠标主机控制的随书源码请见本书配套资料。

11.3.2 键 盘

本节主要介绍支持 USB 键盘的 USB 主机开发,在 USB 库中已经编写好 USB 键盘驱动,供 USB 主机使用,相关函数包含在 usbhhidkeyboard.h 头文件中。主机包含 6 个功能函数,实现对键盘的数据读取、操作:USBHKeyboardOpen、USBHKeyboardClose、USBHKeyboardInit、USBHKeyboardModifierSet、USBHKeyboardPollRateSet、USBHKeyboardUsageToChar。在 USB 库中,默认报告为 8 个字节。如果有必要,可以打开 USBlib 源码,在 usbhhidkeyboard.c 中修改"#define USBH-

KEYB_REPORT_SIZE   8"为键盘报告的实际字节数,由 USB 库做的 USB 键盘默认为 8 个字节。

```
unsigned long USBHKeyboardOpen(tUSBCallback pfnCallback,
                               unsigned char * pucBuffer,
                               unsigned long ulBufferSize);
unsigned long USBHKeyboardClose(unsigned long ulInstance);
unsigned long USBHKeyboardInit(unsigned long ulInstance);
unsigned long USBHKeyboardModifierSet(unsigned long ulInstance,
                                      unsigned long ulModifiers);
unsigned long USBHKeyboardPollRateSet(unsigned long ulInstance,
                                      unsigned long ulPollRate);
unsigned long USBHKeyboardUsageToChar(unsigned long ulInstance,
                                      const tHIDKeyboardUsageTable * pTable,
                                      unsigned char ucUsageID);
```

(1) unsigned long USBHKeyboardOpen(tUSBCallback pfnCallback,
                                   unsigned char * pucBuffer,
                                   unsigned long ulBufferSize);

此函数用于打开一个键盘实例。pfnCallback,回调函数,一但发生一个新的键盘事件时会调用此回调函数;* pucBuffer,USB 主机与 USB 键盘数据传输的数据缓存。该缓存的大小至少能容纳一个普通的设备描述符报告,如果没有足够的空间,则只有部分报告能够读取,一般定义为 128 即可满足需求;ulSize,USB 主机与 USB 键盘数据传输的数据缓存大小。本函数主要用于打开一个键盘实例。本函数返回的值用于区别其他所有 USB 键盘的"句柄",一个常量对应一个键盘实例,如果没有键盘实例,则返回 0。

USBHKeyboardOpen 的函数代码如下:

```
unsigned long USBHKeyboardOpen(tUSBCallback pfnCallback, unsigned char * pucBuffer,
                               unsigned long ulSize)
{
    // 外部键盘回调函数传递给全局键盘结构体的回调函数,g_sUSBHKeyboard
    // pfnCallback
    g_sUSBHKeyboard.pfnCallback = pfnCallback;
    // 打开 USBHID 实例,用于初始化 HID 设备层
    g_sUSBHKeyboard.ulHIDInstance =
        USBHHIDOpen(USBH_HID_DEV_KEYBOARD, USBHKeyboardCallback,
                    (unsigned long)&g_sUSBHKeyboard);

    return((unsigned long)&g_sUSBHKeyboard);
}
```

第 11 章　USB 主机开发

通过 USBHKeyboardOpen 的函数代码可看出,此函数通过 USBHHIDOpen 初始化 USBHID 驱动层,并返回键盘实例,同时 USBHKeyboardOpen 函数完成键盘实例 g_sUSBHKeyboard 的初始化,其中 g_sUSBHKeyboard 主要保存与键盘实例相关的参数与函数,包含键盘实例的所有数据。

例如:打开一个键盘实例,程序代码如下所示。

```
// USB 键盘接口需要的内存空间
#define KEYBOARD_MEMORY_SIZE    128
// 分配内存单元给键盘数据存储
unsigned char g_pucBuffer[KEYBOARD_MEMORY_SIZE];
// 声明 USB 驱动,并产生 g_sUSBEventDriver
DECLARE_EVENT_DRIVER(g_sUSBEventDriver, 0, 0, USBHCDEvents);
//注册 g_USBHIDClassDriver HID 设备类,同时也注册一个 g_sUSBEventDriver 事件驱动
static tUSBHostClassDriver const * const g_ppHostClassDrivers[] =
{
    &g_USBHIDClassDriver
    ,&g_sUSBEventDriver
};
static const unsigned long g_ulNumHostClassDrivers =
    sizeof(g_ppHostClassDrivers) / sizeof(tUSBHostClassDriver *);
//键盘实例,可以理解为句柄
static unsigned long g_ulKeyboardInstance;
// 注册驱动
USBHCDRegisterDrivers(0, g_ppHostClassDrivers, g_ulNumHostClassDrivers);
// 输出信息
UARTprintf("Host Keyboard Application\n");
// 打开键盘驱动,初始化驱动
g_ulKeyboardInstance = USBHKeyboardOpen(KeyboardCallback, g_pucBuffer,
                           KEYBOARD_MEMORY_SIZE);
```

KEYBOARD_MEMORY_SIZE,定义为 128 个字节,USB 主机与 USB 键盘数据传输的数据缓存大小,该缓存的大小至少能容纳一个普通的设备描述符报告,所以定义 128 个字节完全足够;g_pucBuffer[KEYBOARD_MEMORY_SIZE]定义缓存,用于存放设备描述符报告;DECLARE_EVENT_DRIVER(g_sUSBEventDriver, 0, 0, USBHCDEvents),声明 USBHCDEvents 的事件驱动;g_ppHostClassDrivers,是设备驱动与事件驱动的集合,用于驱动注册时使用;g_ulNumHostClassDrivers,记录设备驱动与事件驱动个数;g_ulKeyboardInstance,键盘实例,可以理解为句柄,是一个 32 位的无符号数;USBHCDRegisterDrivers,注册事件驱动与设备驱动;USBHKeyboardOpen,打开键盘驱动,并返回一个驱动标号(句柄)g_ulKeyboardInstance。

(2) unsigned long USBHKeyboardClose(unsigned long ulInstance);

本函数用于关闭一个已打开的键盘实例。参数 ulInstance,键盘实例的标识值,应该为应用程序调用 USBHKeyboardOpen()函数时返回的值。本函数返回0。

例如:关闭一个已经打开的键盘实例,程序代码如下所示。

```
// USB 键盘接口需要的内存空间
#define KEYBOARD_MEMORY_SIZE    128
// 分配内存单元给键盘数据存储
unsigned char g_pucBuffer[KEYBOARD_MEMORY_SIZE];
// 声明 USB 驱动,并产生 g_sUSBEventDriver
DECLARE_EVENT_DRIVER(g_sUSBEventDriver, 0, 0, USBHCDEvents);
//注册 g_USBHIDClassDriver HID 设备类,同时也注册一个 g_sUSBEventDriver 事件驱动
static tUSBHostClassDriver const * const g_ppHostClassDrivers[] =
{
    &g_USBHIDClassDriver
    ,&g_sUSBEventDriver
};
static const unsigned long g_ulNumHostClassDrivers =
    sizeof(g_ppHostClassDrivers) / sizeof(tUSBHostClassDriver *);
//键盘实例,可以理解为句柄
static unsigned long g_ulKeyboardInstance;
// 注册驱动
USBHCDRegisterDrivers(0, g_ppHostClassDrivers, g_ulNumHostClassDrivers);
// 输出信息
UARTprintf("Host Keyboard Application\n");
// 打开键盘驱动,初始化驱动
g_ulKeyboardInstance = USBHKeyboardOpen(KeyboardCallback, g_pucBuffer,
KEYBOARD_MEMORY_SIZE);

.......

//关闭 g_ulKeyboardInstance 这个键盘实例
USBHKeyboardClose(g_ulKeyboardInstance );
```

(3) unsigned long USBHKeyboardInit(unsigned long ulInstance);

本函数用于当检测到有一个键盘实例时,初始化一个键盘接口,然后向 USB 主机控制器报告此事件。ulInstance,键盘实例的标识值,应该为应用程序调用 USBHKeyboardOpen()函数时返回的值,在调用函数 USBHKeyboardOpen()后接收到 USB_EVENT_CONNECTED 事件时,本函数必须调用,并且每次检测到 USB_EVENT_CONNECTED 事件发生时都要调用一次,不管怎么样,该函数只能在回调函数之外调用。本函数如果返回非零值表明发生了一个错误事件。

USBHKeyboardInit 的函数代码如下：

```
unsigned long USBHKeyboardInit(unsigned long ulInstance)
{
    unsigned char ucModData;
    tUSBHKeyboard * pUSBHKeyboard;
    // 获取键盘实例数据
    pUSBHKeyboard = (tUSBHKeyboard *)ulInstance;
    //HID 主机设为空闲
    USBHHIDSetIdle(pUSBHKeyboard->ulHIDInstance, 0, 0);
    // 获取报告描述符
    USBHHIDGetReportDescriptor(pUSBHKeyboard->ulHIDInstance,
                               pUSBHKeyboard->pucBuffer,
                               USBHKEYB_REPORT_SIZE);

    // 设置启动协议
    USBHHIDSetProtocol(pUSBHKeyboard->ulHIDInstance, 1);
    ucModData = 0;
    //通过报告描述符,设置键盘灯
    USBHHIDSetReport(pUSBHKeyboard->ulHIDInstance, 0, &ucModData, 1);
    return(0);
}
```

通过 USBHKeyboardInit 的函数代码可看出，此函数调用 USBHHIDSetIdle 使 HID 主机处于空闲状态；通过 USBHHIDGetReportDescriptor 获取报告描述符；USBHHIDSetProtocol，设置设备的引导协议状态。通过 USBHHIDSetReport 设置报告描述符，控制键盘灯。从而完成键盘实例的初始化。

例如：初始化键盘实例，程序代码如下所示。

```
unsigned long KeyboardCallback(void * pvCBData, unsigned long ulEvent,
                               unsigned long ulMsgParam, void * pvMsgData)
{
    unsigned char ucChar;
    switch(ulEvent)
    {
        // 检测连接
        case USB_EVENT_CONNECTED:
        {
            UARTprintf("Keyboard Connected\n");
            g_eUSBState = STATE_KEYBOARD_INIT;
            break;
        }
        //键盘移出
```

```c
case USB_EVENT_DISCONNECTED:
{
    UARTprintf("Keyboard Disconnected\n");
    g_eUSBState = STATE_NO_DEVICE;
    g_ulModifiers = 0;
    break;
}
// 按键被按下
case USBH_EVENT_HID_KB_PRESS:
{
    // Caps Lock 状态改变
    if(ulMsgParam == HID_KEYB_USAGE_CAPSLOCK)
    {
        g_eUSBState = STATE_KEYBOARD_UPDATE;
        g_ulModifiers ^= HID_KEYB_CAPS_LOCK;
    }
    else
    {
        // backspace 按键
        if((unsigned char)ulMsgParam == HID_KEYB_USAGE_BACKSPACE)
        {
            ucChar = ASCII_BACKSPACE;
        }
        else
        {
            //获取按键的 ASCII 码
            ucChar = (unsigned char)
                USBHKeyboardUsageToChar(g_ulKeyboardInstance,
                                        &g_sUSKeyboardMap,
                                        (unsigned char)ulMsgParam);
        }
    }
    break;
}
case USBH_EVENT_HID_KB_MOD:
{
    //your code
    break;
}
case USBH_EVENT_HID_KB_REL:
{
    //your code
```

```c
            break;
        }
    }
    return(0);
}

// 应用程序
........
 while(1)
 {
        switch(g_eUSBState)
        {
            // 第一次插入键盘,用于初始化
            case STATE_KEYBOARD_INIT:
            {
                // 键盘初始化
                USBHKeyboardInit(g_ulKeyboardInstance);
                // 改变当前状态为已经连接到键盘
                g_eUSBState = STATE_KEYBOARD_CONNECTED;
                break;
            }
            case STATE_KEYBOARD_UPDATE:
            {
                // 已经有键盘数据到达
                g_eUSBState = STATE_KEYBOARD_CONNECTED;
                //设定键盘的按键状态,g_ulModifiers 为功能键屏蔽位
                USBHKeyboardModifierSet(g_ulKeyboardInstance, g_ulModifiers);
                break;
            }
            case STATE_KEYBOARD_CONNECTED:
            {
                //空闲状态
                break;
            }
            case STATE_UNKNOWN_DEVICE:
            {
                //位置设备
                break;
            }
            case STATE_NO_DEVICE:
            {
                //无设备
```

```
            break;
        }
        default:
        {
            break;
        }
    }
    //定时调用 USB 主控
    USBHCDMain();
}
```

USBHKeyboardInit 用于初始化一个键盘实例,必须在检测到有 USB_EVENT_CONNECTED 事件时调用此函数,即使是 KeyboardCallback 内检测 USB_EVENT_CONNECTED 事件,也不能直接在 KeyboardCallback 键盘回调函数内调用,应该通过修改标志位,在 KeyboardCallback 之外调用 USBHKeyboardInit。例如上面代码中,KeyboardCallback 函数检测到 USB_EVENT_CONNECTED 事件时,修改 g_eUSBState 标志位为 STATE_KEYBOARD_INIT,在主函数中,检测到 g_eUSBState 标志位为 STATE_KEYBOARD_INIT 时,调用 USBHKeyboardInit,初始化键盘实例,之后修改 g_eUSBState 标志位为 STATE_KEYBOARD_CONNECTED,表示键盘连接成功,可以使用。注意,每次 KeyboardCallback 检测到有 USB_EVENT_CONNECTED 事件发生时,都要调用 USBHKeyboardInit 对键盘进行初始化。

通过以上 3 个函数,就可以完成 USB 键盘的初始化与控制,那么 USB 键盘数据(按键被按下、松开等)怎么传递给应用程序呢?

在调用 USBHKeyboardOpen 打开键盘驱动实例时,其第一个参数 KeyboardCallback,用于 USB 键盘事件回调,所有键盘事件与数据都通过此函数传递。此函数的函数原型如下:

```
unsigned long KeyboardCallback(void * pvCBData, unsigned long ulEvent,
                               unsigned long ulMsgParam, void * pvMsgData)
```

其中,pvCBData 为回调函数数据指针,此数据由内部函数给出,通常值为 0;ulMsgParam,一个事件的具体参数;pvMsgData,一个事件特定的数据指针。

```
unsigned long USBHKeyboardModifierSet( unsigned long ulInstance,
                                       unsigned long ulModifiers );
```

这个函数用来设定键盘的按键状态。ulInstance,键盘的实例,通过调用 USBHKeyboardOpen()的返回值。ulModifiers 键盘要修改的位屏蔽,可以使用的值包括:HID_KEYB_NUM_LOCK;HID_KEYB_CAPS_LOCK;HID_KEYB_SCROLL_LOCK;HID_KEYB_COMPOSE;HID_KEYB_KANA。

上述值支持所有键盘,如果键盘不支持,则将忽略。如果函数调用成功则返回

0,否则返回非 0,并指示错误状态。注意:本函数主要用于修改大小写键盘灯状态。

```
unsigned long USBHKeyboardPollRateSet   (unsigned long   ulInstance,
                            unsigned long   ulPollRate);
```

这个函数用来设定键盘的查询速率。ulInstance 指向一个键盘实例;ulPollRate 表示键盘查询速率,以 ms 为单位。这个函数允许应用程序设定键盘查询速率。ulInstance 是从 USBHKeyboardOpen()函数返回的值。ulPollRate 是新的查询速率,以 ms 为单位。这个值初始化时为 0,表示键盘指示在状态更新时改变,任何正数,都会产生一个自动重复的查询。如果调用成功则返回 0,否则返回非 0,并指示错误状态。

```
unsigned long USBHKeyboard UsageToChar    (unsigned long   ulInstance,
                                const   tHIDKeyboard   UsageTable * pTable,unsigned
                                char ucUsageID)
```

这个函数用来映射一个 UsageID 到一个可以打印输出的字符。ulInstance 表示这个键盘的实例;pTable 返回可以打印输出的字符;ucUsageID 表示要映射的 UsageID。输出 pTable 字符,这个字符定义在 tHIDKeyboardUsageTable 结构体中,可以查看文档获取 tHIDKeyboardUsageTable 这个结构体的更多细节问题。这个函数使用当前 shift 和 Caps Lock 的状态码映射一个 UsageID 到 ASCII 码。

例如:获取按键的 ASCII 码,程序代码如下所示。

```
switch(ulEvent)
{
    // 检测连接
    case USB_EVENT_CONNECTED:
    {
        UARTprintf("Keyboard Connected\n");
        g_eUSBState = STATE_KEYBOARD_INIT;
        break;
    }
    //键盘移出
    case USB_EVENT_DISCONNECTED:
    {
        UARTprintf("Keyboard Disconnected\n");
        g_eUSBState = STATE_NO_DEVICE;
        g_ulModifiers = 0;
        break;
    }
    // 按键被按下
    case USBH_EVENT_HID_KB_PRESS:
    {
```

```
        // Caps Lock 状态改变
        if(ulMsgParam == HID_KEYB_USAGE_CAPSLOCK)
        {
            g_eUSBState = STATE_KEYBOARD_UPDATE;
            g_ulModifiers ^= HID_KEYB_CAPS_LOCK;
        }
        else
        {
            // backspace 按键
            if((unsigned char)ulMsgParam == HID_KEYB_USAGE_BACKSPACE)
            {
                ucChar = ASCII_BACKSPACE;
            }
            else
            {
                //获取按键的ASCII码
                ucChar = (unsigned char)
                    USBHKeyboardUsageToChar(g_ulKeyboardInstance,
                                            &g_sUSKeyboardMap,
                                            (unsigned char)ulMsgParam);
            }
        }
        break;
    }
    .......
}
```

本小结主要讲解 USB 键盘的控制及数据读取。本小节的 USB 键盘主机控制的源码请见本书配套资料。

11.3.3 U 盘

本节主要介绍支持 U 盘的 USB 主机开发,在 USB 库中已经编写好 U 盘驱动,供 USB 主机使用,相关函数包含在 usbhmsc.h 头文件中。主机包含 5 个功能函数,实现对 U 盘的数据读取、操作:USBHMSCDriveOpen、USBHMSCDriveClose、USB-HMSCDriveReady、USBHMSCBlockRead、USBHMSCBlockWrite。函数原型如下:

```
unsigned long USBHMSCDriveOpen(unsigned long ulDrive,
                               tUSBHMSCCallback pfnCallback);
void USBHMSCDriveClose(unsigned long ulInstance);
long USBHMSCDriveReady(unsigned long ulInstance);
```

```
long USBHMSCBlockRead(unsigned long ulInstance, unsigned long ulLBA,
                      unsigned char * pucData,
                      unsigned long ulNumBlocks);
long USBHMSCBlockWrite(unsigned long ulInstance, unsigned long ulLBA,
                       unsigned char * pucData,
                       unsigned long ulNumBlocks);
```

(1) unsigned long USBHMSCDriveOpen(unsigned long ulDrive,
tUSBHMSCCallback pfnCallback);

这个函数用来打开一个大容量存储设备实例。在任何设备连接之前,允许驱动器连接和断开。ulDrive 是一个为 0 的驱动器在系统中存在的索引。如果有一个系统中的 USB 集线器中有固定的驱动器数量,这个数字应该大于 0。应用程序必须提供 pfnCallback 回调函数,如设备的枚举和移动存储设备相关的事件。这个函数将返回驱动程序实例编号,如果没有驱动程序,此函数将返回零。

(2) void USBHMSCDriveClose(unsigned long ulInstance);

这个函数被调用时,MSC 驱动器要处于关闭或切换到 USB 设备模式。

(3) long USBHMSCDriveReady (unsigned long ulInstance);

这个函数检查驱动器是否准备好访问。ulInstance 用于确定哪个设备已准备就绪。任何非零的返回表明该设备还没有准备好或者发生错误。

(4) long USBHMSCBlockRead(unsigned long ulInstance,
                           unsigned long ulLBA,
                           unsigned char * pucData,
                           unsigned long ulNumBlocks)

这个函数执行 MSC 设备的读取。这个函数将从 ulInstance 指定的 U 盘实例中读取相关数据。参数 ulLBA 指定了读取设备的逻辑块地址。此功能将只按照 ulNumBlocks 块大小读取,在大多数情况下,每次读取 512 字节的数据。在缓冲 * pucData 中至少包含 ulNumBlocks×512 个字节的数据。该函数返回零表示成功,返回负数表示失败。

(5) long USBHMSCBlockWrite (unsigned long   ulInstance,
                              unsigned long   ulLBA,
                              unsigned char * pucData,
                              unsigned long   ulNumBlocks)

这个函数执行 MSC 设备的写入。这个函数将从 ulInstance 指定的 U 盘实例中写入相关数据。参数 ulLBA 指定了写入设备的逻辑块地址。此功能将只按照 ulNumBlocks 块大小写入,在大多数情况下,每次写入 512 字节的数据。在缓冲 * pucData 中至少包含 ulNumBlocks×512 个字节的数据空间。该函数返回零表示成功,返回负数表示失败。

例如：U 盘初始化，程序代码如下所示。

```c
//FAT 文件系统
static FATFS g_sFatFs;


#define HCD_MEMORY_SIZE            128
unsigned char g_pHCDPool[HCD_MEMORY_SIZE];
unsigned long g_ulMSCInstance = 0;
// 声明 USB 事件驱动接口
DECLARE_EVENT_DRIVER(g_sUSBEventDriver, 0, 0, USBHCDEvents);
// 主机驱动
static tUSBHostClassDriver const * const g_ppHostClassDrivers[] =
{
    &g_USBHostMSCClassDriver
    ,&g_sUSBEventDriver
};
static const unsigned long g_ulNumHostClassDrivers =
    sizeof(g_ppHostClassDrivers) / sizeof(tUSBHostClassDriver *);
#define FLAGS_DEVICE_PRESENT       0x00000001
// Holds global flags for the system.
int g_ulFlags = 0;
//文件夹深度及文件名长度
#define MAX_DIR_DEPTH              10
#define MAX_FILE_NAME_LEN          (8 + 1 + 3 + 1)   // 8.3 + 1
// 文件夹数据结构
struct
{
    FILINFO FileInfo[20];
    unsigned long ulIndex;
    unsigned long ulSelectIndex;
    unsigned long ulValidValues;
    DIR DirState;
    char szPWD[MAX_DIR_DEPTH * MAX_FILE_NAME_LEN];
    u8 ucAuth[20];
}
g_DirData;


//设备状态
volatile enum
{
    STATE_NO_DEVICE,
    STATE_DEVICE_ENUM,
```

```c
    STATE_DEVICE_READY,
    STATE_UNKNOWN_DEVICE,
    STATE_POWER_FAULT
}
g_eState;//DMA 使用
tDMAControlTable g_sDMAControlTable[6] __attribute__ ((aligned(1024)));
//初始化 FileSystem,Mount 到硬盘 0
tBoolean FileInit(void)
{
    if(f_mount(0, &g_sFatFs) != FR_OK)
    {
        return(false);
    }
    return(true);
}
//U 盘回调函数
void  MSCCallback(unsigned long ulInstance, unsigned long ulEvent, void * pvData)
{
    switch(ulEvent)
    {
        case MSC_EVENT_OPEN:
        {
            g_eState = STATE_DEVICE_ENUM;
            break;
        }
        case MSC_EVENT_CLOSE:
        {
            g_eState = STATE_NO_DEVICE;
            FileInit();
            break;
        }
        default:
        {
            break;
        }
    }
}
//事件
void USBHCDEvents(void * pvData)
{
    tEventInfo * pEventInfo;
    pEventInfo = (tEventInfo *)pvData;
```

```c
        switch(pEventInfo->ulEvent)
        {
            case USB_EVENT_CONNECTED:
            {
                //连接成功
                UARTprintf("U disk is Connect!.............\n");
                g_eState = STATE_UNKNOWN_DEVICE;
                break;
            }
            case USB_EVENT_DISCONNECTED:
            {
                //设备移出
                UARTprintf("U disk is DisConnect!............\n");
                g_eState = STATE_NO_DEVICE;
                break;
            }
            case USB_EVENT_POWER_FAULT:
            {
                //电源错误
                g_eState = STATE_POWER_FAULT;
                break;
            }
            default:
            {
                break;
            }
        }
}
//main 函数
int main(void)
{
    FRESULT FileResult;
    FIL file,file1;
    DIR CurrentDir;
    s8 getbuff[10];

    BYTE ucBuff[200] = "/*--------------------------------*/\r\nhello \
    paul! \r\nThis is a File System example! \r\nCopyright @ \
    2010\r\n/*--------------------------------------*/\r\n";\
    char Buff[200];
    BYTE ucCnt = 0;
```

```c
WORD ulSize = 200;
WORD x;
WORD i;
g_eState = STATE_NO_DEVICE;
SysCtlClockSet(SYSCTL_SYSDIV_4 | SYSCTL_USE_PLL | SYSCTL_OSC_MAIN |
            SYSCTL_XTAL_8MHZ);
UARTStdioInit(0);
//DMA 控制
SysCtlPeripheralEnable(SYSCTL_PERIPH_UDMA);
uDMAEnable();
uDMAControlBaseSet(g_sDMAControlTable);
//注册 Host 驱动
UARTprintf("USBHCDRegisterDrivers.............\n");
USBHCDRegisterDrivers(0, g_ppHostClassDrivers, g_ulNumHostClassDrivers);
// 打开 USB 大容量存储设备驱动
g_ulMSCInstance = USBHMSCDriveOpen(0, MSCCallback);
// 初始化电源
USBHCDPowerConfigInit(0, USBHCD_VBUS_AUTO_HIGH);
// 初始化主机控制
USBHCDInit(0, g_pHCDPool, HCD_MEMORY_SIZE);
UARTprintf("Initialize the file system.............\n");
FileInit();
strcpy((s8 *)g_DirData.ucAuth,"root");
while(1)
{
    USBHCDMain();
    switch(g_eState)
    {
        case STATE_DEVICE_ENUM:
        {
            UARTprintf("STATE_DEVICE_ENUM.............\n");
            if(USBHMSCDriveReady(g_ulMSCInstance) != 0)
            {
                SysCtlDelay(SysCtlClockGet()/30);
                break;
            }
            FileResult = f_opendir(&g_DirData.DirState, "/");
            if(FileResult == FR_OK)
            {
                g_DirData.ulIndex = 0;
                g_DirData.ulSelectIndex = 0;
                g_DirData.ulValidValues = 0;
```

```c
            g_eState = STATE_DEVICE_READY;
            UARTprintf("U disk is Connect!..............\n");
        }
        else if(FileResult != FR_NOT_READY)
        {
        }
        memset(g_DirData.szPWD,0,MAX_DIR_DEPTH * MAX_FILE_NAME_LEN);
        g_ulFlags = FLAGS_DEVICE_PRESENT;
    break;
}
//可以在此读写U盘
case STATE_DEVICE_READY:
{
        UARTprintf("Open is ok!..............\n");

        f_open(&file, "test.txt", FA_OPEN_ALWAYS | FA_READ | FA_WRITE);
        f_write(&file, ucBuff,sizeof(ucBuff), &x);

        ......

    break;
}
//没有设备
case STATE_NO_DEVICE:
{
    if(g_ulFlags == FLAGS_DEVICE_PRESENT)
    {
        UARTprintf("U disk is DisConnect!..............\n");
        ucCnt = 0;
        g_ulFlags &= ~FLAGS_DEVICE_PRESENT;
    }
    break;
}
//未知设备插入
case STATE_UNKNOWN_DEVICE:
{
    if((g_ulFlags & FLAGS_DEVICE_PRESENT) == 0)
    {
    }
    g_ulFlags = FLAGS_DEVICE_PRESENT;
    break;
}
```

第 11 章　USB 主机开发

```
            //电源故障
            case STATE_POWER_FAULT:
            {
                break;
            }
            default:
            {
                break;
            }
        }
    }
}
```

　　本小节主要讲解 U 盘的控制及数据读取。以上列出部分源代码,关于底层文件驱动没有在本书列出,更多 U 盘主机控制的源码请见本书配套资料。

11.4　小　　结

　　USB 的硬件与软件互连支持数据在 USB 主机与 USB 设备之间流动。本章主要讲述了 USB 主机接口,该类接口简化了用户程序与设备之间的通信复杂度。通过 3 个实例介绍使用 USB 库对 LM3S 系列处理器进行 USB 主机开发及相关流程。完成本章学习,可以开发简单的鼠标、键盘、U 盘的主机接口,还可在此基础上进一步扩展,完成更复杂的 USB 系统。

第 12 章

USB OTG 开发

简单地说,OTG 就是 On The Go,正在进行中的意思。USB OTG 是 USB On-The-Go 的缩写,是近年发展起来的技术,2001 年 12 月 18 日由 USB Implementers Forum(USB IF)公布,主要应用于各种不同的设备或移动设备间的连接,进行数据交换,特别是 PDA、移动电话、消费类设备。改变如数码照相机、摄像机、打印机等设备间多种不同制式连接器,多达 7 种制式的存储卡间数据交换的不便。USB 技术的发展使得 PC 和周边设备能够通过简单方式、适度的制造成本将各种设备连接在一起,上述提到的应用都可以通过 USB 总线作为 PC 的周边,在 PC 的控制下进行数据交换。但这种方便的交换方式,一旦离开了 PC,各设备间无法利用 USB 口进行操作,因为没有一个设备能够充当 PC 一样的 Host。OTG 技术就是实现在没有 Host 的情况下,实现设备间的数据传送。例如将数码相机直接连接到打印机上,通过 OTG 技术,连接两台设备间的 USB 口,将拍出的相片立即打印出来;也可以将数码照相机中的数据,通过 OTG 发送到 USB 接口的移动硬盘上。

12.1 OTG 介绍

随着 PDA、移动电话、数码相机、打印机等消费类产品的普及,用于这些设备与计算机,或设备与设备之间的高速数据传输技术越来越受到人们的关注,IEEE1394 和 USB 是用于此类传输的两个主要标准。这两个标准都提供即插即用和热插拔功能,都可以向外提供电源,也都支持多个设备的连接。其中 IEEE1394 支持较高的数据传输速度,但相对比较复杂、价格较高,主要用于需要高速通信的 AV 产品;而最初的 USB 标准主要面向低速数据传输的应用,其中 USB 1.1 支持 1.5 Mbps 和 12 Mbps 的传输速率,被广泛用于传输速率要求不高的 PC 机外设,如:键盘、鼠标等。USB 2.0 标准的推出使 USB 的传输速度达到 480 Mbps。而 USB OTG 技术的推出则可实现没有主机时设备与设备之间的数据传输。例如:如图 12-1 所示,数码相机可以直接与打印机连接并打印照片,从而拓展了 USB 技术的应用范围。

第 12 章　USB OTG 开发

图 12-1　数码相机与打印机连接并打印照片

12.1.1　主机通信协议与对话请求协议

　　USB OTG 标准在完全兼容 USB 2.0 标准的基础上,增添了电源管理(节省功耗)功能,它允许设备既可作为主机,也可作为外设操作(两用 OTG)。OTG 两用设备完全符合 USB 2.0 标准,并可提供一定的主机检测能力,支持主机协商协议(HNP)和会话请求协议(SRP)。在 OTG 中,初始主机设备称为 A 设备,外设称为 B 设备。可用电缆的连接方式来决定初始角色。两用设备使用新型 mini-AB 插座,从而使 mini-A 插头、mini-B 插头和 mini-AB 插座增添了第 5 个引脚(ID),以用于识别不同的电缆端点。mini-A 插头中的 ID 引脚接地,mini-B 插头中的 ID 引脚浮空。当 OTG 设备检测到接地的 ID 引脚时,表示默认的是 A 设备(主机),而检测到 ID 引脚浮空的设备则认为是 B 设备(外设)。系统一旦连接后,OTG 的角色还可以更换。主机与外设采用新的 HNP,A 设备作为默认主机并提供 VBUS 电源,并在检测到有设备连接时复位总线、枚举并配置 B 设备。OTG 标准为 USB 增添的第二个新协议称为会话请求协议(SRP)。SRP 允许 B 设备请求 A 设备打开 VBUS 电源并启动一次会话。一次 OTG 会话可通过 A 设备提供 VBUS 电源的时间来确定(注:A 设备总是为 VBUS 供电,即使作为外设)。也可通过 A 设备关闭 VBUS 电源来结束会话以节省功耗,这在电池供电产品中非常重要。例如,在两台蜂窝电话通过连接互相交换信息时,一台连接在 mini-A 端,是 A 设备,默认为主机。另一台是 B 设备,默认为外设。当在不需要 USB 通信时,A 设备可以关闭 VBUS 线,此时 B 设备就会检测到该状态并进入低功耗模式。

12.1.2　OTG 功能的构建

　　构建 OTG 功能时需要在 USB 外设上添加电路,电路中的通用串行总线控制器可以是一个微处理器和 USB SIE（串口引擎），也可以是集成的 μP/USB 芯片或与 USB 收发器相连的 ASIC。为总线提供电源的外部设备需要一路 3.3 V 稳压输出供电电压，以便为逻辑电路和连接在 D+、D− 引脚上的 1 500 Ω 电阻提供电源。通过 D+、D− 引脚上的上拉电阻可向主机发出设备已连接的信号，并指示设备的工作速度。电阻上拉至 D+ 表示全速运行，电阻上拉至 D− 表示低速运行。其他端点（包括 D+ 和 D− 的 15 kΩ 下拉电阻）用于检测上拉电阻的状态。由于 USB 设计需要提供热插拔功能。因此，其 ESD 保护电路主要用于为 D+、D− 和 VBUS 引脚提供保护。为了增加 OTG 的两用功能，必须扩充收发器功能来使 OTG 设备既可作为主机使用，也可以作为外设使用。而要实现上述功能，就需要在电路中添加 D+ 和 D− 端的 15 kΩ 下拉电阻并为 VBUS 提供供电电源。此外，收发器还需要具备以下 3 个条件：

(1) 可切换 D+/D− 线上的上拉和下拉电阻，以提供外设和主机功能。

(2) 作为 A 设备时，需要具有 VBUS 监视和供电电路；作为 B 设备初始化 SRP 时，需要监视和触发 VBUS。

(3) 具有 ID 输入引脚。

作为两用 OTG 设备，ASIC、DSP 或其他与收发器连接的电路必须具备充当外设和主机的功能，并应按照 HNP 协议转换其角色。收发器所需添加的大多数电路用于 VBUS 引脚的管理。作为主机，它必须能够提供 5 V、输出电流可达 8 mA 的电源。ASIC 和控制器还必须包含 USB 主机逻辑控制功能，包括发送 SOF（帧启动）包、发送配置、输入、输出数据包，在 USB 1 msec 帧内确定传输进程、发送 USB 复位信号、提供 USB 电源管理等。

12.1.3　LM3S 的 OTG 功能

该 Stellaris USB 控制器支持 USB 主机/设备/OTG 功能，在点对点通信过程中可运行在全速和低速模式。它符合 USB 2.0 标准，包含挂起和恢复信号。它包含 32 个端点，其中包含两个用于控制传输的专用连接端点（一个用于输入，一个用于输出）以及 30 个由固件定义的端点，并带有一个大小可动态变化的 FIFO，以支持多包队列。可通过 μDMA 来访问 FIFO，将对系统软件的依赖降至最低。USB 设备启动方式灵活，可软件控制是否在启动时连接。USB 控制器遵从 OTG 标准的会话请求协议（SRP）和主机协商协议（HNP）。

Stellaris USB 控制器大部分都支持 OTG 功能，与 USB 有关的引脚有：USB0DM、USB0DP、USB0EPEN、USB0PFLT、USB0RBIAS、USB0VBUS、USB0ID 等 7 个主要引脚，其中 USB0DM 和 USB0DP 提供双向差分数据，与 USB 其他设备或主机进行数据交换，是 USB 各个功能的物理连接线。USB0EPEN 和 USB0PFLT 用于主机模式对电源控制操作，USB0EPEN 可选择性地以主机模式控制外部电源，向 USB 总线供电；USB0PFLT 可选择性地用于主机模式对外部电源的错误状态检测。USB0RBIAS，USB 模拟电路内部 9.1 kΩ 电阻（1%精度），普通贴片电阻即可满足要求。USB0VBUS 和 USB0ID 为 OTG 模式专用，USB0ID 此信号用于检测 USB ID 信号的状态。此时 USB PHY 将在内部启用一个上拉电阻，通过外部元件（USB 连接器）检测 USB 控制器的初始状态（即电缆的 A 侧设置下拉电阻，B 侧设置上拉电阻）；USB0VBUS 此信号用于会话请求协议。USB PHY 可通过此信号检测 VBUS 的电平，并在 VBUS 脉冲期间短时上拉。

12.1.4 OTG 函数

在 USB 库中提供一组工作在 OTG 模式下的函数，主要负责 OTG 控制和通信等。下面就常用的函数及数据类型进行介绍。

tUSBOTGState 用于指示 OTG 模块的工作模式，如下所示。

```
typedef enum
{
    //OTG 主机处于空闲状态
    USB_OTG_MODE_IDLE,
    //等待 ID
    USB_OTG_MODE_WAITID,
    //转换等待
    USB_OTG_MODE_WAIT,
    //工作在 B 端,等待连接
    USB_OTG_MODE_B_WAITCON,
    //工作在 B 模式
    USB_OTG_MODE_B_DEVICE,
    //工作在 A 模式
    USB_OTG_MODE_A_HOST,
}
tUSBOTGState;
```

void USBOTGSetMode(tUSBMode eUSBMode)，此函数用于主机、设备及其他模式切换。其中 eUSBMode 参数是 USB_MODE_HOST、USB_MODE_DEVICE、USB_MODE_NONE 中的任意一个。USBOTGSetMode 函数如下：

```
static void USBOTGSetMode(tUSBMode eUSBMode)
{
    if((g_eDualMode != eUSBMode) || (g_eDualMode == USB_MODE_NONE))
    {
        // 从主机模式切换到未配置模式
        if((g_eDualMode == USB_MODE_HOST) && (eUSBMode == USB_MODE_NONE))
        {
            // 关闭电源
            USBOTGRemovePower(0);
```

```
    }

    //从设备模式切换到未配置模式
    if((g_eDualMode == USB_MODE_DEVICE) && (eUSBMode == USB_MODE_NONE))
    {
        //结束当前会话
        USBOTGSessionRequest(USB0_BASE, false);
    }
    if(eUSBMode == USB_MODE_NONE)
    {
        g_ulWaitTicks = g_ulPollRate;
    }
    // 如果有模式改变时调用的回调函数,处理如下
    if((g_pfnUSBModeCallback) && (g_eDualMode != eUSBMode))
    {
        //当前工作模式传给回调函数
        g_pfnUSBModeCallback(0, eUSBMode);
    }
    //保存新的工作模式
    g_eDualMode = eUSBMode;
}
```

void USBStackModeSet(unsigned long ulIndex, tUSBMode eUSBMode,
                    tUSBModeCallback pfnCallback)

USBStackModeSet 函数允许工作在双模式时,USB 设备与 USB 主机之间切换。ulIndex,USB 模块编号,固定为 0;eUSBMode,切换到某个工作模式,USB_MODE_DEVICE 表示工作在设备模式,USB_MODE_HOST 表示作 USB 主机使用。pfnCallback 是回调函数,每次工作模式改变时调用。工作在双模式时,USB 中断处理函数名应该为:USB0DualModeIntHandler;工作在设备模式下时,USB 中断处理函数名应该为:USB0DeviceIntHandler;工作在主机模式下时,USB 中断处理函数名应该为:USB0HostIntHandler。USBStackModeSet 的函数如下所示:

```
void USBStackModeSet(unsigned long ulIndex, tUSBMode eUSBMode,
                    tUSBModeCallback pfnCallback)
{
    // 存储当前 USB 的工作模式
    g_eUSBMode = eUSBMode;
    //记住回调函数
    g_pfnUSBModeCallback = pfnCallback;
    // 工作在主机或者设备模式时,将当前工作模式传给回调函数
```

```
        if((eUSBMode == USB_MODE_DEVICE) || (eUSBMode == USB_MODE_HOST))
        {
            g_pfnUSBModeCallback(0, eUSBMode);
        }
}
```

void USBDualModeInit(unsigned long ulIndex)函数用于双通道模式初始化。参数 ulIndex 为 USB 模式编号，固定为 0。此函数用于初始化 USB 硬件控制器，为 USB 模块工作在双模式提供条件。

USBDualModeInit 函数如下所示：

```
void USBDualModeInit(unsigned long ulIndex)
{
    //配置端点 0
    USBHostEndpointConfig(USB0_BASE, USB_EP_0, 64, 0, 0,
                          (USB_EP_MODE_CTRL | USB_EP_SPEED_FULL |
                          USB_EP_HOST_OUT));
    // 使能 USB 中断
    USBIntEnableControl(USB0_BASE, USB_INTCTRL_RESET | USB_INTCTRL_DISCONNECT |
                                   USB_INTCTRL_SESSION | USB_INTCTRL_BABBLE |
                                   USB_INTCTRL_CONNECT | USB_INTCTRL_RESUME |
                                   USB_INTCTRL_SUSPEND | USB_INTCTRL_VBUS_ERR);
    //使能全部端点中断
    USBIntEnableEndpoint(USB0_BASE, USB_INTEP_ALL);
    // 初始化 USBTick
    InternalUSBTickInit();
    // 使能 USB 模块中断
    IntEnable(INT_USB0);
    //打开 USB 会话请求
    USBOTGSessionRequest(USB0_BASE, true);
    // 初始化电源控制
    USBHostPwrConfig(USB0_BASE, USBHCDPowerConfigGet(ulIndex));
    //使能电源自动控制
    if(USBHCDPowerAutomatic(ulIndex))
    {
        //打开电源
        USBHostPwrEnable(USB0_BASE);
    }
}
```

从 USBDualModeInit 函数代码中可以看出，此函数主要用于初始化 USB 硬件模块，使其能正常工作在双模式下。

```
void USBOTGModeInit(unsigned long ulIndex, unsigned long ulPollingRate,
                    void * pvPool, unsigned long ulPoolSize);
```

USBOTGModeInit 函数用于初始化 USB 硬件模块,使其能正常工作在 OTG 模式下。参数 ulIndex,指定要初始化的 USB 控制器;ulPollingRate,轮询控制器模式改变的速度,以 ms 表示;pvPool,指向 USB 控制器保存数据的缓冲池;ulPoolSize, USB 控制器保存数据的缓冲池的大小。这个函数初始化 USB 控制器硬件进入一个适合 OTG 模式操作的状态。应用程序必须确保控制器在中间状态并能够接收合适的中断,主机和设备模式通过 OTG 选择。ulPollingRate 参数被用来设定速度。

```
void USBOTGMain(unsigned long ulMsTicks);
```

这个函数是 OTG 控制器驱动的主循环函数。参数 ulMsTicks 是上次调用这个函数到现在的时间差,以 ms 为单位。这个函数必须在应用程序主函数中周期性地调用。USBOTGMain 函数代码如下所示:

```
void USBOTGMain(unsigned long ulMsTicks)
{
    if(ulMsTicks > g_ulWaitTicks)
    {
        g_ulWaitTicks = 0;
    }
    else
    {
        g_ulWaitTicks -= ulMsTicks;
    }
    switch(g_eOTGModeState)
    {
        case USB_OTG_MODE_IDLE:
        {
            g_eOTGModeState = USB_OTG_MODE_WAITID;
            // 初始化会话请求,并检查 ID 信号
            USBOTGSessionRequest(USB0_BASE, true);
            break;
        }
        case USB_OTG_MODE_WAIT:
        case USB_OTG_MODE_WAITID:
        {
            //超时检查
            if((g_ulWaitTicks == 0) && (g_ulPollRate != 0))
            {
                //结束会话
                USBOTGSessionRequest(USB0_BASE, false);
```

```c
        //进入空闲模式
        USBOTGSetMode(USB_MODE_NONE);
        // 自动电源控制
        if(USBHCDPowerAutomatic(0) == 0)
        {
            //禁止电源输出
            InternalUSBHCDSendEvent(USB_EVENT_POWER_DISABLE);
        }
        // 进入 IDLE 模式
        g_eOTGModeState = USB_OTG_MODE_IDLE;
    }
    break;
}
case USB_OTG_MODE_A_HOST:
{
    //调用主机控制主程序
    USBHCDMain();
    break;
}
case USB_OTG_MODE_B_WAITCON:
case USB_OTG_MODE_B_DEVICE:
default:
{
    break;
}
}
}
```

从 USBOTGMain 函数中可以看出，USBOTGMain 函数只支持 OTG 模式与 USB 主机模式，不支持设备模式。USBOTGMain 的函数调用图如图 12-2 所示，可以看出，USBOTGMain 只调用了 USBHCDMain 函数实现主机功能，并进行枚举，没有设备工作模式。

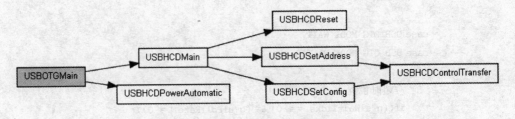

图 12-2 USBOTGMain 函数调用图

12.2 OTG B 开发

OTG B 设备开发,与 USB 设备开发类似。OTG 实际上是 USB 的一种扩展,能支持设备与主机功能,但在某一时刻只能完成其中一种功能。通过 OTG 协议可实现主从模式切换。OTG B 设备开发,实际是把 OTG 功能当作设备使用。

注意以下几点:
① OTG B 设备开发时,与前面几章的 USB 设备开发过程一样;
② 启动代码中的 USB 中断服务函数应该为 USB0DeviceIntHandler;
③ 硬件连接上,ID 信号应为高电平(或者悬空),经过试验验证,ID 为高电平时最稳定,USB 处于 OTG B 设备模式,与 USB 设备开发一样。
④ USB OTG 处理器的 OTG B 程序与 USB 设备程序是兼容的。

12.3 OTG A 开发

OTG A 开发,实际是 USB 主机开发,但又与 USB 主机开发有所区别,读者应该注意以下几点:
① 启动代码中的 USB 中断服务函数应该为 USB0OTGModeIntHandler;
② 应该周期调用 USBOTGMain 函数;
③ USBOTGMain 的入口参数必须提供 GetTickms(),以获取上次调用这个函数到现在的时间差;
④ 初始化 USB 堆栈工作在 USB_MODE_OTG 模式,如 USBStackModeSet(0, USB_MODE_OTG, ModeCallback)。

12.4 OTG 开发实例

前面已经介绍了 OTG 的基本工作模式,即 OTG A 与 OTG B,其中 OTG B 即 B 设备开发与普通 USB 设备开发模式一样,并且程序代码兼容,但要从外部提供 ID 信号引脚为高电平。在 OTG A 主机模式中,USB 开发有点不一样,在第 12.3 节已经介绍。本节主要以一个实例(OTG 键盘主机)介绍 USB A 设备与 USB 主机开发的异同点。

(1) 声明足够的内存空间,用于 OTG 数据缓存;注册驱动,便于 OTG 枚举设备。

```
// 定义主机控制单元内存空间的大小
#define HCD_MEMORY_SIZE         128
unsigned char g_pHCDPool[HCD_MEMORY_SIZE];
```

```c
// 键盘内存单元大小
#define KEYBOARD_MEMORY_SIZE    128
// 分配内存单元给键盘数据存储
unsigned char g_pucBuffer[KEYBOARD_MEMORY_SIZE];
// 声明 USB 事件驱动接口
DECLARE_EVENT_DRIVER(g_sUSBEventDriver, 0, 0, USBHCDEvents);
// 注册 HID 类驱动
static tUSBHostClassDriver const * const g_ppHostClassDrivers[] =
{
    &g_USBHIDClassDriver,
    &g_sUSBEventDriver
};
static const unsigned long g_ulNumHostClassDrivers =
    sizeof(g_ppHostClassDrivers) / sizeof(tUSBHostClassDriver *);
```

(2) 使用 tick 定时器,实现基本时钟定时计数,为 USBOTGMain 函数提供上次调用这个函数到现在的时间差,以 ms 为单位,这点有别于 USB 主机开发。

```c
// Tick 时钟
#define TICKS_PER_SECOND 100
#define MS_PER_SYSTICK (1000 / TICKS_PER_SECOND)
unsigned long g_ulSysTickCount;
unsigned long g_ulLastTick;
// SysTick 中断处理函数
void SysTickIntHandler(void)
{
    g_ulSysTickCount++;
}
```

(3) 设置系统内核参数与 USB 硬件使能。

```c
// 设置主频
SysCtlClockSet(SYSCTL_SYSDIV_4 | SYSCTL_USE_PLL | SYSCTL_OSC_MAIN |
               SYSCTL_XTAL_16MHZ);
// 初始化变量
g_eUSBState = STATE_NO_DEVICE;
eLastMode = USB_MODE_OTG;
g_eCurrentUSBMode = USB_MODE_OTG;
//使能 USB 模块
SysCtlPeripheralEnable(SYSCTL_PERIPH_USB0);
//配置 Tick 时钟
SysTickPeriodSet(SysCtlClockGet() / TICKS_PER_SECOND);
SysTickEnable();
SysTickIntEnable();
```

```c
// 使能总中断
IntMasterEnable();
// 配置 UART 端口
GPIOPinTypeUART(GPIO_PORTA_BASE, GPIO_PIN_0 | GPIO_PIN_1);
UARTStdioInit(0);
// 配置 USB 引脚
GPIOPinTypeUSBDigital(GPIO_PORTA_BASE, GPIO_PIN_6 | GPIO_PIN_7);
GPIOPinTypeUSBDigital(GPIO_PORTB_BASE, GPIO_PIN_0 | GPIO_PIN_1);
// 初始化 USBstack
USBStackModeSet(0, USB_MODE_OTG, ModeCallback);
// 注册驱动
USBHCDRegisterDrivers(0, g_ppHostClassDrivers, g_ulNumHostClassDrivers);
// 打开键盘实例
g_ulKeyboardInstance = USBHKeyboardOpen(KeyboardCallback, g_pucBuffer,
                                        KEYBOARD_MEMORY_SIZE);
// USB 电源配置
USBHCDPowerConfigInit(0, USBHCD_VBUS_AUTO_HIGH | USBHCD_VBUS_FILTER);
// 配置为 OTG 模式
USBOTGModeInit(0, 2000, g_pHCDPool, HCD_MEMORY_SIZE);
UARTprintf("Host Keyboard Application\n");
```

在这一步中，USBStackModeSet 函数初始化 USB 堆栈工作在 USB_MODE_OTG 模式；USBHKeyboardOpen 函数调用与 USB 主机开发时调用一样；USBOTGModeInit 函数用于初始化 USB 硬件模块，使其能正常工作在 OTG 模式下。参数 ulIndex，指定要初始化的 USB 控制器；ulPollingRate，轮询控制器模式改变的速度，以 ms 表示；pvPool，指向 USB 控制器保存数据的缓冲池；ulPoolSize，USB 控制器保存数据的缓冲池的大小。这点与 USB 主机开发有很大区别。

（4）周期性地调用 USBOTGMain 函数，实现 USB OTG 功能。

```c
//USB OTG 主控
USBOTGMain(GetTickms());
// 状态是否改变
if(g_eCurrentUSBMode != eLastMode)
{
    eLastMode = g_eCurrentUSBMode;
    switch(eLastMode)
    {
        case USB_MODE_HOST:
            pcString = "HOST";
            break;
        case USB_MODE_DEVICE:
            pcString = "DEVICE";
```

```
                    break;
                case USB_MODE_NONE:
                    pcString = "NONE";
                    break;
                default:
                    pcString = "UNKNOWN";
                    break;
            }
            UARTprintf("USB mode changed to %s\n", pcString);
        }
        switch(g_eUSBState)
        {
            // 键盘第一次连接到主机
            case STATE_KEYBOARD_INIT:
            {
                // 初始化键盘
                USBHKeyboardInit(g_ulKeyboardInstance);
                g_eUSBState = STATE_KEYBOARD_CONNECTED;
                break;
            }
            case STATE_KEYBOARD_UPDATE:
            {
                g_eUSBState = STATE_KEYBOARD_CONNECTED;
                USBHKeyboardModifierSet(g_ulKeyboardInstance, g_ulModifiers);
                break;
            }
            ……
        }
```

通过以上4步就能完成 USB OTG 开发,在必要时可以动态切换工作模式。综上所述:OTG 开发分为两部分,即 OTG A 与 OTG B,OTG A 设备开发与 USB 主机开发有所区别,要注意它们之间的微小差别;OTG B 就是 B 设备开发,和 USB 设备开发模式一样,并且程序代码兼容。这两种模式最初都由 ID 引脚的电平决定,ID 为低电平时,工作在主机(A)模式下;ID 为高电平(或者悬空),工作在设备(B)模式下。完成第一次默认状态配置后,通过会话请求进行主从切换。

12.5 OTG 开发小结

本章主要介绍 USB OTG 技术,USB OTG 由 USB Implementers Forum(USB IF)公布,主要应用于各种不同的设备或移动设备间的连接,进行数据交换。特别是 PDA、移动电话、消费类设备。能过本章的学习,读者可以掌握 USB OTG 的基本概念,对 USB OTG 开发有初步了解,能独立完成简单的 OTG 系统开发。

第 13 章

USB 设备工程实例

通用串行总线(Universal Serial Bus,USB)是一种快速、灵活的总线接口。与其他通信接口相比,USB 接口的最大特点是易于使用,这也是 USB 的主要设计目标。作为一种高速总线接口,USB 适用于多种设备,比如数码相机、MP3 播放机、高速数据采集设备等。易于使用还表现在 USB 接口支持热插拔,并且所有的配置过程都由系统自动完成,无需用户干预。

USB 接口支持 1.5 Mbps(低速)、12 Mbps(全速)和高达 480 Mbps(USB 2.0 规范)的数据传输速率,扣除用于总线状态、控制和错误监测等的数据传输,USB 的最大理论传输速率仍达 1.2 Mbps 或 9.6 Mbps,远高于一般的串行总线接口。USB 接口芯片价格低廉,一个支持 USB 1.1 规范的 USB 接口芯片价格大多在人民币(2002年)20~40 元之间,这也大大促进了 USB 设备的开发与应用。

Stellaris 的 USB 控制器支持 USB 主机/设备/OTG 功能,其价格大约为 30 元人民币,随着使用量的增加,其价格也有所降低。该系列处理器除 USB 功能外还集成了其他功能,比如:I^2C、SSI、PWM、EPI、CAN、UART、10/100 以太网、I^2S、ADC、CCP 等数十种功能,该处理器使用 ARM Cortex-M3 处理器内核,80 MHz 运行速度,100 DMIPS 性能,所有程序开发都使用现成的 C 语言库,所以选择 Stellaris 的USB 控制器开发 USB 系统是比较合理的选择。

13.1 USB 设备开发流程

USB 设备是 USB 系统的主体,USB 是因便携式设备需要而出现的,所以 USB 设备开发非常重要。使用 Stellaris 的 USB 控制器开发 USB 设备的流程图如图 13-1 所示,首先根据具体设计要求确定 USB 设备类型,明确 USB 设备的功能;通过 USB 功能选择合适的通信协议,人机接口类的使用 HID,比如键盘、鼠标等,大容量存储设备使用 Mass Storage 设备类等;确定好 USB 协议后,参考相关手册完成 MCU 应用程序开发。当然这只是设备开发,与设备开发密切相关的是上位机的设计,这也是设备开发的一部分,包括驱动程序开发和用户应用程序开发,其中驱动一般由 USB 处理器公司提供,可不用考虑;用户应用程序也并非一定要开发者用代码实现,比如HID 设备为标准驱动,在系统中已经集成好所有的驱动,由系统自身完成其控制,不用人为操作。

使用 Stellaris 的 USB 控制器开发 USB 设备的注意事项如下：

(1) 几乎所有 Stellaris 的 USB 控制器都有设备功能,在开发时,启动代码中的 USB 中断服务程序的名字应该为 USB0DeviceIntHandler,在 USB 库中已经集成好此中断服务函数,不必自己再编写,但一定要在中断向量表中加入该中断服务函数名。

(2) Stellaris 的 USB 控制器有些版本出现 bug,需要开发人员避开,如 LM3S3XXX 与 LM3S5XXX 的 A0 版本

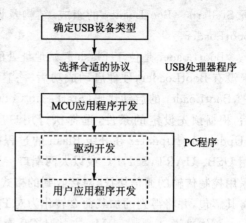

图 13-1 USB 设备开发流程图

有一个 bug。图 11-1 为 PB0 与 PB1 硬件连接图,在 A0 版本中,USB 处理器工作在主机和设备模式下时,PB0 与 PB1 不能当 GPIO 使用,因为在此版本中,PB0 与 PB1 为主机与设备提供电平信号。当 USB 处理器工作在主机模式下时,PB0 应该连接到低电平;工作在设备模式下时,PB0 应该连接到高电平。同时,PB0 引脚与电压信号之间应该连接一个电阻,其典型值为 10 Ω。PB1 必须连接到 5 V(4.75～5.25 V)。如果 USB 处理器不是 A0 版本,PB0 与 PB1 可以用作 GPIO 功能,但建议读者不要将其用作 GPIO。

(3) OTG 用作设备开发时,ID 引脚应该连接到高电平;启动代码中的 USB 中断服务程序名字也应该为 USB0DeviceIntHandler。

(4) 使用 USB 功能时,处理器的主频不要低于 20 MHz,建议设置为 50 MHz,并且内核电压最好调节到 2.75 V,提高处理器的工作电压,让系统更稳定可靠。

为了保证 USB 设备的正常工作,其硬件电路要严格按照官方提供的数据手册进行开发,以保证硬件电路的正确性,否则在 USB 系统开发时,会增加调试难度。

13.2 USB 设备之 USB BootLoader

现今,嵌入式系统越来越受到人们的重视。随着系统复杂程度的提高,小型化和网络化也成为嵌入式系统发展的必然趋势。如何利用现成的通信网络,安全、快捷地对各个节点单片机进行在线软件升级(ISP),成为嵌入式系统发展的一项重要课题,对工业控制、航空航天、通信等领域意义重大。而实现这一功能,需要一段核心代码的支持,这段代码就是 BootLoader。Luminary 设计并生产的 Stellaris 系列单片机。基于先进的 ARM Cortex-M3 内核,芯片提供的高效性能、广泛的集成功能,适用于各种关注成本并明确要求具有过程控制以及连接能力的应用方案。不可多得的是,

Luminary 官方提供了 BootLoader 的全部源代码,大大减小了开发难度。本节将分析 Stellaris BootLoader 的组成、结构及设计思路,讲解关键技术,以更好地应用 BootLoader。

BootLoader 是位于 Flash 起始地址处的一小段代码,占据空间默认为 2 KB。如果没有 BootLoader,硬件启动成功后,将直接运行用户应用程序(Application)。反之,BootLoader 的启动代码(Start-up Code)将先被执行,进行一系列的初始化操作后,根据预先设定的条件,选择执行用户应用程序(Application)或升级控制程序(Updater)。Updater 在升级 Flash 的过程中,需要与上位机通信,通信的端口可选用 USB、UART、SSI、I²C 或以太网端口。为了保证数据的无差错传输,BootLoader 采用控制传输的通信协议,对接收到的格式正确且校验成功的数据包,Updater 能够将其解包,并将得到的加载命令转化为对 Flash 底层寄存器的操作。

总体来说,Cortex-M3 支持 4 GB 存储空间,如图 13-2 所示被划分成若干区域,不像其他的 ARM 架构,它们的存储器映射由半导体厂家说了算,Cortex-M3 预先定义好了"粗线条的"存储器映射。通过把片上外设的寄存器映射到外设区,就可以简单地以访问内存的方式来访问这些外设的寄存器,从而控制外设的工作。结果,片上外设可以使用 C

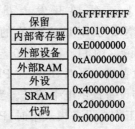

图 13-2 Cortex-M3 存储器映射图

语言来操作。这种预定义的映射关系,也使得对访问速度可以做高度的优化,而且对于片上系统的设计而言更易集成(还有一点很重要,即不用每学一种不同的单片机就要熟悉一种新的存储器映射)。Cortex-M3 的内部拥有一个总线基础设施,专用于优化对这种存储器结构的使用。在此之上,Cortex-M3 甚至还允许这些区域之间"越权使用"。比如说,数据存储器也可以被放到代码区,而且代码也能够在外部 RAM 区中执行(但是会变慢不少)。处于最高地址的系统级存储区,是 Cortex-M3 用于藏"私房钱"的——包括中断控制器、MPU 以及各种调试组件。所有这些设备均使用固定的地址。通过把基础设施的地址定死,使得至少在内核水平上,为应用程序的移植扫清了障碍。

由于 Cortex-M3 的地址空间中,0~0.5 GB 被映射为 Flash 空间,0.5~1 GB 被映射为 SRAM 空间。由于 SRAM 是易失性存储器,故系统上电时,SRAM 中并没有内容,系统必须从 Flash 开始启动。Flash 空间起始地址处必须存放向量表。向量表是异常产生时获取异常处理函数入口的一块连续内存,每一个异常都在向量表固定的偏移地址处(偏移地址以字对齐),通过该偏移地址可以获取异常处理函数的入口指针。向量表中前 4 个字分别为:栈顶地址、复位处理函数地址、NMIISR 地址、硬故障 ISR 地址。一张向量表至少由这 4 项组成。在程序代码开始运行后,向量表的基地址也可以改变。通过软件设置 NVIC 中的向量表偏移寄存器(NVIC_

VTABLE,0xE000ED08),可以在任意32字对齐处建立向量表。

Stellaris系列单片机硬件启动原理如下：硬件复位时，NVIC_VTABLE复位为0，向量表默认位于Flash空间起始地址处(0x00000000)。内核读取向量表第1个字设置主堆栈(SP_main)，读取第2个字设置PC指针，之后跳转到复位处理函数中运行。自此，系统的控制权交由软件接管。BootLoader启动需先执行以下一系列操作：配置向量表、初始化存储器、复制BootLoader代码到SRAM、从SRAM中执行代码。BootLoader的作用之一，是提供运行时修改Flash的功能。而由于Stellaris系列单片机具有单周期的Flash读写能力，因此默认的代码段本身就位于Flash中。这样，如果内核直接从Flash中加载修改其自身的指令，则既容易造成时序上的混乱，又有可能因Flash中某些关键指令被修改而导致整个系统崩溃。

为了解决这个问题，Stellaris系列处理器采用以下方案：在SRAM中建立BootLoader的映像，即把BootLoader复制到SRAM中，然后从SRAM中加载指令，如图13-3所示。这样，指令的加载源(SRAM)与修改操作的目标(Flash)相分离，一定程度上保证了软件升级的可靠性和安全性。这样的存储器映射允许修改Flash的全部代码。此外，BootLoader也提供了保护机制，用于保护Flash中的BootLoader代码本身，以及Flash空间顶部的一段存储区(保存即使Flash升级也不需要擦除的代码)。除非相应的配置选项使能，否则，这两段代码不可随意修改。但是有一个缺点，Flash下载地址填写为0时，BootLoader将会被清除。

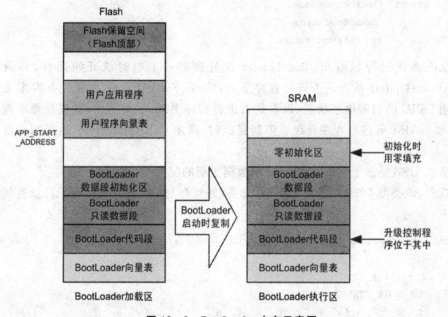

图13-3 BootLoader内存示意图

在Keil版本的BootLoader中有这样一段代码，用于BootLoader复制到SRAM中，请看下面一段代码：

```
;**************************************************************
;复位处理
;**************************************************************
    export   Reset_Handler
Reset_Handler
    ;初始化处理器
    bl       ProcessorInit
    ;跳转到 SRAM 中的 Reset 处理函数处
    ldr      pc, = Reset_Handler_In_SRAM
;**************************************************************
;SRAM 中的 Reset 处理函数
;**************************************************************
    align    4
    area     ||.text||, code, readonly, align = 2
Reset_Handler_In_SRAM
    ;初始化
    if       :def:_BL_HW_INIT_FN_HOOK
    import   $_BL_HW_INIT_FN_HOOK
    bl       $_BL_HW_INIT_FN_HOOK
    endif
    ;检查是否有更新需求,并调用应用程序 CallApplication
    import   CheckForceUpdate
    bl       CheckForceUpdate
    cbz      r0, CallApplication
```

从上面代码可以看出,BootLoader 在处理器一上电时就开始运行,并通过 CheckForceUpdate 检查是否进行程序下载,CheckForceUpdate 是通过查看定义更新使用 GPIO 的引脚电平状态,确定是否进行程序更新,如果更新,就根据要求选择 USB 和 UART 等进行程序升级。更新完成时,调用 CallApplication,运行用户应用程序。

那么 USB 是怎么与 PC 通信,完成数据更新的呢?

首先,该类型 USB 系统属于 USB 设备,并且有与 PC 通信的功能。其设备描述符如下:

```
const unsigned char g_pManufacturerString[] =
{
    2 + (14 * 2),
    USB_DTYPE_STRING,
    'L', 0, 'u', 0, 'm', 0, 'i', 0, 'n', 0, 'a', 0, 'r', 0, 'y', 0,
    ' ', 0, 'M', 0, 'i', 0, 'c', 0, 'r', 0, 'o', 0
};
const unsigned char g_pProductString[] =
```

```c
{
    (23 + 1) * 2,
    USB_DTYPE_STRING,
    'D', 0, 'e', 0, 'v', 0, 'i', 0, 'c', 0, 'e', 0, ' ', 0, 'F', 0, 'i', 0,
    'r', 0, 'm', 0, 'w', 0, 'a', 0, 'r', 0, 'e', 0, ' ', 0, 'U', 0, 'p', 0,
    'g', 0, 'r', 0, 'a', 0, 'd', 0, 'e', 0
};
const unsigned char g_pSerialNumberString[] =
{
    (3 + 1) * 2,
    USB_DTYPE_STRING,
    '0', 0, '.', 0, '1', 0
};
const unsigned char * const g_pStringDescriptors[] =
{
    g_pLangDescriptor,
    g_pManufacturerString,
    g_pProductString,
    g_pSerialNumberString
};
```

以上字符串描述符主要用于枚举,让 USB 主机识别该设备的语言、厂商、设备编号等。

```c
//****************************************************************
//DFU 设备描述符
//****************************************************************
const unsigned char g_pDFUDeviceDescriptor[] =
{
    18,                             // Size of this structure
    USB_DTYPE_DEVICE,               // Type of this structure
    USBShort(0x110),                // USB version 1.1 (if we say 2.0, hosts assume
                                    // high-speed - see USB 2.0 spec 9.2.6.6)
    USB_CLASS_VEND_SPECIFIC,        // USB Device Class
    0,                              // USB Device Sub-class
    0,                              // USB Device protocol
    64,                             // Maximum packet size for default pipe
    USBShort(USB_VENDOR_ID),        // Vendor ID (VID)
    USBShort(USB_PRODUCT_ID),       // Product ID (PID)
    USBShort(USB_DEVICE_ID),        // Device Release Number BCD
    1,                              // Manufacturer string identifier
    2,                              // Product string indentifier
    3,                              // Product serial number
```

第 13 章　USB 设备工程实例

```
        1                               // Number of configurations
};
```

DFU 设备描述符定义该设备为 USB 1.1 规范的设备，最大数据包为 64 个字节，并定义了 VID、PID 和设备编号，第 1 个字符串描述符为厂商字符串，第 2 个字符串描述符为产品描述符，第 3 个字符串描述符为产品串号。

```c
//***************************************************************
// DFU 设备的配置描述符
//***************************************************************
const unsigned char g_pDFUConfigDescriptor[] =
{
    // Configuration descriptor header
    9,                              // Size of the configuration descriptor
    USB_DTYPE_CONFIGURATION,        // Type of this descriptor
    USBShort(27),                   // The total size of this full structure
    1,                              // The number of interfaces in this
                                    // configuration
    1,                              // The unique value for this configuration
    0,                              // The string identifier that describes this
                                    // configuration
#if USB_BUS_POWERED
    USB_CONF_ATTR_BUS_PWR,          // Bus Powered
#else
    USB_CONF_ATTR_SELF_PWR,         // Self Powered
#endif
    (USB_MAX_POWER / 2),            // The maximum power in 2 mA increments
    // Interface descriptor
    9,                              // Length of this descriptor
    USB_DTYPE_INTERFACE,            // This is an interface descriptor
    0,                              // Interface number
    0,                              // Alternate setting number
    0,                              // Number of endpoints (only endpoint 0 used)
    USB_CLASS_APP_SPECIFIC,         // Application specific interface class
    USB_DFU_SUBCLASS,               // Device Firmware Upgrade subclass
    USB_DFU_PROTOCOL,               // DFU protocol
    0,                              // Description in string ID number 4
    // Device Firmware Upgrade functional descriptor
    9,                              // Length of this descriptor
    0x21,                           // DFU Functional descriptor type
    (DFU_ATTR_CAN_DOWNLOAD |        // DFU attributes
     DFU_ATTR_CAN_UPLOAD |
```

```
    DFU_ATTR_MANIFEST_TOLERANT),
    USBShort(0xFFFF),                    // Detach timeout (set to maximum)
    USBShort(DFU_TRANSFER_SIZE),         // Transfer size 1KB
    USBShort(0x0110)                     // DFU Version 1.1
};
```

本设备配置包含了配置描述符头、接口描述符、其他功能描述符。配置描述符头规定了本设备只有一个接口描述符,定义了供电方式;接口描述符定义了该设备为 UDF 类型;功能描述符定义该设备可以下载、上传等功能,传输 1 KB 数据、使用 1.1 版本。

```
void USB0DeviceIntHandler(void)
{
    unsigned long ulTxStatus, ulGenStatus;
    // Get the current full USB interrupt status
    ulTxStatus = HWREGH(USB0_BASE + USB_O_TXIS);
    ulGenStatus = HWREGB(USB0_BASE + USB_O_IS);
    // Received a reset from the host
    if(ulGenStatus & USB_IS_RESET)
    {
        USBDeviceEnumResetHandler();
    }
    // USB device was disconnected
    if(ulGenStatus & USB_IS_DISCON)
    {
        HandleDisconnect();
    }
    // Handle end point 0 interrupts
    if(ulTxStatus & USB_TXIE_EP0)
    {
        USBDeviceEnumHandler();
    }
}
```

USB 中断处理函数,获取 USB 中断标志,根据中断标志进行相应的数据处理,比如:枚举准备、进行枚举、断开连接等。

当枚举成功后,PC 机的上位机程序与 USB 设备进行通信,完成数据交换,将要下载的程序传输到处理器内部,通过处理器的 Flash 自编程功能完成程序下载。整个数据传输只是一个简单的数据通信,在此不作具体介绍。

13.3　USB 设备开发总结

为简化 USB 设备的开发过程,USB 提出了设备类的概念。所有设备类都必须支持标准 USB 描述符和标准 USB 设备请求。如果有必要,设备类还可以自行定义其专用的描述符和设备请求,分别被称为设备类定义描述符和设备类定义请求。另外,一个完整的设备类还将指明其接口和端点的使用方法,如接口所包含端点的个数、端点的最大数据包长度等。

HID 设备类就是设备类的一类,HID 是 Human Interface Device 的缩写,即人机交互设备,例如键盘、鼠标及游戏杆等。不过 HID 设备并不一定要有人机接口,只要符合 HID 类别规范的设备都是 HID 设备。

USB 协议制定时,为了方便不同设备的开发商基于 USB 进行设计,定义了不同的设备类来支持不同类型的设备。虽然在 USB 标准中定义了 USB_DEVICE_CLASS_AUDIO——AUDIO 设备。但是很少有此类设备问世。目前称为 USB 音箱的设备,大都使用 USB_DEVICE_CLASS_POWER,仅仅将 USB 接口作为电源使用。完全基于 USB 协议的 USB_DEVICE_CLASS_AUDIO 设备,采用一根 USB 连接线,在设备中不同的端点实现音频信号的输入、输出包括相关按键的控制。

AUDIO 设备是专门针对 USB 音频设备定义的一种专用类别,它不仅定义了音频输入、输出端点的标准,还提供了音量控制、混音器配置、左右声道平衡,甚至包括对支持杜比音效解码设备的支持,功能相当强大。不同的开发者可以根据不同的需求对主机枚举自己的设备结构,主机则根据枚举的不同设备结构提供相应的服务。

USB 通道的数据传输格式有两种,而且这两种格式还是互斥的。有消息和流两种格式,对于流状态,不具有 USB 数据的格式,遵循的规则就是先进先出。对于消息通道,它的通信模式符合 USB 的数据格式,一般由 3 个阶段组成,分别是建立阶段、数据阶段和确认阶段。所有通信的开始都是由主机方面发起的。

USB 协议制定时,为了方便不同设备的开发商基于 USB 进行设计,定义了不同的设备类来支持不同类型的设备。Bulk 也是其中一种,Bulk 设备使用端点 Bulk 传输模式,可以快速传输大量数据。

批量传输(Bulk)采用的是流状态传输,批量传送的一个特点就是支持不确定时间内进行大量数据传输,能够保证数据一定可以传输,但是不能保证传输的带宽和传输的延迟。而且批量传输是一种单向的传输,要进行双向传输必须使用两个通道。本章将介绍双向 Bulk 的批量传输。

USB 的 CDC 类是 USB 通信设备类(Communication Device Class)的简称。CDC 类是 USB 组织定义的一类专门给各种通信设备(电信通信设备和中速网络通信设备)使用的 USB 子类。根据 CDC 类所针对通信设备的不同,CDC 类又被分成以下不同的模型:USB 传统纯电话业务(POTS)模型、USB ISDN 模型和 USB 网络模型。通常一个 CDC 类又由两个接口子类组成通信接口类(Communication Interface Class)和数据接口类(Data Interface Class)。通信接口类对设备进行管理和控制,而数据接口类传送数据。这两个接口子类占有不同数量和类型的终端点(Endpoints),不同 CDC 类模型,其所对应的接口的终端点需求也是不同的。

USB 的 Mass Storage 类是 USB 大容量存储设备类（Mass Storage Device Class）。专门用于大容量存储设备，比如 U 盘、移动硬盘、USB CD‐ROM、读卡器等，在日常生活中经常用到。USB Mass Storage 设备开发相对简单。

USB 的 Composite 类是 USB 复合设备类，一个 USB 设备具有多种设备功能，比如一个 USB 设备同时具有鼠标和键盘功能。单一的 USB 设备开发相对简单，但在很多时候使用的 USB 设备具有多种功能。Composite 类可以满足这种要求。

USB 设备之所以会被大量应用，主要具有以下优点：

① 可以热插拔。这就让用户在使用外接设备时，不需要重复"关机将并口或串口电缆接上再开机"这样的动作，而是直接在计算机工作时，就可以将 USB 电缆插上使用。

② 携带方便。USB 设备大多以"小、轻、薄"见长，对用户来说，同样 20 GB 的硬盘，USB 硬盘比 IDE 硬盘要轻一半，在想要随身携带大量数据时，当然 USB 硬盘会是首选了。

③ 标准统一。大家常见的是 IDE 接口的硬盘，串口的鼠标键盘，并口的打印机和扫描仪，可是有了 USB 之后，这些应用外设统统可以用同样的标准与计算机连接，这时就有了 USB 硬盘、USB 鼠标、USB 打印机等。

④ 可以连接多个设备。USB 在计算机上往往具有多个接口，可以同时连接几个设备，如果接上一个有 4 个端口的 USB HUB 时，就可以再连上 4 个 USB 设备（注：最高可连接 127 个设备）。

本章主要介绍了 USB 设备的开发流程，简单介绍了 USB BootLoader 的基本概念与原理和 USB 设备开发，并讲解了各种设备的优缺点。通过本章的学习，使读者对 USB 设备开发有了一个更高层次的理解。

第 14 章

USB 主机开发实例

USB 互连支持数据在 USB 主机与 USB 设备之间的流动。主机上的客户软件（software client）与设备的功能部件（function）之间的通信必须通过主机接口（host interface）实现。

14.1 USB 主机开发流程

如图 14-1 所示，主机与设备都被划分成不同的层次。主机上垂直的箭头是实际的信息流。设备上对应的接口是基于不同端点实现的。在主机与设备之间的所有通信最终都是通过 USB 电缆进行的，然而，在"客户-功能部件"层存在"主机-设备"的逻辑信息流。主机上的客户软件和设备功能部件之间的通信是基于实际的应用需求及设备所能提供的能力。客户软件与功能部件之间的透明通信的要求，决定主机和设备下层部件的功能以及它们的接口（interface），图 14-1 展示了 USB 通信模型之间基本的信息流与互连关系。

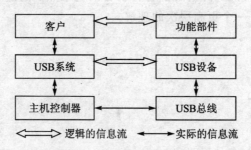

图 14-1 通信模型层次关系图

USB 主机有 3 个主要组成部分：
- 主机控制器驱动（Host Controller Driver）；
- USB 驱动（USB Driver）；
- 主机软件（host software）。

USB 控制器的存在，方便地将各种不同的 USB 控制器实现映射到 USB 系统，客户可以不必知道设备到底接在哪个主机控制器上就能同设备进行通信。USB 驱动提供了基本的面向客户的主机界面。在 HCD 与 USB 之间的接口称为主机控制器驱动接口(Host Controller Driver Interface,HCDI)。这层接口不能被客户直接访问,所以也不是由USB具体来完成的。一个典型的HCDI是由支撑各种不同USB控制器的操作系统来定义的。

USBD提供I/O请求包(I/O Request Packets)形式的数据传输,以某一特定通道来传输数据。另外,USBD为它的客户提供了一个容易被支配及配置的抽象的设备。作为这种抽象的一部分,USBD拥有标准通道对设备进行一些标准的控制。该标准通道实现了USBD与抽象设备之间的逻辑通信。

在有些操作系统中,提供了额外的非USB系统软件以支持设备的配置及设备驱动程序的加载。在这样的操作系统中,设备驱动程序应使用提供的主机软件接口而不是直接访问USBDI。

客户层描述的是直接与USB设备进行交互所需要的软件包。当所有的设备都已连上系统时,这些客户就可以直接通过设备进行通信。一个客户不能直接访问设备的硬件。所以,USB主机需要提供如下功能:

> 检测USB设备的连接与断开。
> 管理主机与设备之间的标准控制流。
> 管理主机与设备之间的数据流。
> 收集状态及一些活动的统计数字。
> 控制主机控制器与USB设备的电气接口,包括提供有限的能源。

对于特定的主机平台与操作系统下实现的接口请参照相关的操作系统手册。所有的集线器都通过状态改变通道报告其状态的改变,其中包括设备的连接与断开等。USBD的一类特殊客户,即集线器驱动器拥有这些状态改变通道,接收这些状态的改变。对于像设备连接这种状态改变,集线器驱动器将加载设备的驱动程序。在有些系统中,这种集线器驱动程序是操作系统提供的主机软件的一部份,它用来管理设备。

在所有控制中,USB控制器都必须提供基本相同的功能。USB控制器对主机及设备来讲都必须满足一定的要求。下面是USB控制器所提供的功能:

① 状态处理(State Handling)。作为主机的一部份,主机控制器报告及管理它的状态。

② 串行化与反串行化。对于从主机输出的数据,USB控制器将协议及数据信息从它的原始形状转换为字位流。而对于主机接收的数据,USB主机控制器进行反向操作。

③ 帧产生(Frame Generation)。主机控制器以每1 ms为单位产生SOF标志包。

④ 数据处理。主机控制器处理从主机输入输出数据的请求。

⑤ 协议引擎。主机控制器支持USB具体规定的协议。

⑥ 传输差错控制。所有的主机控制器在发现和处理已定义的错误时展现相似

第14章 USB主机开发实例

的行为。

⑦ 远程唤醒。所有的主机控制器都应具有将总线置于挂起状态及在远程唤醒事件下重新启动的能力。

⑧ 集线器。集线器提供了标准的将多个USB设备连接到主机控制器的功能。

⑨ 主机系统接口。主机控制器在主机系统控制器之间建立一个高速的数据通道。

对于使用Stellaris的USB控制器开发USB主机时,其流程如图14-2所示,USB主机有两种方式可以实现USB host和OTG,可以使用OTG开发USB主机,所以在USB主机开发时,必须明确使用什么模式进行USB主机开发;嵌入式USB主机系统与PC中的USB主机开发不一样,嵌入式USB主机系统的资源有限,不能将所有驱动都集成到USB处理器中,并且所有的设备驱动并不一致,所以是对特定设备开发特定的驱动;完成以上工作后,必须进行设备控制、数据传输等MCU应用程序开发。

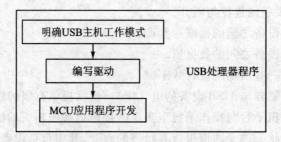

图14-2 USB主机开发流程图

使用Stellaris的USB控制器进行USB主机开发时的注意事项如下所示:

(1) 几乎所有Stellaris的USB控制器都有主机功能,在开发时,启动代码中的USB中断服务程序的名字应该为USB0HostIntHandler,在USB库中已经集成好此中断服务函数,不必自己再编写,但一定要在中断向量表中加入该中断服务函数名。

(2) Stellaris的USB控制器有些版本出现bug,需要开发人员避开,如LM3S3XXX与LM3S5XXX的A0版本有一个bug,当USB处理器工作在主机模式下时,PB0应该连接到低电平(连接到地线)。

(3) OTG用作主机开发时,ID引脚应该连接到低电平(连接到地线);启动代码中的USB中断服务程序的名字也应该为USB0OTGModeIntHandler。使用USBOTGMain函数,OTG控制器驱动的主循环函数,这个函数必须在应用程序主函数中周期性地调用。

(4) 使用USB功能时,处理器的主频不要低于20 MHz,建议设置为50 MHz,并且内核电压最好调节到2.75 V,提高处理器的工作电压,让系统更稳定可靠。

为了保证USB主机的正常工作,其硬件电路要严格按照官方提供的数据手册进行开发,保证硬件电路的正确性,不然在USB系统开发时,会增加调试难度。

14.2 USB 主机之音频输入输出

USB 音频主机是对 USB 音频设备的控制、读取多媒体信息的一种 USB 主机，这种主机包含了数据传输、音频信息解码及音量控制等。

(1) 完成系统的基本信息配置。

```
unsigned long ulTemp;
// 设置主频为 50 MHz
ROM_SysCtlClockSet(SYSCTL_SYSDIV_4 | SYSCTL_USE_PLL | SYSCTL_OSC_MAIN |
                   SYSCTL_XTAL_16MHZ);
ROM_SysTickPeriodSet(SysCtlClockGet() / TICKS_PER_SECOND);
ROM_SysTickEnable();
ROM_SysTickIntEnable();
// 配置中断优先级
ROM_IntPriorityGroupingSet(4);
ROM_IntPrioritySet(INT_USB0, USB_INT_PRIORITY);
ROM_IntPrioritySet(FAULT_SYSTICK, SYSTICK_INT_PRIORITY);
ROM_IntPrioritySet(INT_ADC0SS3, ADC3_INT_PRIORITY);
// 打开中断
ROM_IntMasterEnable();
g_ulFlags = 0;
g_ulSysTickCount = 0;
g_ulLastTick = 0;
```

首先配置系统内部主频为 50 MHz，保证其主频能够顺利处理音频数据，完成整个系统控制；调整中断优先级，让 USB 数据能实时响应，并且不影响其他模块使用。

(2) 初始化音频设备。

```
void USBSoundInit(unsigned long ulFlags, tEventCallback pfnCallback)
{
    // 使能音频控制端口
    SysCtlPeripheralEnable(SYSCTL_PERIPH_GPIOA);
    // 配置 USB 相关引脚
    GPIOPinTypeUSBDigital(GPIO_PORTA_BASE, GPIO_PIN_6 | GPIO_PIN_7);
    //打开 DMA
    SysCtlPeripheralEnable(SYSCTL_PERIPH_UDMA);
    uDMAEnable();
    uDMAControlBaseSet(g_sDMAControlTable);
    //初始化 USB 堆栈为 USB_MODE_OTG 的主机模式
    USBStackModeSet(0, USB_MODE_OTG, 0);
    //注册主机驱动
```

```c
    USBHCDRegisterDrivers(0, g_ppHostClassDrivers, g_ulNumHostClassDrivers);
    // 打开音频实例
    g_ulAudioInstance = USBHostAudioOpen(0, AudioCallback);
    //电源初始化
    USBHCDPowerConfigInit(0, USBHCD_VBUS_AUTO_HIGH | USBHCD_VBUS_FILTER);
    // OTG 初始化
    USBOTGModeInit(0, 2000, g_pHCDPool, HCD_MEMORY_SIZE);
    //保存回调函数
    g_sAudioState.pfnCallbackEvent = pfnCallback;
}
```

通过使用 USBSoundInit 初始化音频设备，包括初始化音频控制引脚、DMA；配置 USB 处理器工作模式为 OTG 的主机模式；注册驱动并打开驱动实例等，完成音频设备一系列的初始化工作。

(3) 控制并读取音频数据。

```c
void USBMain(unsigned long ulTicks)
{
    //定时调用 OTG 控制函数
    USBOTGMain(ulTicks);
    //根据 USB 当前状态判断数据处理
    switch(g_sAudioState.eState)
    {
        case STATE_DEVICE_READY:
        {
            if(HWREGBITW(&g_sAudioState.ulEventFlags, EVENT_CLOSE))
            {
                HWREGBITW(&g_sAudioState.ulEventFlags, EVENT_CLOSE) = 0;
                g_sAudioState.eState = STATE_NO_DEVICE;
                // 调用回调函数
                if(g_sAudioState.pfnCallbackEvent)
                {
                    g_sAudioState.pfnCallbackEvent(SOUND_EVENT_DISCONNECT, 0);
                }
            }
            break;
        }
        // 没有设备
        case STATE_NO_DEVICE:
        {
            if(HWREGBITW(&g_sAudioState.ulEventFlags, EVENT_OPEN))
            {
```

```
            g_sAudioState.eState = STATE_DEVICE_READY;
            HWREGBITW(&g_sAudioState.ulEventFlags, EVENT_OPEN) = 0;
            GetVolumeParameters();
            // 回调
            if(g_sAudioState.pfnCallbackEvent)
            {
                g_sAudioState.pfnCallbackEvent(SOUND_EVENT_READY, 0);
            }
        }
        break;
    }
    // 未知设备
    case STATE_UNKNOWN_DEVICE:
    {
        break;
    }
    // 电源异常
    case STATE_POWER_FAULT:
    {
        break;
    }
    default:
    {
        break;
    }
    }
}
```

此函数是 USB 音频主机的主函数,负责整体调度和数据处理,根据 USB 的不同状态回调相关函数,完成控制功能。

```
static void USBAudioOutCallback(void * pvBuffer, unsigned long ulEvent)
{
    // If a buffer has been played then schedule a new one to play
    if((ulEvent == USB_EVENT_TX_COMPLETE) &&
       (HWREGBITW(&g_ulFlags, FLAGS_PLAYING)))
    {
        // Indicate that a transfer was complete so that the non-interrupt
        // code can read in more data from the file
        HWREGBITW(&g_ulFlags, FLAGS_TX_COMPLETE) = 1;
        // Increment the read pointer
        g_pucRead += g_ulTransferSize;
        // Wrap the read pointer if necessary.
```

第 14 章　USB 主机开发实例

```
            if(g_pucRead >= (g_pucAudioBuffer + g_ulBufferSize))
            {
                g_pucRead = g_pucAudioBuffer;
            }
            // Increment the number of bytes that have been played
            g_ulBytesPlayed += g_ulTransferSize;
            // Schedule a new USB audio buffer to be transmitted to the USB
            // audio device
            USBSoundBufferOut(g_pucRead, g_ulTransferSize, USBAudioOutCallback);
        }
    }
```

音频设备的回调函数,通过 USBMain 进行回调。本函数用于音频数据处理,把音频数据通过 USB 传输到 USB 音频设备中,播放所传送的音乐。

以上 3 步就是 USB 音频开发的主要流程,其与 USB 主机开发类似,比如 HID、MSC 类主机基本一样,并且在 TI 的 USB 库中已经建立了完善的驱动程序,开发人员直接调用即可,如果遇到陌生的 USB 设备,可以仿照 TI 提供的 USB 库进行驱动开发。

14.3　USB 主机开发总结

USB 主机开发相对简单,在使用 USB 库编程时更加简单,USB 主机开发时,首先确定主机要读取设备的类型,如在 USB 库中没有该类设备驱动,则开发人员可按照 USB 库中设备驱动编写方式自行编写;编程开发时,首先配置 USB 处理器的内核;再根据情况编写、修改驱动;注册设备驱动,一个函数就能实现;通过 USB 库函数来运行主机。嵌入式 USB 主机驱动是为 USB 设备定做的,没有一种驱动能完全与各种设备结合,这需要开发人员进一步了解协议,进行设备驱动开发。

通过本章的学习,读者可以对 USB 主机开发有所了解,能使用 USB 库开发简单的 USB 主机,控制自带的 USB 设备类型,比如 HID、MSC 等。在 USB 主机开发中,主要是设备驱动开发,驱动是 USB 主机识别控制 USB 设备的一组信息(包括数据结构、控制函数等)。

第 15 章

USB 系统开发总结

USB 是一种方便人们生活而产生的新型总线,结构简单,仅仅只有 4 根线就能完成数据通信,并且传输速度并不低。由于 USB 的不断发展,便携式设备越来越离不开 USB。没有一种手机不具备 USB 功能,没有一种 MP3 没有 USB 接口,没有哪种计算机没有 USB 卡口。当然,由于 USB 开发的复杂性,在开发 USB 系统时或多或少会遇到一些开发、调试的问题。本章主要归纳了开发时常见的问题,并详细介绍其解决方法。

15.1 常见问题

15.1.1 概念问题

(1) 为什么要开发 USB 这种通信协议?

开发通用串行总线架构(USB)的目的主要基于以下 3 方面考虑。

① 计算机与电话之间的连接:显然用计算机来进行计算机通信将是下一代计算机的基本应用。机器和人们的数据交互流动需要一个广泛而又便宜的连通网络。然而,由于目前产业间的相互独立发展,尚未建立统一标准,而 USB 则可以广泛地连接计算机和电话。

② 易用性:众所周知,PC 机的改装极不灵活。对用户友好的图形化接口和一些软硬件机制的结合,加上新一代总线结构使得计算机的冲突大量减少,且易于改装。但以终端用户的眼光来看,PC 机的输入/输出,如串行/并行端口、键盘、鼠标、操纵杆接口等,均还没有达到即插即用的特性,USB 正是在这种情况下问世的。

③ 端口扩充:外围设备的添加总是被相当有限的端口数目限制着。缺少一种双向、价廉、与外设连接的中低速总线,限制了外围设备(诸如电话/传真/调制解调器的适配器、扫描仪、键盘、PDA)的开发。现有的连接只可对极少设备进行优化,对于 PC 机的新功能部件的添加需定义一个新的接口来满足上述需要,USB 就应运而生了。它是快速、双向、同步、动态连接且价格低廉的串行接口,可以满足 PC 机发展的现在和未来的需要。

(2) USB 与 IEEE1394 的相同点主要有哪些?

两者都是一种通用外接设备接口;两者都可以快速传输大量数据;两者都能连接多个不同设备;两者都支持热插拔;两者都可以不用外部电源。

(3) USB 与 IEEE1394 的不同点有哪些?

① 两者的传输速率不同。USB 的最高速度可达 5 Gbps,但由于 USB 3.0 尚未普及,目前主流的 USB 2.0 只有 480 Mbps,并且速度较稳定;相比之下,IEEE1394 目前的速度虽然只有 800 Mbps,但很稳定,故在数码相机等高速设备中还保留了 IEEE1394 接口,但大量采用 USB 接口。

② 两者的结构不同。USB 在连接时必须至少有一台计算机,并且必须需要 HUB 来实现互连,整个网络中最多可连接 127 台设备。IEEE1394 并不需要计算机来控制所有设备,也不需要 HUB,IEEE1394 可以用网桥连接多个 IEEE1394 网络,也就是说在用 IEEE1394 实现了 63 台 IEEE1394 设备之后也可以用网桥将其他的 IEEE1394 网络连接起来,达到无限制连接。

③ 两者的智能化不同。IEEE1394 网络可以在其设备进行增减时自动重设网络。USB 是以 HUB 来判断连接设备的增减的。

④ 两者的应用程度不同。现在 USB 已经被广泛应用于各个方面,几乎每台 PC 机主板都设置了 USB 接口,USB 2.0 也会进一步加大 USB 应用的范围。IEEE1394 现在只被应用于音频、视频等多媒体方面。

(4) 常用的 USB 控制芯片有哪些?其各种芯片的特点是什么?

从芯片的构架来划分,市面上所有的 USB 控制器芯片可以分为不需要外接微处理器的和需要外接微处理器的两类芯片。不需要外接微处理器的芯片又可以分为 USB 接口专用芯片和嵌入通用微控制器内核的芯片。

USB 接口专用芯片内部装有特定指令集的微控制器,如 Cypress 公司的 USB M8 系列和 enCoRe USB 系列芯片。它们所能实现的功能有限,但是因为具有专门为 USB 应用优化的指令集,所以实现 USB 通信非常方便。

内嵌通用微控制器的 USB 控制芯片,一般是在通用微控制器的基础上扩展了 USB 功能,其优点是开发者熟悉这些通用微控制器的结构和指令集,相关资料丰富,易于进行开发。如 Cypress 基于 8051 的 EZ-USB 系列,Microchip 基于 PIC 的 16C7x5,Motorola 基于 68HC08 系列的 68HC08JB8,Atmel 基于 AVR 的 AT76C711 等 USB 控制芯片。

需要外接微控制器的芯片,只处理与 USB 相关的通信工作,而且必须由外部微控制器对其控制才能正常工作,这些芯片必须提供一个串行或并行的数据总线与微控制器进行连接。此外,还需要一个中断引脚,当数据收到或发送完,这个中断引脚会向微控制器发出中断请求信号。其优点是芯片价格便宜,而且便于用户使用自己熟悉的微控制器进行开发。比如 USBD12、max3420 等。

总之,USB 芯片是一种集成了 USB 协议的微处理器,它能自动对各种 USB 事件作出响应,以处理 USB 总线上的数据传输。USB 芯片按功能可以分为 USB 主控制器芯片、USB集线器芯片和USB功能设备芯片3大类。在对USB控制器芯片性能进行分析时,主要研究数据传输速度、功耗、电源、程序/数据存储器容量、封装以及何种USB规范等通用的技术指标。另外,针对不同的功能类型,还会有不同的要求。下面将对最新有代表性的USB控制芯片进行特性比较。

USB主控制器芯片负责实现主机与USB设备之间的物理数据传输,它是构成USB主机不可或缺的核心部件。另外,虽然有的芯片是主机/设备控制器芯片,但是一般只将其作为主机控制器芯片使用,如SL811HS。

USB集线器芯片负责将一个USB上行端口转化为多个下行端口,它是构成USB集线器不可或缺的核心部件。它所需要关心的性能指标与USB主机的要求不完全相同,支持的下行端口的数目是一个很重要的指标。如Cypress公司的CY7C66113、Alcor Micro公司的AU9254 A21和Philips公司的ISP1251。

本书选用Stellaris的USB控制器支持USB主机/设备/OTG功能,在点对点通信过程中可运行在全速和低速模式。它符合USB 2.0标准,包含挂起和恢复信号。它包含32个端点,其中包含两个用于控制传输的专用连接端点(一个用于输入,一个用于输出)以及30个由固件定义的端点,并带有一个大小可动态变化的FIFO,以支持多包队列。可通过μDMA来访问FIFO,将对系统软件的依赖降至最低。USB设备启动方式灵活,可软件控制是否在启动时连接。USB控制器遵从OTG标准的会话请求协议(SRP)和主机协商协议(HNP)。

(5) USB接口定义是怎样的?

USB是一种常用的PC接口,它只有4根线,两根电源线和两根信号线,故信号是串行传输的,USB接口也称为通用串行口,USB 2.0的速度可以达到480 Mbps。可以满足各种工业和民用需要。USB接口的输出电压和电流是:+5 V和500 mA。实际上有误差,最大不能超过+/−0.2 V,也就是4.8~5.2 V。USB接口的4根线一般是按图15-1所示分配的,需要注意的是千万不要把正负极弄反了,否则会烧掉USB设备、USB主机或者计算机的南桥芯片。图15-1为USB接口定义图,黑线为GND,地线;红线为VCC,电源正;绿线为DATA+,DP;白线为DATA−,DM。一般的排列方式为从左到右是红白绿黑,红色为USB电源,标有VCC、Power、5 V、5 VSB字样;白色为USB数据线(负),标有DATA−、USBD−、PD−、USBDT−字样;绿色为USB数据线(正),标有DATA+、USBD+、PD+、USBDT+字样;黑色为地线,标有GND、Ground字样。

图15-2为USB A型公头接口图,是USB设备常用的一种接口,比如U盘、鼠标、键盘等都使用此接口。

第 15 章 USB 系统开发总结

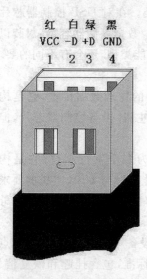

图 15-1 USB 接口定义图

图 15-2 USB A 型公头接口

图 15-3 为 Mini B 型 5Pin 接口。这种接口可以说是目前最常见的一种接口了,这种接口由于防误插性能出众,体积也比较小巧,所以正在赢得越来越多的厂商青睐,现在这种接口广泛出现在读卡器、MP3、数码相机以及移动硬盘上。

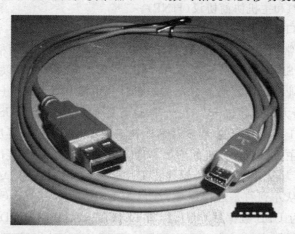

图 15-3 Mini B 型 5Pin 接口

图 15-4 为 Mini B 型 4Pin 接口,这种接口常见于以下品牌的数码产品:奥林巴斯的 C 系列和 E 系列,柯达的大部分数码相机,三星的 MP3 产品(如 Yepp),SONY 的 DSC 系列,康柏的 IPAQ 系列产品……

Mini B 型 4Pin 还有一种形式,那就是 Mini B 型 4Pin Flat。顾名思义,这种接口比 Mini B 型 4Pin 要更加扁平,在设备中的应用也比较广泛。

第 15 章　USB 系统开发总结

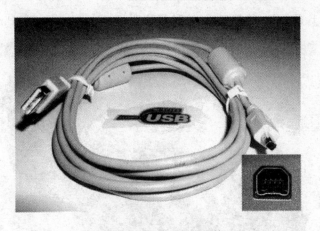

图 15-4　Mini B 型 4Pin 接口

图 15-5 为 MINI B 型 8Pin 接口图，这种接口和前面的普通型比起来，就是将原来的 D 型接头改成了圆形接头，并且为了防止误插在一边设计了一个凸起。

这种接头常见于一些 Nikon 的数码相机，CoolPix 系列比较多见。虽然 Nikon 一直坚持用这种接口，但是在一些较新的机型中，例如 D100 和 CP2000 也采用了普及度较高的 Mini B 型 5Pin 接口。

图 15-5　MINI B 型 8Pin 接口

图 15-6 为 MINI B 型 8Pin 2×4 接口，这种接口也是一种比较常见的接口，例如 iRiver 的著名 MP3 系列，其中号称"铁三角"的 180TC，以及该系列的很多其他产品采用的均是这种接口。这种接口的应用范围也还算广，不过 iRiver 自从 3XX 系列

第15章 USB系统开发总结

全面换成 Mini B 型 5Pin 的接口后，这种规格明显没有 Mini B 型 5Pin 抢眼了。

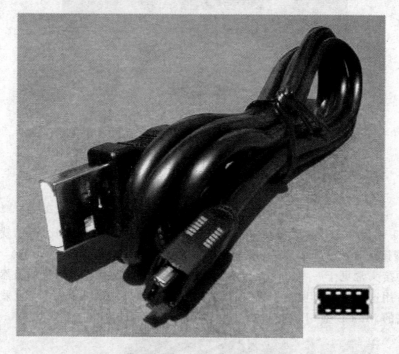

图 15-6　MINI B 型 8Pin 2×4 接口

(6) USB 3.0 较前两个版本有什么优点？

英特尔公司和业界领先的公司一起携手组建了 USB 3.0 推广组，旨在开发速度超过当今 10 倍的超高效 USB 互连技术。该技术是由英特尔，以及惠普(HP)、NEC、NXP 半导体以及德州仪器(Texas Instruments)等公司共同开发的，应用领域包括个人计算机、消费及移动类产品的快速同步即时传输。随着数字媒体的日益普及以及传输文件的不断增大——甚至超过 25 GB,快速同步即时传输已经成为必要的性能需求。

USB 3.0 具有向后兼容标准，并兼具传统 USB 技术的易用性和即插即用功能。该技术的目标是推出比目前连接水平快 10 倍以上的产品，采用与有线 USB 相同的架构。除对 USB 3.0 规格进行优化以实现更低的能耗和更高的协议效率之外，USB 3.0 的端口和线缆能够实现向后兼容，以及支持未来的光纤传输。

新规范提供了十倍于 USB 2.0 的传输速度和更高的节能效率，可广泛用于 PC 外围设备和消费电子产品。制定完成的 USB 3.0 标准已经移交给该规范的管理组织 USB Implementers Forum(简称 USB IF)。该组织将与硬件厂商合作，共同开发支持 USB 3.0 标准的新硬件。USB 3.0 最大的优点是传输速度快。

(7) Stellaris 的 USB 处理器支持高速传输吗？

目前，所有 Stellaris 的 USB 处理器支持全速和低速模式，并符合 USB 2.0 标准。暂时不支持高速模式。

15.1.2 开发问题

(1) 在使用 Stellaris 所提供的 USB 处理器进行 USB 设备开发时，什么不能正常枚举？

Stellaris 所提供的 USB 处理器中，LM3S3XXX 与 LM3S5XXX 的 A0 版本有一个 bug，如图 11-1 所示，为 PB0 与 PB1 硬件连接图，在 A0 版本中，USB 处理器工作在主机和设备模式下时，PB0 与 PB1 不能当 GPIO 使用，因为在此版本中，PB0 与 PB1 为主机与设备提供电平信号。当 USB 处理器工作在主机模式下时，PB0 应该连接到低电平；工作在设备模式下时，PB0 应该连接到高电平。同时，PB0 引脚与电压信号之间应该连接一个电阻，其典型值为 10 Ω。PB1 必须连接到 5 V(4.75～5.25 V)。如果 USB 处理器不是 A0 版本，PB0 与 PB1 可以用作 GPIO 功能。

图 15-7 为 LM3S3XXX 与 LM3S5XXX 系列的 USB 功能引脚连接图，USB 处理器工作在主机模式下时，USB 功能引脚连接与设备模式下一样。①USB0RBIAS，连接方式固定，为 USB 模拟电路内部必须的 9.1 kΩ 电阻(1% 精度)，普通贴片即可满足其需求；②USB0DP 和 USB0DM，USB0 的双向差分数据引脚，连接方式固定，分别连接到 USB 规范中的 D+ 和 D- 中，在使用时，请特别注意 D+ 和 D- 的连接方式。USB0EPEN，USB 主机电源输出使能引脚，主机输出电源使能信号，高电平有效，用于使能外部电源。如图 15-8 所示，通过 USB 处理器的 USB0EPEN 引脚输出高电平使能 VBUS 电源。USB0PFLT，主机模式下的外部电源异常输入引脚，指示外部电源的错误状态，低电平有效。当 TPS2051 的 VBUS 电源输出电流小于 1 mA 或者大于 500 mA 时，从 OCn 引脚输出低电平，开漏输出，通过 USB 处理器的 USB0PFLT 上拉，读取 OCn 的电平变化，并产生中断。注意，USB0EPEN 和 USB0PFLT 主要工作在主机模式下，为设备提供 VBUS 电源。当然这两个引脚可以不使用，直接通过主板上的 5 V 提供 VBUS 电源，这会缺少 USB 电源欠压过流保护，但电路简单。开发人员可以根据自己设计的需要，自行剪裁硬件，但是 USB0RBIAS、USB0DP、USB0DM 必不可少，如果是 A0 版本处理器，工作在主机和设备模式下时，PB0 与 PB1 不能当 GPIO 使用，PB1 接 5 V；主机模式下，PB0 接地；设备模式下，PB0 接 5 V。

对 LM3S3XXX、LM3S5XXX、LM3S9XXX 系列处理器，有一部分具有 USB OTG 功能，与传统不具有 USB OTG 功能的处理器比较，PB0 作为 USB0ID，此信号用于检测 USB ID 信号的状态。此时 USBPHY 将在内部启用一个上拉电阻，通过外部元件(USB 连接器)检测 USB 控制器的初始状态(即电缆的 A 侧设置下拉电阻，B 侧设置上拉电阻)。

第15章 USB系统开发总结

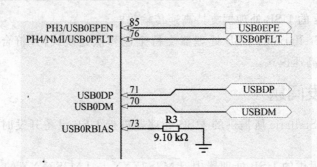

图15-7 USB功能引脚连接图

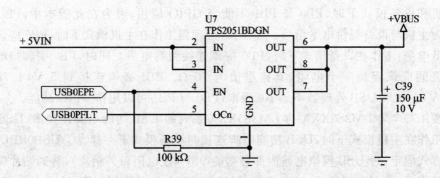

图15-8 VBUS控制电路

注意:对于具有 USB OTG 功能的 USB 处理器来说,也有一个重要的 bug 要注意,根据官方提供的数据手册可以看出,只要处理器不工作在 USB OTG 模式下,USB0VBUS 和 USB0ID 引脚可配置为 GPIO 使用,但在 B1 版本中(其他版本也有此 bug)无论 USB 处理器工作在什么模式,USB0ID 必须归 USB 使用,不能用于 GPIO 功能,如果工作在主机模式,USB0ID 连接低电平(地);工作在设备模式,USB0ID 连接高电平(5 V)。同时 USB0VBUS 必须提供 5V 输入。这个 bug 是怎么产生的呢? 由于 USB 控制状态寄存器(USBGPCS)的 DEVMODOTG 位工作不正常,造成无法通过寄存器来实现 USB 工作模式的切换,必须通过外面的 USB0ID 引脚来实现,相当于使用 USB OTG 来让 USB 处理器工作在主机或者设备模式下。

所以 USB 处理器不能正常工作,请检查芯片是不是有 bug 的芯片,是否有回避这些 bug。

(2) 为什么之前能正常下载程序,现在不能下载,并提示不能初始化设备?

可能是因为芯片被锁,调试器不能连接到处理器,必须解锁。那么死锁是怎么发生的呢? 除了 5 个 JTAG/SWD 引脚(PB7 和 PC[3:0])之外,所有 GPIO 引脚默认都是三态引脚(GPIOAFSEL=0,GPIODEN=0,GPIOPDR=0,且 GPIOPUR=0)。JTAG/SWD 引脚默认为 JTAG/SWD 功能(GPIOAFSEL=1,GPIODEN=1 且 GPIOPUR=1),如果 JTAG 引脚在设计中用作 GPIO 功能,在上电的一瞬间,由于处理器的运行速度很快,程序代码立即将 JTAG 引脚变成它们的 GPIO 功能,那么在 JTAG 引脚功能切换前调试器(下载器)将没有足够的时间去连接和中止控制器(如图 15-9 所示)。这会将调试器(下载器)锁在元件外,这就是死锁的原因。一但死锁,芯片几乎没法再次使用。通过使用一个基于外部或软件的触发器来恢复 JTAG 功能的软件程序就可以避免这种情况发生。在 IO 数量足够的情况下,不使用 JTAG 引脚的 GPIO 功能,就可以避免芯片死锁的发生。

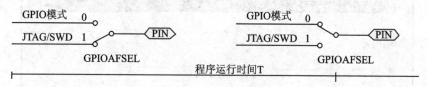

图 15-9 死锁过程

(3) 怎么使用 LMFlashProgrammer 解锁已经被死锁的处理器?

① 安装 LMFlashProgrammer 软件。

② 将 JTAG 连接到开发板上,并上电。打开 LMFlashProgrammer 软件,如图 15-10 所示,为 LMFlashProgrammer 的主界面。

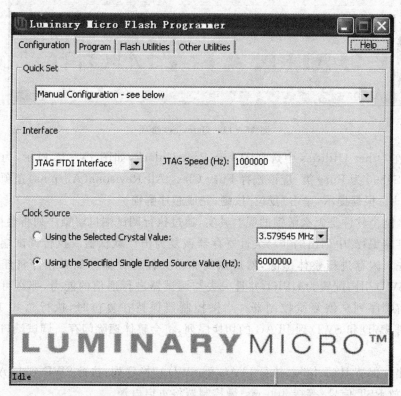

图 15-10 打开 LMFlashProgrammer 软件

第 15 章　USB 系统开发总结

③ 根据下载器选择合适的选项,如图 15 - 11 所示,在 LMFlashProgrammer 主界面中的 Quick set 中选择 Stellaris Development Board。

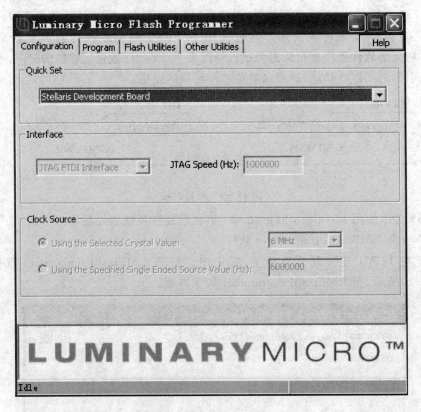

图 15 - 11　Quick Set 选项

④ 在 Other Utilities 中选择 Debug Port Unlock,如图 15 - 12 所示,为解锁界面,LM3S9b96 为 Fury 类,应该选择 Fury Class All Revisons(All pins)选项。并单击 Unlock。根据提示,单击"复位"按键,完成芯片解锁。

那么这个软件是怎么解锁的呢?其实,通过执行调试端口解锁序列将会把"非易失性寄存器编程"中列出的非易失性寄存器恢复为出厂默认值。执行调试端口解锁序列将先对闪存进行整体擦除,然后恢复非易失性寄存器。如果软件将任意一个 JTAG/SWD 引脚配置为 GPIO,并且失去与调试器进行通信的能力,那么可以用调试端口解锁序列来恢复微控制器。当微控制器保持在复位时,执行总共 10 次 JTAG 到 SWD 和 SWD 到 JTAG 的切换序列,将会整体擦除闪存。调试端口解锁序列为:

发出并保持 RST 信号;执行 JTAG 到 SWD 切换序列;执行 SWD 到 JTAG 切换序列;释放 RST 信号;等待 400 ms;微控制器将重启电源。

(4) USB 库的不同版本有没有区别? 向后兼容吗?

第 15 章　USB 系统开发总结

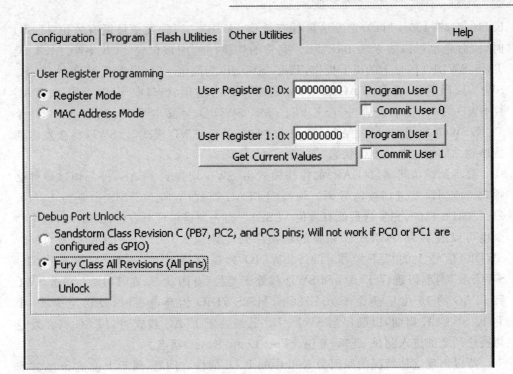

图 15-12　解锁界面

每一个版本的 USB 有所不同,比如现在的 USB 库已经能支持 CM4 了,USB 库功能更强大,向后兼容。但也有极少数函数、结构体不兼容,比如:tUSBDCompositeDevice 这个结构在 revision 8049 中比更低级版本多了 unsigned long * pulDeviceWorkspace,所以使用 USB 的 revision 8049 版本时要注意,定义复合设备时要定义 pulDeviceWorkspace 的空间,不然会工作不正常。

(5) 在 Keil 中使用 ICDI 调试时,为什么在 SWD 转换为 JTAG 模式时,会提示不能初始化目标设备,如图 15-13 所示。

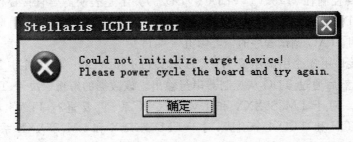

图 15-13　ICDI 提示

为了将调试访问端口(DAP)的操作模式由 JTAG 切换到 SWD,外部调试硬件必须向微控制器发送切换前导码。用于切换到 SWD 模式的 16 位 TMS 命令为

第 15 章　USB 系统开发总结

b1110.0111.1001.1110,先发送最低有效位(LSB)。当先发送最低有效位(LSB)时,该命令也可以表示为 0xE79E。完整的切换序列应该包含 TCK/SWCLK 以及 TMS/SWDIO 信号上的以下操作:TMS/SWDIO 为高电平时,发送至少 50 个 TCK/SWCLK 周期,以确保 JTAG 和 SWD 都处于复位/空闲状态;在 TMS 上发送 JTAG 到 SWD 的 16 位切换命令 0xE79E;TMS/SWDIO 为高电平时,发送至少 50 个 TCK/SWCLK 周期,以确保当 SWJ-DP 已经处于 SWD 模式时,SWD 将在发送切换序列之前进入总线复位状态。

要将调试访问端口(DAP)的操作模式由 SWD 切换到 JTAG,外部调试硬件必须向微控制器发送切换前导码。切换到 JTAG 模式的 16 位 TMS 命令为 b1110.0111.0011.1100,先发送最低有效位(LSB)。当先发送最低有效位(LSB)时,这个命令也可以表示为 0xE73C。完整的切换序列应该包含 TCK/SWCLK 以及 TMS/SWDIO 信号上的以下操作:TMS/SWDIO 为高电平时,发送至少 50 个 TCK/SWCLK 周期,以确保 JTAG 和 SWD 都处于复位/空闲状态;在 TMS 上发送 SWD 到 JTAG 的 16 位切换命令 0xE73C;TMS/SWDIO 为高电平时,发送至少 50 个 TCK/SWCLK 周期,以确保当 SWJ-DP 已经处于 JTAG 模式时,JTAG 将在发送切换序列之前进入测试-逻辑-复位(Test Logic Reset)状态。

所以出现这种情况的原因为 Keil 中的 ICDI 插件有问题,属于上位机 bug,并非 ICDI 调试器硬件问题。可以上电重新复位,消除此错误。所以每次从 SWD 转换为 JTAG 模式时,都应该让目标板重新上电复位一次。

(6) 最小系统中 VDD25(VDDC)引脚有什么用,LDO 引脚有什么用呢?

如图 15-14 所示,为 LM3S 处理器的功率控制图,从图中可以看出,处理器 VDD 与 VDDA 需要外部提供 3.3 V 电源,其中 VDD 为 IO 提供电源,而 VDDA 为内部模拟器件提供模拟电源。注意:VDDA 必须用 3.3 V 供电,否则微控制器不能正常工作,VDDA 为设备的所有模拟电路供电,包括时钟电路。VDDC 是 CPU 处理器和 PLL 的电源,由 LDO 输出,通过滤波输入到 VDDC,所以在倍频时,内核 LDO (内核电压)要控制在 2.75 V。VDDC(VDD25)并不是表明需要直接提供 2.5 V 电压,而是与 LDO 连接。

(7) 使用 JTAG 调试程序时需要哪几个端口?

JTAG 接口由 4 个引脚组成:TCK、TMS、TDI 和 TDO。数据通过 TDI 串行发送至控制器,然后通过 TDO 从控制器串行输出。该数据的解析取决于 TAP 控制器的当前状态。对于 LM3S5XXX 和 LM3S3XXX 系列,要进行 JTAG 调试时,需 TCK、TMS、TDI 和 TDO 4 个引脚;对于 LM3S6XXX 和 LM3S8XXX 系列处理器,JTAG 调试时,需 TCK、TMS、TDI、TDO 和 RST 等共 5 个引脚。在 JTAG 规范中,RST 引脚是可以不用的,因为 RST 引脚主要用于复位 JTAG 模块,使用相关命令也可复位 JTAG 模块,进行调试。为什么 LM3S6XXX 和 LM3S8XXX 系列处理器要使用 RST 引脚才能正常调试,可能是这两种处理器内部的 JTAG 模块不支持命令复

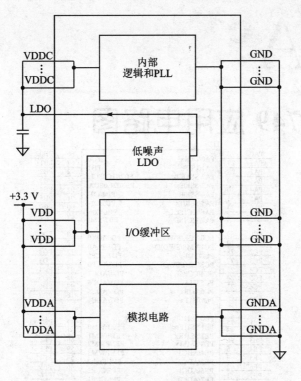

图 15-14 功率控制图

位。为了弥补这种缺陷,请读者将 TCK、TMS、TDI、TDO、RST 引脚都与 JTAG 调试器连接。

(8) LM3S8XXX 等 Cortex-M3 处理器数据手册中说其每个 IO 引脚可以承受 5 V 电压,为什么加上 5 V 电压时,其处理器出现死机现象?

在官方数据手册中,确实有说每个 IO 能承受 5 V 电压,实际最多可以承受 4 V。这是处理器本身的 bug,新生产的处理器已经解决了此类 bug。为了让系统更安全稳定,不建议直接使用处理器的 IO 引脚处理 5 V 数字信号。

15.2 本章小结

USB 结构简单但开发复杂,本章就 Stellaris 的 USB 处理器使用中常遇到的问题做了基本介绍,阐述了其产生的基本原因,提供了解决此类问题的方案。通过本章的学习,读者可以避免一些开发疑问,轻松学习 USB 开发。

附录 A

LM3S5749 应用电路图

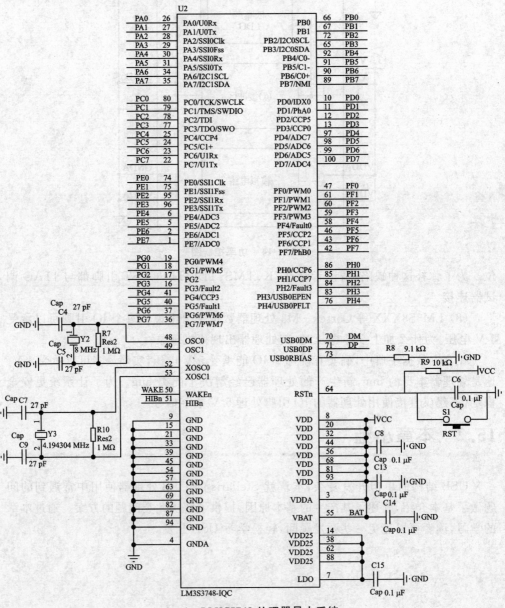

A-1 LM3S5749 处理器最小系统

附录 A LM3S5749 应用电路图

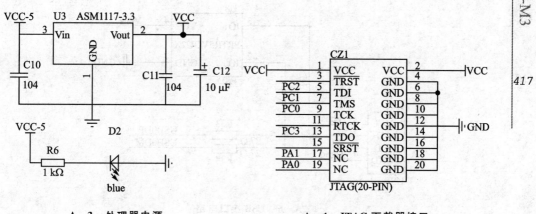

A-2 I/O 扩展接口

A-3 处理器电源

A-4 JTAG 下载器接口

附录 A LM3S5749 应用电路图

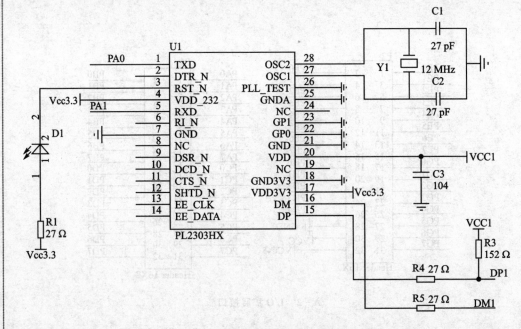

A-5 USB 转 UART 电路

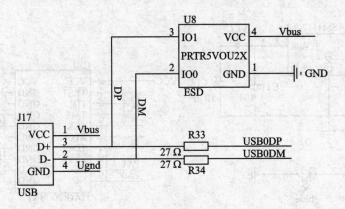

A-6 USB 接口电路

附录 A　LM3S5749 应用电路图

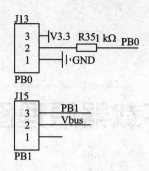

A-7　A0 版本 bug 处理电路

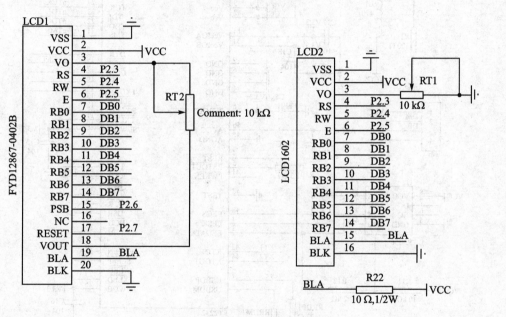

A-8　LCD1602 与 12864 接口电路

附录 B

LM–Link 下载器原理图

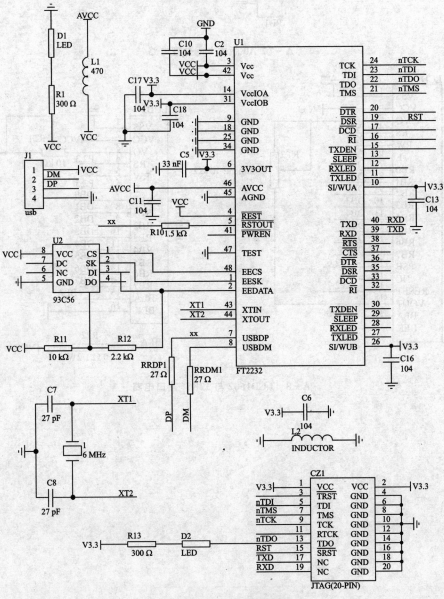

B–1　LM–Link 下载器原理图

附录 B LM-Link 下载器原理图

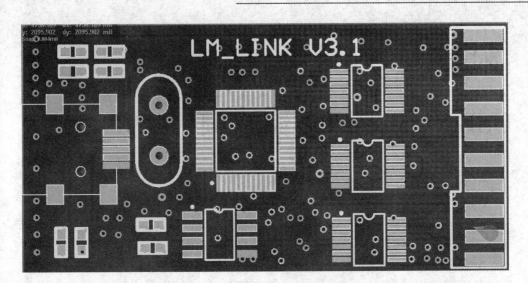

B-2 LM-Link 下载器 3D 图

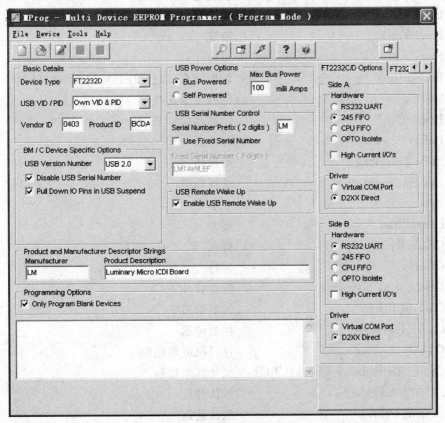

B-3 ft2232d 配置

附录 C

USB 常见术语及缩略词

ACK	确认信号
Active Device	正在使用的设备
Asynchronous Data	异步数据
Asynchronous RA	异步自适应速率
Asynchronous SRC	异步抽样转换率
Audio Device	音频设备
AWG♯(American Wire Gauge)	美国电线标准
Babble	帧传输中的总线动作
BandWidth	带宽
Bit	比特
Bit Stuffing	数据填充，以使 PLL 可以提取时钟信号
b/s	每秒多少比特
B/s	每秒多少字节
Buffer	缓冲区
Bulk Transfer	批量传送
Bus Enumeration	总线标准
Byte	字节
Capabilities	能力
Characteristics	特征
Client	客户
Configuring Software	配置软件
Control PIPE	控制管道
Control Transfer	控制传送
CTI	计算机电信组织
Cyclic Redundancy Check(CRC)	循环冗余校验
Default Address	默认地址
Default PIPE	默认通道
Device	设备

附录C USB常见术语及缩略词

Device Address	设备地址
Device EndPoint	设备端口
Device Resource	设备资源
DownStream	下行
Driver	驱动
DWORD	双字
Dynamic Insertion and Removal	动态插入和拆除
EEPROM	电擦写可编程只读存储器
End User	终端用户
EndPoint	端口
EndPoint Address	端口地址
EndPoint Direction	端口指向
EndPoint Number	端口号
EOF	帧结束
EOP	包结束
External Port	外设端口
False EOP	错误的包结束标志
Frame	帧
Frame Pattern	帧结构
Full-duplex	全双工
Function	功能,功能部件
Handshake Packet	握手包
Host	主机
Host Controller(HCD)	主机控制器
Host Resources	主机资源
Hub	集线器
Hub Tier	Hub 层
Interrupt Request(IRQ)	中断请求
Interrupt Transfer	中断传送
I/O Request Packet(IRP)	输出/输入请求包
Isochronous Data	同步数据
Isochronous Device	同步设备
Isochronous Sink EndPoint	同步接收端点
Isochronous Source EndPoint	同步源端
Isochronous Transfer	同步传送
Jiffer	抖动

附录 C USB 常见术语及缩略词

kb/s	传送速率每秒几千比特
kB/s	传送速率每秒几千字节
LOA	有始无终的总线协议
LSb	最低比特
LSB	最低字节
bcdUSB 2B BCD	USB 规划发布号
LSB	最低字节
Mb/s	传送速率每秒几兆比特
MB/s	传送速率每秒几兆字节
Message PIPE	消息通道
MSb	最高比特
MSB	最高字节
NAK	不确认
NRZI	非归零翻转码
Object	对象
Packet	数据包
Packet Buffer	数据包缓冲区
Packet ID(PID)	数据包标识符
Phase	相位
Phase Locked Loop（PLL）	锁相环
Physical Device	物理部件
PIPE	管道
Polling	查询
Port	口、端口
Power On Reset（POR）	电源复位
Programmable Data Rate	可编程数据速率
Protocol	协议
Rate Adapter（RA）	自适应速率
Request	请求，申请
Retire	取消，终止
Root Hub	根集线器，主机 Hub
Root Port	根集线器下游端口
Sample	抽样，取样
Sample Rate(Fs)	抽样速率
Sample Rate Conversion（SRC）	抽样转换率
Service	服务

Service Internal	服务间歇
Service Jitter	抖动参数
Service Rate	指定端口每单位时间的服务项目
SOP	包开始
Stage	控制传输的某个阶段
Start-Of-Frame(SOF)	帧开始
Stream PIPE	流通道
Synchronization Type	同步类型
Synchronous RA	同步的 RA
Synchronous SRC	同步的 SRC
SPI	系统可编程接口
TDM	时分复用
TimeOUt	超时
Token Packet	标志包
Transaction	处理事务
Transfer	传送
Transfer Type	传送类型
Turn-around Time USB	传输中包与包之间的间隔时间,防传输冲突
USBD	USB 驱动器
Universal Serial Bus Resources	USB 提供的资源
UpStream	上行
Virtual Device	虚拟设备
Word	字(16)

参考文献

[1] 屈召贵. 嵌入式系统原理及应用[M]. 成都:电子科技大学出版社,2011.

[2] [英]Joseph Yiu 著. ARM Cortex - M3 权威指南[M]. 宋岩译. 北京:北京航空航天大学出版社,2009.

[3] 周立功. Cortex - M3 开发指南——基于 LM3S8000[M]. 广州周立功单片机发展有限公司,2007.

[4] 周立功. ARM 微控制器基础与实战[M]. 北京:北京航空航天大学出版社,2003.

[5] Compaq、Intel、Microsoft、NEC, Universal Serial Bus Specification,1998.

[6] USB - IF, Device Class Definition for Human Interface Devices (HID),1996—2010.